THE FONTANA HISTORY OF
THE ENVIRONMENTAL SCIENCES

PETER J. BOWLER is Professor of History and
Philosophy of Science at the Queen's University of Belfast. He has taught at universities in
Canada, Malaysia and the United Kingdom, and
is the author of a number of books on the history
and impact of the theory of evolution.

D1464667

FONTANA HISTORY OF SCIENCE
GENERAL EDITOR: ROY PORTER

already published:

Environmental Sciences   PETER J. BOWLER
Chemistry   W. H. BROCK

forthcoming:

Technology   DONALD CARDWELL
Mathematics   IVOR GRATTAN-GUINNESS
Physics   RUSSELL MACCORMMACH
Astronomy   JOHN NORTH
Biology   ROBERT OLBY
Medicine   ROY PORTER
Science in Society
LEWIS PYENSON & SUSAN SHEETS-PYENSON
Human Sciences   ROGER SMITH

FONTANA HISTORY OF SCIENCE
(Editor: Roy Porter)

THE FONTANA HISTORY OF
THE
ENVIRONMENTAL
SCIENCES

Peter J. Bowler

FontanaPress
An Imprint of HarperCollinsPublishers

Fontana Press
An Imprint of HarperCollins*Publishers*,
77–85 Fulham Palace Road,
Hammersmith, London W6 8JB

A Fontana Press Original 1992

1 3 5 7 9 10 8 6 4 2

A catalogue record for this book is
available from the British Library

ISBN 0 00 686184 9

Set in Linotron Meridien

Printed in Great Britain by
HarperCollins*Manufacturing*, Glasgow

PREFACE TO
THE FONTANA HISTORY OF SCIENCE

Academic study of the history of science has advanced dramatically, in depth and sophistication, during the last generation. More people than ever are taking courses in the history of science at all levels, from the specialized degree to the introductory survey; and, with science playing an ever more crucial part in our lives, its history commands an influential place in the media and in the public eye.

Over the past two decades particularly, scholars have developed major new interpretations of science's history. The great bulk of such work, however, has been published in detailed research monographs and learned periodicals, and has remained hard of access, hard to interpret. Pressures of specialization have meant that few survey works have been written that have synthesized detailed research and brought out its wider significance.

It is to rectify this situation that the Fontana History of Science series has been set up. Each of these wide-ranging volumes examines the history, from its roots to the present, of a particular field of science. Targeted at students and the general educated reader, their aim is to communicate, in simple and direct language intelligible to non-specialists, well-digested and vivid accounts of scientific theory and practice as viewed by the best modern scholarship. The most eminent scholars in the discipline, academics well-known for their skills as communicators, have been commissioned.

The volumes in this series survey the field and offer powerful overviews. They are intended to be interpretative, though not primarily polemical. They do not pretend to a timeless, definitive quality or suppress differences of viewpoint, but are meant to be books of and for their time; their authors offer their own interpretations of contested issues as part of a wider, unified story and a coherent outlook.

Carefully avoiding a dreary recitation of facts, each volume develops a sufficient framework of basic information to ensure that the beginner finds his or her feet and to enable student readers to use such books as their prime course-book. They rely upon chronology as an organizing framework, while stressing the importance of themes, and avoiding the narrowness of anachronistic 'tunnel history'. They incorporate the best up-to-the-minute research, but within a larger framework of analysis and without the need for a clutter of footnotes – though an attractive feature of the volumes is their substantial bibliographical essays. Authors have been given space to amplify their arguments and to make the personalities and problems come alive. Each volume is self-contained, though authors have collaborated with each other and a certain degree of cross-referencing is indicated. Each volume covers the whole chronological span of the science in question. The prime focus is upon western science, but other scientific traditions are discussed where relevant.

This series, it is hoped, will become the key synthesis of the history of science for the next generation, interpreting the history of science for scientists, historians and the general public living in a uniquely science-oriented epoch.

ROY PORTER
Series Editor

CONTENTS

PREFACE

This book is a history of the environmental sciences in the broadest sense of that term. It includes all the sciences that deal with our physical and organic environments, ranging from geography and geology to ecology and evolution theory. It is not meant, however, as a collected history of all the special disciplines involved. An account that included details of every branch of natural history would have little interest to the general reader, and would also be impossibly long. Instead, I have concentrated on major areas of science and on important theoretical innovations. Even so, to survey such a broad range of topics is a daunting task, and it has been necessary to select those issues which I thought would most interest the non-specialist reader. If that choice betrays my origins as a historian of evolution theory, I offer my apologies. I have tried to build on my own strengths and add material as widely as possible. This has meant that the choice of topics outside my own areas of expertise has been influenced by the availability of good secondary literature.

The main purpose of the book is to show how modern historians try to understand the development of sciences that influence the way we think and behave. The history of science no longer provides a dry catalogue of factual discoveries. It is a discipline that attempts to put science into its social context, to understand the cultural and professional factors that influence the way scientists explain their observations. The results are controversial because many scientists think of their work as the simple accumulation of factual knowledge. They may actually prefer a catalogue of discoveries to a sociological analysis that implies that sometimes their choice of research topic or theoretical model is influenced by external factors ranging from economic pressures to religious beliefs.

It must also be conceded that historians of science have allowed the direction of their own research to be influenced by a preconceived sense of which issues are the 'big' ones. The result is that some topics – the Darwinian revolution, for instance – have been written about almost endlessly, while others have been left on one side. This situation is changing, partly as a result of new attitudes in society as a whole. The history of ecology is at last becoming fashionable, although in other areas (meteorology and oceanography, for example) one can count the number of sophisticated studies on the fingers of one hand.

As far as I know this is the first comprehensive history of the 'environmental sciences', and the fact that it has been commissioned in this format reflects a sense that the general public has now begun to think of 'the environment' as an important problem in which science is deeply involved. One facet of the history of these sciences is the changing attitude of western civilization to the environment, and I have tried to include this in my account. At one level the 'environmental sciences' represent an artificial category – a collection of distinct specializations that have no unity beyond that thrust upon them by the public concern over the environment. Many environmentalists argue that, unless we think of our planet as a system of interlocking natural processes, we may destroy it altogether through our interference. Science often seems part of the problem: its professional fragmentation symbolizes the materialistic trend in modern thought, the desire to divide Nature up into separate units, each of which can be studied in isolation and exploited for short-term profit. Efforts to reintroduce a sense of the unity of Nature for environmentalist purposes have been rejected as the expression of an unworkable idealism.

This trend towards materialism and professional fragmentation is part of the story told below, but we shall also see that science has passed through phases when the study of broad natural interactions has been actively encouraged owing to the influence of rival philosophies such as romanticism. We now have more practical reasons for reintroducing a sense of the unity of Nature, but this may prove equally powerful as a means of persuading scientists to rethink their tendency to compartmentalize everything. I hope that a survey illustrating the wide range of motivations that have influenced scientists in the course of time will encourage the hope that a new, more responsible science of the environment is not ruled out by the very nature of science itself.

February 1992
PETER J. BOWLER
The Queen's University of Belfast

ILLUSTRATIONS

FIGURES

1

The Problem of Perception

The environmental sciences have now become a matter of acute concern. We all know that modern technology has encouraged massive exploitation of the world's resources, with all the resulting problems of pollution and environmental exhaustion. Science has clearly played a role in this exploitation – yet, at the same time, it is the scientists (or at least some of them) who are warning us of the dangers. A historical study of how those sciences dealing with the environment have developed may throw some light on our predicament. If science is simultaneously part of both the problem and the solution, we need to know how our attempt to understand and exploit the world has taken on its modern form. To what extent do these sciences reflect the underlying values of western culture – indeed, to what extent is the very notion of a rational study of Nature part of a purely western world view? Can there be a scientific study of the natural world that does not to some extent reflect the values of those who fund and conduct the research? Have those values changed through time and, if so, does the very concept of an environmental 'science' depend on the nature of the society that sustains it? These are questions of vital interest in the modern world, but they may be illuminated by a historical analysis of how the environmental sciences have developed.

The scope of the problem is evident from the fact that the very term 'environmental sciences' has a modern context that would not have been recognized by the scientists of earlier generations. Twentieth-century science has become highly specialized, and the research programmes of disciplines often differ in character even when – to the outsider – they seem to be dealing with closely related topics. Environmentalist critics argue that science has encouraged a fragmented image of Nature in which the details of everything are studied, but there is no scope for an overview that might help us to understand the problems of the earth as a whole. The unity of the 'environmental sciences' is not created by the sciences themselves; it is imposed by the public's growing awareness of the threat posed to the environment by our own activities.

The history of the environmental sciences may thus seem a somewhat artificial category. In the twentieth century, at least, it must encompass a wide range of studies that have little real interaction. Our growing awareness that the environment as a whole is something worthy of study provides a kind of unity through hindsight, but this is external to the sciences themselves. The growth of environmentalism provides no historical focus because most sciences have traditionally been associated with the desire to understand Nature in order to dominate and exploit it. Even ecology began as a science devoted to understanding natural relationships for the purpose of improving our ability to control them.

It is possible, however, that history may supply its own form of conceptual unity. The specialization of scientific disciplines that created the fragmentation of modern perspectives at the research level has built up over the last couple of centuries. Geology emerged as a coherent field of study around 1800, ecology about a century later. Earlier scientists did not divide the study of Nature into such rigid categories, and were thus in a better position to appreciate the links between what have now become distinct research topics. Even when independent disciplines began to emerge, individual scientists often worked in several of them simultaneously, and could appreciate how they might interact. Much of Charles Darwin's early research was in the area of geology, not evolutionary biology. There have been several episodes in the history of science where active efforts were made to encourage the kind of interactive studies that modern environmentalists demand.

Too much of the history of science has concentrated on exciting episodes such as the 'Darwinian revolution', leaving other topics unexplored even when they were of direct interest to the central figures in the debates we have chosen to highlight. Historians were fascinated by the image of a 'war' between science and religion, and focused their attention on those episodes where the conflict seemed most apparent. We are now beginning to realize the damage that this preconception has done to our interpretation of the past. The growth of environmentalism provides historians with a new incentive, a new level of hindsight that directs our attention to hitherto ignored areas of science. And as we study these areas, we see both the range and the unity of earlier efforts to understand the environment. If our current desire to create a unified group of environmental sciences forces the scientists themselves to take a broader perspective, this will merely revive a sense of unity that has been lost in the period of increased specialization. But historians, too, have been given a new agenda that is forcing them to look at areas of science once dismissed as too technical to be of interest to non-specialists.

Another way of illustrating the problems confronting a historian of the environmental sciences is to expose the relationship between science and changing cultural values. One could write a history of western attitudes towards Nature that paid very little attention to the character of science itself. The unit of analysis would be how the people of a given century thought about the world in which they lived, and the historian could appeal to literature and the visual arts as well as to natural history and the disciplines that we accept as part of 'science'. It is true that a rigid separation of scientific and non-scientific ways of looking at Nature only came about during the last few centuries.

But part of our story must surely be how the environmental *sciences* emerged. We need to know how and why western culture began to allow its vision of Nature to be determined by professionals who were able to set themselves up as the source of authority on such matters.

Modern science reflects the attitudes of western culture towards the natural world. Other cultures perceive Nature in different ways, and we need to know which particular aspects of the classical and Judeo-Christian traditions have shaped the origins of those areas of study that gave rise to the environmental sciences. To do this, we must go outside the realm of science itself to identify the social and cultural milieu within which the scientific hypotheses were conceived and articulated. This milieu changed through time and thus permitted the emergence of different ideas about Nature that could be used in the framing of scientific theories. There was never a single, unified western culture that served as the basis for all scientific thought. Differences in religious, philosophical and social background have always made room for debate over how Nature was to be interpreted. Such debates occur both within science itself and between science and those other areas of knowledge that also hope to influence our view of the world. Although science has, on the whole, come to exert greater influence in modern times, there are still energetic resistance movements such as the creationists.

We now take it for granted that there are differences between scientific and non-scientific ways of looking at Nature. The emergence of disciplines charged with the *rational* investigation of Nature is an important aspect of the development of western thought. The historian needs to ask if this element of rationality is characteristic of the west's more exploitative attitude towards Nature, as contrasted with the viewpoint of other cultures. The emergence of the environmentalist or 'Green' movement, with its emphasis on the use of science to pinpoint the problems of the modern world, suggests that, by itself, rationality is a two-edged sword. One can apply the principle of rational investigation to support either an exploitative or a conservationist view of the environment. When we see people who call themselves 'scientists' speaking on both sides in debates over environmental issues, we are forced to confront the possibility that science must be something more than a value-free search for factual information. However rational the scientific method, it is used to test hypotheses that are generated by human beings living within a particular culture and society. The history of the environmental sciences must thus encompass both the emergence of science and changing cultural attitudes towards Nature.

Even if we concentrate on the emergence of science, there are problems created by the peculiar nature of the environmental sciences. The earliest triumphs of science were made through the application of mathematics and the experimental method in areas such as physics and chemistry. Natural history and geology have always been regarded as 'softer' sciences in which methodological rigour was more difficult to attain. It is sometimes argued that 'natural history' is mere fact-gathering, involving no attempt to explain the observations and hence lacking the characteristics of true science. Only when this approach was replaced by the more rigorous attitude of the biologist and the geologist did something justifiably termed 'environmental *science*' emerge. In fact, the transition from natural history to biology and geology was a complex process, and some areas have remained largely untouched by explanatory schemes such as the theory of evolution. The entomologist who specializes in the description and classification of insect species may still find little of value in modern evolutionism. Yet 'mere' classification is itself a theoretical process, and, although our attention will naturally be drawn to the creation of the great explanatory theories, we cannot afford to ignore the continued efforts of those whose primary aim is classification.

In recent decades the theories of geologists and biologists

have become more amenable to expression in terms of mathematics. But the environmental sciences have methodological requirements of their own that should not be analysed in terms of categories devised for the study of the physical sciences. The naturalist and the geologist address themselves to questions of a kind that are simply not asked by the physicist or the chemist. If presented with a piece of rock, a chemist would analyse it to determine the substances of which it is composed – but the geologist would try to find out how it had been formed and how it came to be placed at a particular spot on the earth's crust. The forces that sustain the natural world are governed by physical law, but the earth and its inhabitants constitute a unique collection of structures that have been created by the operations of such laws over a vast period of time. To understand how the system works, we need to know both its structure and the historical 'laws' that govern how it can change. In principle, those laws may be compounded from the laws of physics and chemistry, but their operation within a single isolated system gives them a historical character that is unique to the environmental sciences.

One of the most important developments within the cultural framework of modern science is the emergence of this awareness that Nature has a history that determines its present structure. Our modern preoccupation with the fragility of the environment is in part a product of our knowledge that the world can and does change through time. The environmental sciences have thus undergone at least two major reorientations in the course of the last couple of centuries, the first of which has laid the foundations for the second. In the nineteenth century, the emergence of the historical view of Nature was the chief focus for controversy because it challenged the traditional image of a divinely created universe. The twentieth century's growing need to understand the damage that humankind is doing to the environment is, of course, a product of the growing power of our technology. Yet, without the sense of history, our approach to the problems of overdevelopment would be quite different. Owing to the emergence of the evolutionary viewpoint, we know that species are shaped by their environment – and that the environment is constantly changing. We know that species that cannot adjust have always faced extinction. Unlike our ancestors of a few generations ago, we are forced to confront the possibility of a future in which the world might be changed beyond recognition by either natural or artificial factors.

EXPANDING HORIZONS

Let us put on one side, for the moment, the problem of defining what we mean by a 'science' of the environment. Until comparatively recently, no one would have thought it appropriate for the rational investigation of the world to have been separated off into an intellectual category of its own. Even when sciences began to emerge in the eighteenth and nineteenth centuries, it seemed obvious that their subject matter was less accessible to rigorous investigation than the laboratory-scale problems of the physicist and chemist. Traditionally, the sciences dealing with the environment seemed to interact far more directly with ordinary people's perception of the world. However hard one tried to be scientific, to give an account of how our world was created by a natural process involved the changing of deep-seated views hallowed by both philosophical and religious traditions.

Culture and Nature

It may be difficult for the scientist to accept that certain aspects of the way in which we perceive the world are dependent upon the culture within which we are raised. The basic categories that must be used in an accurate description of the natural world seem self-evident. We take it for granted that the whole system is governed by natural law, for instance, not by magical or spiritual forces. Yet the categories we use are unique to the scientific version of western culture. Other cultures perceive Nature in different ways, and thus seek to classify and explain the world in terms that seem meaningless to us. These alternative conceptual schemes are all too easily dismissed as primitive or pre-scientific. Yet representatives of other cultures may say that we ourselves are missing something in our analysis of Nature. Modern environmentalists sometimes admit the superiority of the more reverent attitude towards Nature displayed by so-called 'primitives'.

The local environment may exert a strong influence on the way in which its inhabitants perceive the natural world. A tribe inhabiting deep jungle may have no word for 'horizon' in its vocabulary, while a people whose life is conducted on open plains will have a sense of space built firmly into its way of thought. Eskimo languages contain many words to denote different kinds of snow and ice that are barely distinguishable to a stranger. The concept of a 'wilderness' makes no sense to a tribe of hunter-gatherers that has lived there for generations. Yet to a farmer the wilderness represents a threat or a challenge, while to the modern city-dweller it offers the hope of release from an artificial environment. A medieval European saw mountains as terrifying places of desolation – only with the coming of Romanticism did we begin to see them as beautiful. The fact that such value-judgements have changed even within our own culture must alert us to the possibility that our modern perception of Nature is not based on self-evident or universal categories.

Natural limitations may also impose restrictions on our ability to comprehend the world, although those restrictions have been and are still being removed. The ability of many early civilizations to conceptualize the world was limited by their technical inability to explore it. Many parts of the world remained shrouded in mystery, leaving ample room for speculation. The environmental sciences are a product of the growing ability of the European nations to explore across the whole world. Now that we live in a global village, science is still engaged in a process of exploration that can produce major changes in our perspective. The modern explanation of continental drift, for instance, was only worked out after the development of techniques to study the ocean floor (see chapter 9).

Exploration beneath the sea represents an extension of the trend by which European civilization expanded the range of its geographical knowledge. Yet any culture with a wide understanding of the world around it needs some way of representing that knowledge, and the creation of a map depends upon a particular choice of the features to be included. The environmental thinker Yi-Fu Tuan reports that, when men and women from the same Eskimo tribe were asked to draw a map of their local area, the maps differed significantly because the men and women had different priorities in their choices of what to represent[1]. The schematic maps used to denote transit systems such as the London underground often bear only a loose relationship to the real layout of the city, but they serve their purpose better than a more 'accurate' representation.

Maps also help to define the relationship between the centre and the periphery of a culture. Chinese maps invariably placed China at the centre of the world, while medieval European maps (even those drawn in Britain) put Jerusalem at the centre. Lest we scoff too readily at such obvious distortions of reality, it should be remembered that the establishment of the prime meridian at Greenwich reflected the cartographical perspective of a British empire based on ocean trade. By contrast, many Americans prefer to think of the United States as 'God's country'.

It is even more difficult to create a representation of the world that includes its living inhabitants. Naturalists have always sought to describe and classify the animals and plants that make up the organic environment. Early systems of classification often reflected the fact that many species are useful to humankind. But as Europeans spread around the world, they were confronted by a bewildering array of new species. This placed a strain upon traditional systems and led indirectly to the creation of modern biological taxonomy (the science of classification). The urge to classify things according to a rational system lies deeply rooted in the human mind, yet there are important choices to be made when establishing the basis for a taxonomic system. To those not afflicted by the bug, the process may seem of little more interest than stamp collecting – yet in the realm of modern science there are few topics that can raise hackles more effectively than a disagreement over the principles of biological classification. Even though biologists now agree that species should be classified according to their degrees of relationship, they disagree fundamentally over how the relationships should be worked out.

The emergence of a system that at least took it for granted that classification should be based solely on physical resemblances constitutes a major step in the emergence of modern science. Other cultures, and even the pre-scientific versions of western culture, did not take so restricted an approach to the question. The philosopher and historian Michel Foucault first became interested in the relationship between language and the representation of the world when he read an account of a Chinese encyclopedia in which animals were divided into the following categories[2]:

(a) belonging to the Emperor, (b) embalmed, (c) tame, (d) sucking pigs, (e) sirens, (f) fabulous, (g) stray dogs, (h) included in the present classification, (i) frenzied, (j) innumerable, (k) drawn with a very fine camel hair brush, (l) *et cetera*, (m) having just broken the water pitcher, (n) that from a long way off look like flies.

The possibility that another culture should think it natural to adopt a system that seems so arbitrary to us led Foucault to investigate the changing approaches to classification within the western scientific tradition.

Foucault's example is a particularly bizarre one, and in most cases the relationship between culture and classification involves a subtle interaction between what we would call 'real' and 'imaginary' factors. Because many animals and plants have a practical value as food or medicine, most non-European cultures distinguish the species in their environment in a way that corresponds very closely to the categories of the modern biologist. But they also recognize species that are of no practical value, and all are endowed with mythical properties allowing them to play a role as cultural symbols. Anthropologists have explored the complex interaction between primitive classifications of animals and plants and the cultures from which they emerge. The factors that are included within a classification scheme tell us a great deal about the mental processes of the people who think the scheme is a 'natural' one. Our own culture has drawn upon various traditions in which animals played a symbolic role. The ancient Greeks gave many animals human characteristics – the lion was bold, the fox cunning and so on. Medieval bestiaries continued this tradition, and also took it for granted that a description should include each species' heraldic significance (see chapter 2). Although modern science was supposed to have purged itself of such anthropomorphism, it has sometimes proved difficult for naturalists to escape their cultural background altogether.

Particular attention was often paid to those animals which did not seem to fit into the accepted categories. The anthropologist Mary Douglas suggests that the dietary restrictions imposed on the Jews by the Book of Leviticus are based on anomalies within a primitive scheme of classification[3]. Biblical commentators have often been puzzled by the apparently arbitrary division of animals into 'clean' and 'unclean' species, but Douglas argues that the ancient Hebrews associated holiness with wholeness, i.e. with those things that fitted completely into the divine plan. God had given them sheep and goats, so cud-chewing, cloven-hoofed animals were clean. The pig is cloven-hoofed but does *not* chew the cud; it did not fit into the system and was thus an abomination because the legalistic minds of the priests could not stomach anomalies. Such examples are not restricted to the pre-scientific era. The emergence of modern biology was associated with an extended controversy over whether or not humankind should be classified along with the other animals. Religious thinkers argue that the decision to treat the human race as just another product of animal evolution provides clear evidence of the materialistic tendencies at the heart of modern science.

Religious and philosophical beliefs have traditionally played an important role in shaping how different cultures perceive the world in which they live. Many so-called 'primitive' cultures see Nature as both a physical and a spiritual system, hence their belief that the world can be manipulated by magical techniques or by the calling up of spirits. Traces of such an attitude can still be found even within modern western society, as in the tendency to attribute a mystical significance to sites such as Stonehenge. Societies that subsist by hunting and gathering tend to see Nature as a complex whole with which humankind must harmonize – a viewpoint that the modern Green movement would like to revive. Farming communities are more likely to clothe Nature with female attributes, so the earth-mother becomes the source of all productivity. Feminists and others complain that western science is a fundamentally masculine project, which visualizes Nature as a passive, female entity to be explored and conquered at will. Other critics point out that the image of Nature as a machine is also one that fosters an attitude of control and domination. These are valid points, but the origins of the rival images lie deep in the foundations of western thought.

Science and Christian Tradition

Modern science prides itself upon the rejection of non-materialistic conceptions of Nature, yet it can be argued that the emergence of the scientific world view was itself a product of the Judeo-Christian tradition. The ancient Hebrews were herdsmen, and the God of the Old Testament is thus modelled on the shepherd who both controls and cares for his flock. In the creation story of the Book of Genesis, God creates the universe as something apart from Himself – He does not dwell within Nature. Some historians of science believe that it is this model for the relationship between the Creator and the universe which lies at the heart of the mechanistic viewpoint. From it stemmed the tendency to visualize Nature as something constructed by God and operating according to laws that He had imposed. Instead of being a mysterious place to be understood in spiritual terms, the world became a machine whose operations could be understood in the light of humankind's God-given reason. Far from being intrinsically antagonistic to religion, the attitudes of modern science may have arisen from a particular view of the relationship between God and Nature. If science has now decided to ignore the divine origin of the universe it investigates, this cannot conceal the Judeo-Christian origins of the belief that Nature is comprehensible in terms of rigidly operating laws.

Does the biblical promise that humankind has been given dominion over the earth lie at the root of the west's desire to exploit the natural world without regard to the ecological consequences? Such a ruthless attitude would be anathema to a culture with a less mechanistic view of things. On the other hand, the Bible can also be interpreted as placing us in the position of stewards looking after God's creation – and the steward is not supposed to destroy the property entrusted to him. Medieval Europeans did not set out to destroy their environment, and the growth of industrial civilization has been accompanied by a lessening of interest in the Christian foundations of western culture. Some thinkers would argue that it is science itself that, by emphasizing a purely materialistic view of Nature, has led to the attitude of indifference to the environment characteristic of modern industry. But if science is itself an aberrant product of Judeo-Christian culture, we must look very carefully indeed at the roots of modern attitudes.

The actual structure of the earliest scientific theories was determined not so much by the Bible as by the legacy of Greek thought. Of particular influence in the area of natural history was the thought of the philosopher Plato, who postulated the existence of an ideal world of which the physical universe is but an imperfect copy. It was relatively easy to adapt this Platonic philosophy to the creation story of the Bible by supposing that the ideal world corresponds to the plan of creation existing in the mind of God. The philosophy of 'essentialism' sees the individual animals that make up the population of a species as merely imperfect copies of an ideal or archetypical form that constitutes the species' real essence. Even well into the modern era, many naturalists and geologists still saw themselves as describing and trying to explain a divinely created universe. Linnaeus, the eighteenth-century Swedish naturalist who founded the modern system of biological classification, was quite convinced that his system offered an accurate representation of the divine plan of creation (see chapter 5). His commitment to the fixity of species reflects not only a biblical influence but also the essentialism that stems from the Platonic view of the world.

Linnaeus also developed an early anticipation of ecology, based on the assumption that God had established a carefully organized 'balance of Nature' in which every species played its part. Many seventeenth- and eighteenth-century naturalists found it impossible to accept the extinction of species, because this would leave a gap in God's creation. This attitude declined with the emergence of the modern sciences of paleontology and evolution theory. Some naturalists dared to ask if systems of classification were merely products of the human passion for order, not a reflection of a real pattern existing in Nature itself. By establishing the reality of extinction, the paleontologists forced everyone to confront the possibility that Nature was indifferent to the fate of species, let alone individuals. Modern Darwinism is a product of the view that Nature does not display a rationally ordered sequence of natural kinds. In the end, science has emancipated itself from the tradition of Platonic essentialism that for so long upheld a basically static view of creation. At the same time, we must bear in mind that many nineteenth-century evolutionists tried to retain an element of order or structure within their view of natural development. They still saw evolution as a purposeful process aimed at the production of a morally significant goal.

Geology and evolution theory were products of a major change in Europeans' perception of the timescale of the material universe. In 1650 Archbishop James Ussher pinpointed the date of creation as 4004 BC. Within such a limited vision of the earth's history, divine creation seemed the only way of explaining how the modern state of affairs had been established. Yet by the time Ussher published his estimate, some scholars were already beginning to wonder if the biblical story of human origins was acceptable. Naturalists began to suggest that the fossils hidden within the rocks indicated a period of earth history antedating the creation of humankind by many thousands, if not millions, of years. Although many geologists at first retained the view that the earth's development must be guided by a divine plan, the destruction of the biblical view of creation nevertheless established the wider framework within which a theory of natural evolution would be articulated.

It may seem paradoxical that sciences assuming a vast timespan for the development of the earth and its inhabitants should have emerged within a culture that was traditionally bound by the biblical creation story. Other religions, especially Hinduism, take it for granted that creation extends over millions of years, and would thus seem more natural foundations upon which to build the modern scientific view of universal history. Yet eastern cultures have traditionally based their view of history upon a conceptual scheme that differs in another important respect from that accepted by Christianity. Whatever the limitations of the timespan implied by the Book of Genesis, the biblical conception of history is tied firmly to the assumption that there is a *direction* built into the spiritual development of humankind. There was a creation, and there will be an end to all material things, the whole system serving as a backdrop for the great drama of human sin and redemption. Most other cultures, even those which accept a universe extending over vast periods of time, have preferred a cyclic model of history in which everything is re-created over and over again. Irreversible change was seen as a frightening prospect. The cyclic model does appear in western thought, but only as a stimulus that has forced the advocates of the developmental viewpoint to extend their timescale. Once again, it turns out that the Judeo-Christian foundations of western culture have played a role in creating the framework within which modern science interprets the world.

THE NATURE OF SCIENCE

Modern science is both a product of, and a transforming influence within, western culture. Whatever the origins of the scientific world view, the actual process of investigation has led us away from the biblical image of a static material universe. Scientists now like to think of themselves as the exponents of an objective method designed to generate factual knowledge about the world. They pride themselves upon the repudiation of all external factors that might influence their investigations. The scientific method guarantees that subjective factors such as philosophical and religious beliefs cannot be incorporated into the framework of knowledge. But can we really accept that science has been able to emancipate itself so completely from the world of human beliefs and feelings? Our perception of the world is almost inevitably dictated by cultural factors, and the rise of materialism itself can be counted as just such a factor. The creation and development of what we now call the 'environmental sciences' raises a series of questions centred on our definition of science and our understanding of how the functioning of science itself shapes our perception of the world.

The Scientific Method

There is a vast literature on the 'philosophy of science' intended to elucidate every aspect of the process by which the supposedly objective framework of modern knowledge has been established. But history shows that ideas about the scientific method have changed considerably over the years. At one time it was fashionable to claim that scientists were pure observers. The method of 'induction' required them to gather facts without preconception so as to guarantee that the regularities they observed (the laws of Nature) were free from any taint of subjectivity. Modern philosophers of science realize that such a fact-gathering exercise could not possibly work, because Nature is so complex that we need some sort of guidance to tell us which facts are relevant to a particular line of enquiry. Scientists proceed by setting up a hypothesis that anticipates the discoveries they hope to make. In other words, they make a prediction about what they expect will happen in a certain situation, and only then turn to observation and experiment in order to test that prediction. This is the 'hypothetico-deductive method' in which science proceeds through the testing of consequences deduced from an imaginary model of the phenomenon.

According to this methodology, the objectivity of science lies in its willingness to subject every aspect of the hypothesis to rigorous testing. If the predictions derived from the hypothesis are not confirmed by observation and experiment, the hypothesis is rejected and a new model sought. Sir Karl Popper has insisted that the chief characteristic of science is that its hypotheses are 'falsifiable', i.e. they are framed in such a way that they can be refuted by straightforward observational tests. Pseudo-sciences such as astrology offer deliberately vague theories that can always be saved from falsification. Popper believes that true scientists should actively seek to expose the weaknesses of their own ideas, welcoming the refutation of one hypothesis so that they can pass on to create something better[4].

The philosophers' efforts to define the nature of the scientific method have been largely conditioned by a fascination with physics as the ideal model for a scientific discipline. Their efforts have been less than satisfactory when applied to biology and the environmental sciences. By the very nature of the subject matter, the environmental scientist is forced to ask a different kind of question to the physicist or chemist. The latter deals with repeatable phenomena that are relatively easily mapped in mathematical terms. He or she can propose hypothetical laws of Nature and test them in experiments designed to reveal immediately whether the hypothesis is valid or not. But the naturalist or geologist is placed in a different position by the fact that the subject requires the study not of universal phenomena but of a particular structure or set of structures. To describe and classify the entities that make up our environment, and to explain how those entities were created by natural processes, involves the construction of explanatory schemes that may include physical laws, but also involves factors that the physicist does not have to consider.

Philosophers of science are gradually waking up to the fact that definitions of the scientific method designed for the physical sciences may simply not be relevant in areas such as biology and geology. If we are to call these areas 'sciences' at all, we have to accept that science is a multifaceted activity that cannot be unified by simply reducing the 'softer' areas to physics and chemistry. If the 'soft' sciences need a methodology that leaves greater room for debate, so be it. Only opponents of science such as the creationists have a vested interest in claiming that fields such as geology and evolutionary biology are not really sciences at all – as measured by some totally inappropriate standard derived from physics. It must be possible to engage in a rational study of the environment and its changes, eliminating those hypotheses that are plainly unable to explain the observed facts, even though we are unable to test our hypotheses in quite the same rigorous manner as the physicist.

Until recently, geologists and evolutionary biologists were unable to formulate their ideas in mathematical terms, and their broader hypotheses cannot be tested by experiment. There was a time when physicists looked down on geology as an inferior science and thought that the geologists must defer to them on disputed issues. Nowadays the two fields get on together in a much more amicable fashion, yet we can still find a noted philosopher of science such as Karl Popper dismissing evolution theory – and by implication all other fields that construct historical models to explain present structures – as unscientific because they do not deal with repeatable phenomena and hence cannot test their hypotheses properly. Admittedly, Popper has since withdrawn this accusation, but his original position illustrates how difficult it was for those thinkers who saw physics as the ideal science to appreciate that this model was inappropriate for other areas[5]. Unless we are to accept the present state of the earth as something to be described rather than explained, we must postulate change through time as the means by which the existing structures were formed. We are thus dependent upon the construction of historical models outlining unique sequences of events.

A major feature of the development of sciences such as geology and evolutionary biology has been the emergence of a determination to include only *natural* forces within those historical models. But this does not detract from the need to see those forces working within a particular structure or environment. The earth and its inhabitants may be governed solely by natural laws, but our planet is quite unlike any of the others and it has been formed as the result of a unique historical process.

At the same time, we must recognize the artificiality of trying to compartmentalize the sciences too rigidly into those that do and do not have a historical dimension. The life sciences include some areas such as anatomy and physiology that study the internal functions of organisms, and others such as ecology and evolutionism that study how those organisms interact with one another through space and time. There has always been a distinction to some extent between the medical areas of biology and those we might call 'natural history'. But there have always been interactions: doctors and naturalists were both interested, for instance, in the collection of plants, and botanists routinely noted the medical uses of their specimens until relatively modern times. Modern Darwinism draws heavily on the 'laboratory' science of genetics to explain the processes going on in the wild. Parts of this book must thus be read in conjunction with the companion volume on the history of biology[6].

Science as a Social Activity

Because the emergence of a historical dimension in science challenged the biblical account of creation, the process had traumatic consequences for the development of western thought. As a result, we take it for granted that these sciences have implications for other areas of culture. Everyone was to some extent affected by the loss of faith in Genesis, and the scientists themselves were no exception. It is a mistake to think in terms of a 'war' between science and religion, because many scientists also had religious beliefs that they wished to take into account. The immense controversies that raged within science over the creation of new fields such as geology and evolutionary biology suggest that the scientists themselves were influenced by the broader implications of what they were doing.

The interaction between science and culture is especially clear in the areas of geology and evolution theory. Historians suspect that Popper's image of the purely objective scientist is altogether too idealistic. We know that scientists often cling on desperately to their own theories even when others think they have been decisively refuted. Great controversies erupt when groups committed to rival theories battle it out for domination of the scientific community. If we are to go beyond an armchair image of what science *ought* to be like, we must take into account the human dimension of the search for knowledge. Even the hypothetico-deductive method allows the scientist to anticipate Nature, and thus opens up the prospect that factors external to science might be used as inspirations in the framing of hypotheses. Theories must be understood as something more than value-free models – they incorporate philosophical or even religious preconceptions that command a degree of loyalty not allowed for in Popper's analysis.

A controversial alternative to Popper's philosophy of science is Thomas S. Kuhn's thesis that science advances through a series of conceptual revolutions[7]. According to Kuhn, most scientific research is 'normal science' done within the conceptual framework specified by a 'paradigm', an underlying theory that defines the kind of question that is meaningful and the kind of answer that will be acceptable. Newcomers to the profession are brainwashed by their education into accepting the logic of the paradigm, and anyone who does not conform is dismissed as a crank. The paradigm only begins to face serious questioning when experimental tests reveal anomalies that do not fit into the scheme. Kuhn insists that, to begin with, most scientists stay loyal to the old paradigm, brushing the anomalies under the carpet rather than accepting them as falsifications. Eventually, though, a crisis state is reached and younger scientists begin to look for an alternative conceptual scheme. When a successful alternative is found, i.e. one that can deal with the anomalies facing the old scheme, it soon gains the status of a new paradigm, imposing its own structure on the normal science of the next few generations.

There are conceptual breaks in the development of the environmental sciences, although many historians are suspicious of Kuhn's claim that the transitions constitute abrupt revolutions. The advent of evolution theory is associated with a 'Darwinian revolution' sparked off by the *Origin of Species*, but it is by no means clear that this conceptual innovation was as abrupt as was once supposed. Without denying the importance of the changing conceptual schemes associated with the rise of evolutionism, modern historical research has highlighted a more complex sequence of events taking place over a long period of time. Kuhn himself admits that revolutions may take decades to complete, but in some cases what has been perceived as a single transition may be composed of several distinct steps.

Kuhn emphasizes that the conceptual schemes that the scientists of earlier periods used to analyse Nature were quite different to those we take for granted today. Although hindsight helps us to pick out those developments that turned out to be major turning points, historians now realize that it is pointless to attempt an analysis of the past in terms of modern ideas. Because scientists think that the growth of scientific knowledge is a rational process, they are tempted to see history as a record of cumulative discoveries. They are thus liable to construct interpretations of the past that are distorted by the need to concentrate solely on those steps that seem to get us closer to our present state of knowledge.

Historians of science often dismiss these accounts as forms of 'Whig history' – a term borrowed from political history and used to denote any account of the past designed to highlight only those events that seem to lead in a desired direction. The Whigs were the political party in eighteenth- and early-nineteenth-century Britain who represented the interests of the rising commercial and industrial classes (ancestors of the later Liberal party). Some historians, of whom Lord Macaulay is the best known, constructed an interpretation of Britain's past in which progress was always brought about by those who shared Whig values. 'Whig history' thus means a view of history manufactured to stress the progressive role of certain ideals that are important in the present.

When applied to the history of science, 'Whig history' means a view of the development of science that interprets all major events as steps leading towards our modern level of knowledge. The great heroes of discovery, Newton, Darwin and the like, proposed theories that were immediately recognized as superior to older ideas by anyone capable of rational thought. Those who resisted the new interpretations, or proposed alternative theories that turned out to be blind alleys, are the 'villains' who tried to hold back the development of science. But just as a conservative thinker can create a view of British history in which the Whigs are the bad guys, so it is possible to show that the scientists who participated in the great debates of the past cannot quite so easily be divided up into heroes and villains. By imposing modern categories onto the past (often in an oversimplified way), we create an artificial model of scientific progress that ignores the often complex process by which the current interpretations of Nature were constructed. The conventional stories told about the 'heroes of discovery' are myths created to enhance the view that science is a cumulative process, and to conceal the fact that new theories are often accepted only after intense controversy within the scientific community.

Serious historical research tends to demythologize the great scientists. It has forced us to recognize that the 'discoveries' of the past – even the ones that were in fact crucial – were at first interpreted in terms that do not correspond to the theoretical framework of modern science. Important advances were mixed in with ideas that we would treat as ludicrous today, while those who opposed

the 'forward-looking' theories were often able to call upon
arguments that were quite plausible by the standards of the
time. To understand what was really going on in any of the
great scientific debates, the historian must make an effort
to appreciate the conceptual scheme that was accepted at
the time, even when that scheme seems bizarre by modern
standards. Only then will we be able to understand what
the scientists of earlier periods thought they were doing,
and their reasons for interpreting their discoveries in the
way they did. Far from belittling the achievements of the
great scientists, the new history of science gives us a clearer
understanding of the obstacles they had to overcome in
order to get their theories accepted.

One consequence of this is a recognition that the great
discoveries of the past were often accepted at least in part
for reasons that the modern scientist – who reinterprets
those discoveries in his or her own terms – would not
find acceptable. Sometimes the right theory was accepted
for the wrong reasons, and sometimes wrong theories were
accepted for reasons that made perfectly good sense in terms
of the currently accepted paradigm. We must not assume
that the advantages of 'good' theories were immediately
recognized, because in their original form those theories
often contained elements that had to be weeded out before
the important core could be recognized. Nor should we seek
to blame scientists who made what we would now regard as
bad choices because their judgement was shaped by preconceptions that hardly anyone at the time could have escaped.
To gain a realistic understanding of how new theoretical
schemes were invented and then incorporated into the way
scientists think, we must try to avoid oversimplifying the
past by pretending that earlier ideas were merely immature
versions of our own way of thinking.

An important lead offered by Kuhn's scheme is its emphasis on the role played by the scientific community. The
followers of a paradigm have a professional loyalty to the
scheme they have inherited, and a scientific revolution

depends upon the creation of a new elite within the community that takes over control of the sources of funding, the means of publication and the educational establishment. In recent years, sociologists of science have begun to place a great deal of emphasis on the role of professionalization and the way in which the scientific community functions during the process of innovation. In some cases it seems likely that potential revolutions have been nipped in the bud because the supporters of a new idea were not able to play the game of scientific politics adequately. The same theory has sometimes had different degrees of success in different countries, because each national community has its own cultural and professional interests.

The functioning of the community is even more important in cases where the emergence of a new theory requires the creation of an entirely new discipline. Originally there was no such thing as a science of geology, and the study of the earth itself was parcelled out among a number of other disciplines. Geology emerged as a distinct science in the late eighteenth and early nineteenth centuries – but it was structured in a somewhat different way in each of the countries that contributed to its creation. More recently, the emergence of geophysics marks a significant break with the kind of geology that was done in the nineteenth century, a break that is associated with the promotion of new conceptual models of the earth's structure.

The paradigms of Kuhn's scheme are theories that impose a fundamental structure upon our view of the world. They define research schools and thus represent professional loyalties that may shape the whole of a scientist's career. Because they model Nature at such a deep level, they involve components that represent philosophical and sometimes even religious commitments. The replacement of the Genesis story of creation with the modern theories of geological and biological evolution is an obvious example of a 'revolution' (however long-winded) involving major philosophical and religious reorientations. But the metaphor of a 'war' between science and religion can be seriously misleading. Many early scientists had deep religious convictions, and it would be a mistake to assume that they were thereby constrained to accept false theories. In some cases it turns out that 'conservative' scientists made important contributions, while more radical thinkers went too far and proposed ideas that could not be accepted in their entirety.

Historians of science are increasingly unwilling to invoke non-scientific factors only to explain why earlier thinkers made *wrong* choices (as defined by modern standards). Those who promoted what later turned out to be the 'right' theory often did so for reasons that would no longer be accepted today. The reinterpretations imposed by later conceptual developments often mean that the original form of even a successful theory would be unacceptable, or even unrecognizable, to modern scientists. Theories come to dominate scientific thinking only after a period of debate in which various rival hypotheses are put forward. Since we must now accept that neither side in such a debate had access to the whole truth as confirmed by our modern understanding, it becomes possible to admit that even the proponents of the ultimately successful theory were motivated by at least some non-scientific factors. Whatever the role of empirical research in determining which hypothesis was ultimately accepted, the scientists who promote new ideas have usually had their imaginations stimulated by their wider beliefs about the nature and purpose of the universe. Whether those wider beliefs can actually play a role alongside the empirical evidence in determining which theory is *accepted* is a matter of intense debate.

Science and Ideology

This debate is especially active over the question of whether or not scientific theories reflect the social and political framework of the groups that articulate them. Exponents of the 'sociology of knowledge' have long maintained that what is accepted as knowledge within any community reflects the interests of those who dominate that community. This is obvious enough in the case of religious 'knowledge', where the commands of God or the gods often endorse the values of the priests who interpret those commands. But in recent decades sociologists have begun to realize that the theoretical component within the supposedly objective scientific form of 'knowledge' allows the possibility that social values are being reflected even here.

If new theories are inspired by religious or philosophical beliefs, then the fact that these beliefs have an ideological component means that science itself may be promoting models of the world that lend support to particular interest groups within our society. The hypothetico-deductive model allows for non-rational factors in the creation of a new hypothesis, although it implies that the hypothesis is only accepted as long as it continues to pass empirical tests. Kuhn's notion of a paradigm extends this non-rational element to the whole scientific community, implying that all who accept a theory have committed themselves to a particular world view with all its attendant implications. The sociology of science builds upon this to argue that there can be no value-free knowledge. All theories represent models of Nature designed to legitimize a particular set of social values.

The scientists themselves vigorously repudiate this assault upon their supposed objectivity. They insist that — whatever their source of inspiration — new theories only succeed because they offer a better way of understanding how Nature actually works. They claim that the sociologists are promoting a relativistic view of knowledge, in which one theory is as good as any other because theories are accepted solely for ideological reasons. This is self-evident nonsense, they argue, because science could hardly have given us control over the natural world unless its theories offered improved predictive power. It has to be said that the sociologists have found it difficult to explain why science continues to offer better control over Nature. Few historians of science accept a totally relativistic view of knowledge: they agree that theories would not succeed unless they offered some improvement in the modelling of Nature, but they also find it difficult to see the theories as completely value-free models that stand or fall purely by their empirical consequences.

The development of the environmental sciences offers perhaps the clearest evidence that social values do influence what we regard as 'knowledge' of the external world. There are several reasons why this should be so. We may be reluctant to dismiss biology and geology as 'soft' sciences, but the fact that they deal with broad issues defining our beliefs about the origin and purpose of the world allows them to interact far more readily with cultural, and hence with social, forces. There are several levels at which this kind of interaction may take place. Scientific ideas about the origin and development of the world have emerged from traditional religious explanations. If religious beliefs are to some extent conditioned by social interests, then it would be most unlikely that their scientific replacements would be entirely free from the same influence. The claim that Darwin's theory of the 'survival of the fittest' reflects the competitive ethos of Victorian capitalism is the most frequently quoted example of a modern theory that seems to derive its inspiration from a political source (see chapters 7 and 8).

Science also plays a role in helping us to deal with the environment at a practical level, and it thus becomes caught up within the framework of industrial and commercial interests. Geology has both challenged the Genesis story of creation and provided the mining industry with the information it needs to exploit the earth's mineral resources. It would be easy to imagine that these two levels of interaction with society are so far apart that they can be treated separately. But, in fact, it would be a mistake to compartmentalize things in this way. The vast increase in the knowledge of geology and paleontology that made modern theories of the origin of the earth and its inhabitants possible would not have occurred without the incentive provided by the desire to exploit the earth's natural resources. And the new sciences that so obviously challenged the Genesis story of creation were seized upon by the emerging commercial and professional classes to argue that social progress was merely a continuation of natural evolution – the movement known as 'social Darwinism'. If there is an ideological component within science, that component forms a seamless web in which practical and more general social interests combine to define what kind of world view a particular group regards as acceptable.

Modern discussions of the ideological component in the environmental sciences are more likely to be based on our current preoccupation with the overexploitation and pollution of the natural world. Many scientists get their living by making their expertise available to industry, and it is rare to find a scientist whose estimation of environmental risks runs at variance with the interests of his or her employer. The Green movement often accuses industrial scientists of producing biased research and is striving to create an alternative science more in tune with its own concerns. The Greens may claim that this is the only pure science – but we must beware of assuming that it is necessarily purer than the science paid for by industry, since environmentally conscious scientists have their own political interests to defend. Once we admit that science reflects values, then everyone's science must be included in the analysis.

As we search for the origins of modern environmental theories, we must beware of the oversimplified labels that are often attached to ideas. The theory of evolution is often branded as 'materialistic' and seen as a source of modern attitudes of exploitation. Yet a recent study suggests that the emergence of modern ecological consciousness can be traced back to non-materialistic trends in nineteenth-century evolutionism[8]. The same study notes a strong element of ecologism in the thought of some German Nazis, hardly the sort of intellectual ancestors to be welcomed by modern opponents of ruthless exploitation. We must be prepared to acknowledge the complexity of the philosophical and ideological factors that have shaped the emergence of modern attitudes. Few great ideas are inherently 'good' or 'bad' – they acquire these labels as they are exploited by people with good or bad intentions.

There is never a single coherent philosophy or ideology determining the scientific perspective of any historical period. We are very much aware of the ideological tensions within the modern world, and past societies were as complex as our own. Modern historians of science are acutely aware of the conflicting social interests of the various parties making up the scientific community at all stages in its development. Instead of looking for a single 'spirit of the age', they study the particular ideological background of relatively small groups within the scientific community as they engage in debates with their rivals. By focusing on social differences within the scientific community, we gain better evidence for the case that choice of theory is at least in part conditioned by social environment.

Ideology does not determine the direction of scientific thinking in the crude way postulated by an earlier generation of sociologists. Choice of theory can now be seen to depend on a host of personal, social and professional factors, which impose themselves upon the available evidence and shape the individual scientist's effort to interpret the natural world. Science is not a purely objective study of the world, but scientific knowledge cannot be dismissed as a mere figment of the scientists' imaginations, a socially constructed set of opinions designed to reinforce a particular ideology. Scientific discovery is a process that mediates between the scientists' creative thinking – stimulated by a host of cultural factors – and their efforts to observe and interpret the external world. Precisely because there are so many different interests trying to influence the development of science, it becomes possible to eliminate the more obviously inappropriate suggestions in the course of debate. Successful theories cannot be treated as hard facts, and their success may well depend to some extent upon social processes operating within the scientific community. But the advocates of each hypothesis must defend their interpretation against all comers, and this ensures that there is a steady improvement in the sophistication of the models through which we interpret Nature.

The history of the environmental sciences may offer us valuable lessons as we turn to confront the very real problems of the modern world. As we look to science for the information we need to control pollution and other threats to the environment, we must bear in mind the lessons that can be learned from a study of how science functioned in the past. We know that we should not judge the environmental sciences by standards derived from other disciplines. We can see that science is not a totally disinterested search for knowledge, because the quest for rational explanations has always begun from inspirations that were provided by forces lying outside science itself. We know that a scientist's professional and other interests will to some extent determine how he or she will react to a new idea. Science has made a determined bid to become the chief source of knowledge in modern society, and it is all the more important that we should be aware of the ideologically loaded basis for its assumed air of objectivity. The sheer sophistication of modern scientific techniques must not blind us to the fact that those techniques can be used for different purposes depending upon the interests of those who apply them. The more we know about science itself, the better we shall be able to evaluate the advice it offers us as to how we should deal with current problems.

The Ancient and Medieval Worlds

Hunters, herdsmen and farmers have always had a fund of practical knowledge about the world around them. But a particular environment or way of life can shape the 'world view' of a society, so that knowledge of animals and plants is absorbed into a philosophy of Nature that becomes the exclusive property of shamans or priests. The earliest civilizations developed formalized religions defining the relationship between natural and supernatural forces. In Egypt, the regular flooding of the Nile was seen as a gift from the gods, designed to ensure the continuity of human life. In the western tradition it is customary to see Greece as the source of the intellectual foundations that would – via synthesis with Christianity – go on to become the source of inspiration for medieval Europe. Included in the legacy of Greece are the first attempts to create a rational understanding of the world. Some Greek geographers accepted that the Nile flooded because of seasonal rains in the African interior. These attempts to replace supernatural with natural explanations form the basis of the modern scientific tradition.

This conviction that Greek thought represents the foundation of both modern philosophy and modern science has ensured that vast efforts have been made to clarify our understanding of this source. The historian of science has to pick his or her way through a bewildering array of rival interpretations offered by scholars who study the surviving texts in an effort to reconstruct the philosophies of Plato, Aristotle and many other lesser figures. It is still possible for major reinterpretations to be offered even today.

Scholars disagree over the true meaning of the texts they interpret, a procedure made all the more divisive by the fact that the texts themselves are often fragmentary or inconsistent. Historians of ideas may be more interested in what a particular thinker's followers *thought* he said, even though modern scholars are convinced that the followers had misunderstood their teacher. The misinterpretations are themselves part of the process by which ideas are developed, and may turn out to be even more important than the original.

In the orthodox picture of how western thought developed, ancient Greece is seen as a brilliant starting point whose influence was blighted for a millennium or more by the more prosaic attitude of the Romans and by the dogmatism of the Christian churches. The medieval period used to be presented as a 'dark age' in which ancient knowledge was abandoned in favour of primitive superstition. Only with the revival of learning in the Renaissance were the foundations of modern science re-established. This negative image of the middle ages was very much a product of rationalist attitudes that emerged in the eighteenth century and still influence many scientists. The idea that religion and science must be at war with one another has gained considerable currency, and there is thus a tendency for scientists to think that an age dominated by religion is one in which science could not flourish.

Given such an attitude, anyone sympathetic to science, or to rational thought in general, will be inclined to look down on an age when the Greek spirit of free enquiry was subverted by Christian dogmatism. And yet modern historians have become convinced that this approach does not do justice to the medieval period. Whatever the problems caused by the collapse of Rome, medieval Europe was the scene of considerable technical and social innovation. Nor was the medieval attitude to Nature quite so strongly conditioned by religious superstition – although it did contain elements quite alien to our modern way of thought. Scholars were intensely aware of the natural world and tried to build upon the legacy of Greece in an effort to understand what was going on around them.

Evaluating ancient and medieval science is all the more difficult because our modern level of knowledge almost inevitably conditions our choice of what is significant. Scientists are impressed by accurate observation, and will thus tend to pick out and praise those writers from earlier times whose work shows evidence of this ability. Scientists also have a very clear idea of which theories turned out to be 'right', and will again be inclined to emphasize those ancient writings which seem to anticipate modern knowledge. All too often this can degenerate into a search for the 'precursors' of modern theories, a search that seems to rest on the assumption that at least some ancient thinkers must have had glimpses of truths that seem so obvious today. This approach can lead to passages being taken out of context or misinterpreted in order to show that Aristotle, for instance, was *really* an evolutionist. Ancient thought is cut and stretched to fit a Procrustean bed defined by our modern categories of analysis.

This approach becomes all the more dangerous when we begin to look for the origins of modern disciplines such as biology or geology. Terms such as 'biology' and 'geology' did not come into use until the nineteenth century, and their introduction was part of the process by which the disciplinary specializations of modern science were established. To look for ancient Greek 'geology' is to make an artificial construct out of ideas and observations that were actually embedded in a very different framework. Some ancient and medieval naturalists may have reported observations of the earth's crust, but they could not anticipate the emergence of a discipline that would take this field as its sole object of study. The study of the natural environment was undertaken by geographers who were interested in the different regions opened up by exploration, by naturalists who sought to understand the wide range of different creatures that inhabited the earth, and to some extent by medical men who were concerned about the effect that the environment might have on health. Thinkers from all of these backgrounds were influenced by prevailing controversies over the relationship between the earth, humankind and God.

The rational element in Greek thought was unable to banish completely the influence of earlier myths and folk beliefs. The earliest Greek thinkers proposed cosmologies that, although based on natural causes, built upon themes derived from traditional creation myths. Medical teachers drew upon the extensive fund of practical knowledge and folk remedies already available, trying (not always successfully) to winnow out the good from the bad. Natural history faced a similar problem, and Aristotle's pioneering efforts to develop a rational understanding of the animal kingdom nevertheless reflect traditional anthropocentric beliefs. It was taken for granted that humankind represented the model through which all other creatures should be understood.

One approach to history attempts to identify those components of ancient thought which promoted or held back the rise of modern science. The evolutionist Ernst Mayr, for instance, sees the whole history of biology as a struggle against the essentialist philosophy established by Plato[1]. This treated the animals and plants we observe as merely imperfect copies of ideal or archetypical patterns defined and maintained on a supernatural plane of existence. Mayr is quite right to argue that modern Darwinian evolutionism required the breaking down of this essentialist view of Nature. But to reduce the whole history of science to a conflict between good and bad ideas is to interpret the past in modern terms. This may be worth while for the scientist anxious to stress the importance of certain modern ideas, but it is a dangerous policy for the historian who wants to understand the motivations of earlier naturalists.

Many historians of science would argue that we should adopt a more sympathetic view of past ideas, even when hindsight tells us that those ideas turned out to be wrong. Where the scientist seeks to use history as a means of highlighting the key steps that led towards modern knowledge, the historian needs to suspend the judgement of hindsight in order to understand the context within which earlier scientists functioned. We must recognize that people who thought along very different lines to ourselves could nevertheless use their preconceptions to construct workable models of reality. If those models subsequently became *un*workable, we must try to see how alternative ways of thinking were introduced without falling into the trap of assuming that the old hypotheses were simply disproved by factual discoveries.

GREECE AND ROME

The ancient world of the Mediterranean basin did not constitute a coherent, unified culture, nor did it remain static from the origins of Greek thought through into the age of Roman expansion. New ideas and new values were introduced, and the fortunes of alternative philosophies fluctuated in the course of time. The great age of Greek civilization began in the fifth century BC, by which time the Greek cities (which were politically independent) had begun to trade widely around the Mediterranean coast, founding colonies that developed into prosperous cities in their own right. In the history of ancient philosophy, this is known as the pre-Socratic era, the period leading up to the work of Socrates, the teacher of Plato. The fourth century BC was the most brilliant age of Greece, when the cities combined to fight off the Persian empire and established a flourishing culture that nourished Plato and Aristotle, the two most influential thinkers of ancient times. Aristotle was the tutor to Alexander the Great, whose father had unified Greece by force of arms and who now used this power to conquer the Persian empire.

In the Hellenistic age, Greek culture spread out across

the Mediterranean and across Asia, encountering alien influences from the east. Alexander's empire collapsed after his death in 323 BC, and over the next two centuries the separate provinces were absorbed by the expanding power of Rome. Greece itself became a Roman province in the first century BC. By the time of Christ, the empire was established across the Mediterranean basin in the form it would preserve until its collapse in the fifth century AD. The Romans gave stability to the ancient world and continued the intellectual traditions established by the Greeks, although the rise of Christianity paved the way for a new spiritual and intellectual era. Even after the collapse of Roman power in the west, the eastern empire based on Constantinople (modern Istanbul) survived for a thousand years, although its later influence was curtailed by the rising power of Islam.

Interpreting the scientific thought of this developing culture is rendered difficult by the confused and often fragmentary nature of the records that have survived. Many thinkers are known only from reports of their work transmitted by others, who may often be hostile to the views they are reporting. Even where extensive material is available, problems are created by the very nature of the surviving works and by confusions over what is authentic. Plato wrote dialogues in which different views are put into the mouths of Socrates and other speakers, leaving the reader to decide which, if any, of these views were shared by Plato himself. Some ideas are put forward in the form of myths, again leaving the reader to decide how much is to be taken seriously. A number of works were attributed to Aristotle that are now known to be spurious. Even within the accepted works of Plato and Aristotle, there are significant changes of emphasis, and scholars are forced to decide which are the earlier writings and which the later. Given all this confusion, it is small wonder that, even today, historians are still advancing new interpretations of what these thinkers actually intended to say.

From this welter of fact and opinion we need to extract an outline of the various philosophies of Nature advanced in the ancient world. Greek natural philosophy emerged from a tradition of seeking rational explanations of how the world came to be as we see it today. Certain basic ideas were eventually established that would underlie many later developments in European thought. The doctrine of the four elements (earth, air, fire and water) was formulated along with parallel views about the composition of the human body. Eventually, it was accepted that the earth was a sphere at the centre of a cosmos that also had a spherical structure, the sun, moon and planets being carried around embedded in a transparent ether.

Some Greek thinkers proposed 'cosmogonies' or hypotheses to explain the development of the whole material universe. These ideas seem to anticipate the motivation (and very occasionally the content) of modern theories of geological and biological evolution. But for all their efforts at rational interpretation, the Greek cosmogonists were nevertheless drawing some of their inspiration from traditional creation myths. Many thinkers preferred a cyclic rather than an evolutionary view of change, while Aristotle held that the universe is eternal. There was a prevailing belief that the cosmos was established according to an orderly pattern, and that human life functioned within the same pattern as the wider world. This was consistent with the Christian concept of divine creation. A few Greek scholars suggested radical alternatives to this anthropocentric viewpoint. The astronomer Aristarchus suggested that the earth is not the centre of the whole universe, while atomists such as Epicurus and Lucretius disputed the claim that the order of Nature was divinely constructed.

Rival Cosmologies

Greek philosophy began with the Ionian school, the first member of which was Thales of Miletus (*c.* 640–546 BC).

Miletus was a Greek trading centre on the coast of Asia Minor, and the outburst of critical thinking that marks the origins of western philosophy may have been stimulated by the more cosmopolitan and more materialistic atmosphere of a society based on trade. But the Ionian philosophers were concerned with issues that had their origins in the creation myths of antiquity. They wanted to apply rational thought to the question of the origin and underlying nature of the universe. Was there some fundamental unity underlying the apparently ceaseless changes exhibited by Nature, and, if so, what caused the superficial rearrangements that we recognize as change?

Thales suggested that water was the basic element of which all other things are composed. He believed that the earth floated in water, and explained earthquakes as the result of waves rocking it. In Greek mythology, the god of the sea, Poseidon, was also responsible for earthquakes. Thales' explanation was thus a rationalization of earlier beliefs. His follower Anaximander suggested that thunderbolts were caused by the wind, and lightning by splitting clouds. Anaximander claimed that all matter originated from a basic substance, the 'boundless', and proposed a theory to explain the origin of the cosmos. He thought that the universe began as a seed that separated itself from the boundless and grew into its present state. Living things were generated when the sun acted on moisture. Human beings had come from some earlier animal form, a supposition required by the fact that infants produced without parents would not have been able to survive. The poet Xenophanes (*c.* 570–475 BC) argued that the earth was generated from water, offering fossils as evidence that the whole surface had once been like mud. But where Anaximander thought that the whole earth would eventually dry up, Xenophanes adopted a cyclic world view in which the earth will eventually dissolve back into water.

The philosopher Parmenides argued that the mind, not the senses, should be used to uncover the underlying principles of Nature. He argued that, since there must be a single underlying substance, all the changes revealed by the senses must be illusory. To avoid this implication, Empedocles (c. 500–430 BC) developed the idea that there must be more than one element, change being made possible by the rearrangement of the elements making up ordinary things. The four elements were derived from common-sense experience and were defined as earth, air, fire and water. This view was to be taken up by Aristotle (who added a fifth element for the heavens, the ether) and became the basis of most European thought about the fundamental constitution of matter until the emergence of modern chemistry.

Empedocles was also concerned with the processes by which the world was brought into its present state. He postulated two antagonistic forces, love and strife, but went beyond this to describe a historical cycle of change for the world as a whole. The present order of things is dominated by strife. In the beginning the earth had generative powers that allowed it to produce living structures. These were formed essentially at random – individual organs (arms, legs, etc.) wandered around and combined in various ways to produce animals. Most of these were monstrously deformed or unable to reproduce. They soon died out, leaving only the happier combinations as the founders of the modern species. This speculation has been hailed as an anticipation of the modern theory of evolution by natural selection. In fact, there are significant differences, since Empedocles postulated only an episode of creation in the earth's youth, not an ongoing process within modern species. His emphasis on development through trial and error can be seen as a genuine alternative to the view that the universe is a structure with a rational design imposed on it by supernatural forces. But like Xenophanes, Empedocles adopted a cyclic view, holding that the present world would be destroyed before a new wave of creation dominated by love would begin.

This view of the cosmos as an unstructured system was developed further by the atomists, especially Democritus (*c.* 470–370 BC). Instead of seeing Greek atomism as the forerunner of modern atomic theory, we should recognize its purpose as an antidote to the general belief that the world is somehow designed by wise and benevolent gods. In this world view, the universe consists of the void (empty space) through which indivisible particles of matter (the atoms) move under the influence of purely physical forces. Atoms can combine together to produce temporarily stable structures, but these invariably break up in time because there is no guiding principle to guarantee the stability of any particular form. Democritus supposed that the earth is but one world among many produced by this universal cycle of creation and destruction, and was quite prepared to accept that each world was unique. The earth itself will ultimately grow old and die. Nature is driven by mechanical forces in a totally inevitable manner, but, because there is no guiding force, the resulting combinations seem purely random to observers who cannot perceive the underlying causes.

When developed by Epicurus (342–271 BC), atomism became the foundation for an atheistic philosophy that would provide a major alternative to the more conventional world view. The Epicurean approach survived most effectively through its incorporation into the *De Natura Rerum* (*On the Nature of the Universe*) by the Roman poet Lucretius (*c.* 100–55 BC). Like all ancient poets, Lucretius had a great love of Nature and an interest in rural life. But he refused to accept that the earth was a system designed for people. Large parts are unsuited to cultivation, while even in the best areas much effort is needed to raise a crop. Many animals are a menace to the human race. Lucretius followed Epicurus in depicting the earth as a mother that gave rise to the earliest animals by spontaneous generation. He took it for granted that smaller animals are still produced in this manner today, but speculated that in the past even the larger animals could have been born directly from the earth[2]:

> Even now multitudes of animals are formed out of the earth with the aid of showers and the sun's genial warmth. So it is not surprising if more and bigger ones took shape and developed in those days, when the earth and ether were young ... There was a great superfluity of heat and moisture in the soil. So, whenever a suitable spot occurred, there grew up wombs, clinging to the earth by roots. These, when the time was ripe, were burst open by the maturation of the embryos, rejecting moisture now and struggling for air.

But as in Empedocles' theory, this is seen as an experimental process – many of the early productions were monstrous and could not survive. Only the strong and the swift went on to become the originators of the modern species:

> In those days the earth attempted also to produce a host of monsters, grotesque in build and aspect – hermaphrodites, halfway between the sexes yet cut off from either, creatures bereft of feet or dispossessed of hands, dumb, mouthless brutes, or eyeless and blind, or disabled by the adhesion of their limbs to the trunk, so that they could neither do anything nor go anywhere nor keep out of harm's way nor take what they needed. These and other such monstrous and misshapen births were created. But all in vain. Nature debarred them from increase.

The earliest humans were formed in the same way and (contrary to the tradition of a golden age) they originally lived like the beasts of the field by gathering nuts and fruit. But again this is no simple evolutionary system, since the Epicureans believed that the earth, having been born, must grow old and die. Lucretius appealed to the evidence of weathering and erosion to show that the surface of the land was subject to decay.

A rival philosophy of life was developed in a body of medical writings attributed to the physician Hippocrates. These present an essentially materialistic view of the body's workings, exemplified in the decision to treat the 'sacred disease' (epilepsy) as just another bodily malfunction. Building upon the concept of the four elements, Hippocrates and his followers postulated four bodily 'humours', black bile, yellow bile, blood and phlegm. The body was healthy when the humours were in a natural state of balance, and the physician's job was to help Nature restore the balance if it had become disturbed. This medical philosophy had implications for the study of the environment because the text *Airs, Waters, Places* shows how health can be influenced by the surrounding conditions. Different locations give different medical problems, while the inhabitants of different regions have acquired physical characteristics that are the cumulative result of generations of exposure to their native environments.

The Hippocratic tradition adopted an essentially materialistic view of the body's functions, but took for granted the power of the body to restore itself to a natural state of balance and health. It assumed that natural systems are not random collections of atoms – they are harmoniously designed, self-sustaining and self-correcting entities. This viewpoint, natural enough in a physician, was more compatible with a philosophy that supposed that Nature is a purposeful system designed according to a rational pattern. The belief that the world was designed by gods goes back to the civilizations of Egypt and Mesopotamia, and the materialistic tendencies of the pre-Socratic philosophers were never able to break the hold of this essentially religious view of the world. The philosopher Anaxagoras used astronomical regularities to argue that the structure of the cosmos was established by Mind, i.e. by a designing intelligence. The same theme was developed in the second half of the fifth century BC by Diogenes of Apollonia, who extended the argument to include the structure of the terrestial world. It is an indication of the uncertainty surrounding many of the ancient writers that modern scholars cannot even be sure which of the ancient Apollonias (one in Crete and one on the Black Sea coast) was Diogenes' birthplace.

A clear expression of the view that Nature must have been designed by a superintending intelligence occurs in the *Memorabilia* of Xenophon, a pupil of Socrates. His arguments anticipate those which would be used by Christian exponents of natural theology right through into the nineteenth century (see chapters 5 and 7). The earth is seen as a physical system intended for the sustenance of humankind, its climate, and the animals and plants it supports, being carefully designed to benefit us in our daily lives. In its original form, this argument highlights both the beauty and the usefulness of the natural world, interpreting those characters as explicable only in terms of design by an intelligent supernatural creator who cares about human welfare. The basic assumption that the beauty and complex structure of living things can be explained only in terms of supernatural design would be transmitted almost unchanged from the Greeks through into the Christian world.

This teleological view of Nature (teleology is explanation in terms of the purpose a thing is intended to fulfil) was to become a characteristic of the Stoic philosophy. Zeno, the founder of this school, saw Nature as a craftsman-like force that generated order and purpose in the world. Panaetius (*b. c.* 185 BC) and later Posidonius both used geographical knowledge to extend the design argument, exploiting the view that humankind is a part of Nature and exploring the relationship between human life and the environment. The harmony of all natural relationships betokens their divine origin, and the human mind is able to understand and copy the designs that are built into Nature. The invention of the ship's rudder, for instance, is based upon observations of the tails of fish. Here we have a sense of the natural world as a beautiful and complex system designed by the gods for our benefit, which is dignified by our presence in it and our ability to understand the complexities of what has been prepared for us.

This theme was taken up in the second book of *De Natura Deorum* (*On the Nature of the Gods*), written by the Roman orator Cicero (106–43 BC) shortly before he was executed after choosing the wrong side in the civil wars following the assassination of Julius Caesar. Cicero's book is a dialogue in which a Stoic, Balbus, develops the theme that Nature cannot be a collection of accidental occurrences. Balbus emphasizes a comparison between human and divine workmanship – divine craftsmanship is, of course, superior, but we understand the purpose of the universe by analogy with the purposefulness of our own machines. The animals that can be domesticated were obviously designed for human use: the strong neck of the ox is intended for the yoke. In this way human control of Nature is made to seem part of the divine plan, not an interference. The wild animals too have been provided with all the physical structures appropriate to their habits. Each species is carefully adapted to its way of life, the predators with claws and teeth, etc. The earth itself is designed to sustain life, with the rains providing water and the winds refreshment. The whole system was designed for humankind – even those creatures which we cannot use are part of a system that only we can understand.

Another Roman writer, Seneca, argued that rivers are designed for navigation and that mineral veins had been carefully positioned in the earth for our use. The Stoic viewpoint is also visible in the encyclopedic *Natural History* compiled by Pliny the Elder (AD 23–79). It may be that Stoicism was the source of Christianity's assumption that the earth has been created for humankind, reinforcing a particularly anthropocentric interpretation of the Old

Testament view than humankind was given dominion over Nature.

Plato and Aristotle

The two greatest thinkers of the Athenian period established intellectual traditions that were to survive into the Christian era and form the basis from which modern science began. Plato (429–347 BC) was a student of Socrates who founded a school known as the Academy in an olive grove on the outskirts of Athens. Aristotle (384–322 BC) began as a disciple of Plato, but broke away to found his own philosophy and his own school, the Lyceum, after serving for a time as tutor to the young Alexander the Great. They extended the teleological view of Nature in different ways, providing it with new conceptual foundations that established themselves as distinct philosophical traditions. Of the two, it was Aristotle who was most profoundly concerned with natural history and the attempt to explain the nature of living things.

Plato's overall philosophy has often been regarded as hostile to scientific investigation of Nature. He held that the whole purpose of philosophy was to lead the mind beyond the material world revealed by the senses to the world of pure ideas. True knowledge consisted of recognizing the necessary connections between ideas, and could not be derived from the study of material things. The objects revealed to us by our senses are merely inferior copies of the eternal ideas. It is certainly necessary for the scientist to make abstract models of Nature; the astronomer, for instance, must construct mathematical theories to explain the motions of the planets. But those theories must be tested against the observed motions, and Plato sometimes implies that even this concession to the world of the senses is unnecessary. Plato said little on the subject of natural history, but it is easy to see how his concept of an ideal world could become the basis for an essentialist view of species. If real animals and plants are merely copies of ideal forms that exist in a supernatural world, then the essence of a species is defined by its ideal pattern. Fixity is guaranteed by the stability of the ideal form, since no amount of change among the individuals inhabiting the material world can ever affect the supernatural pattern upon which they have been modelled.

The ultimate end of Plato's philosophy was to subordinate the world of the senses to the world of ideas. Up to a point, the ideal world can be seen as an abstract, refined version of the material world of common sense, but Plato also implies that the ideas have their basis in the absolute, the ultimate good – which would later be identified with the God of Christianity. In fact, Plato's God has significantly different characteristics, but the link was made plausible by the fact that Plato himself proposed a cosmogony to explain the origins of the material world.

In the *Timaeus* he expounds a 'myth' of creation that nevertheless seems to represent his own views on the relationship between the ideal and material worlds. The creative force is a demiurge that imposes order on the recalcitrant world of matter. The demiurge does not create matter, nor can it evade the limitations inherent to the nature of matter. It can only use the material substratum to give physical expression to the forms that exist in the ideal world. Material Nature is thus a compromise – but Plato insists that it is the best possible compromise. The physical universe is, at bottom, an expression of the divine mind. In particular, Plato implies that the demiurge could not refuse existence to any idea that could be conceived on the mental plane. All possible forms must have a real existence; there can be no gaps in Nature, no missing species whose form exists only in the ideal world. This principle of 'plenitude' was to have a profound effect upon later thought about Nature.

Although the *Timaeus* deals with the creation of the world, its approach distances Plato from the cosmogonical theories proposed by the pre-Socratics. The story has been idealized to such an extent that it can scarcely be called a *historical* account of how the world was formed. What matters to Plato is the relationship between the ideal and the material world – the actual order in which the events of creation take place is not really important. To the extent that the *Timaeus* does advance a philosophy of change through time, it is based on a cyclic concept in which the world is periodically destroyed by catastrophes.

Plato's story of the destruction of Atlantis – a mythical continent that was supposed to have existed in the Atlantic Ocean – is part of this cyclic approach. The story of how the Greeks invaded Atlantis to destroy the military threat posed by its great civilization is told to the traveller Solon by some Egyptian priests[3]:

> At a later time there were earthquakes and floods of extraordinary violence, and in a single dreadful day and night all your fighting men were swallowed up by the earth, and the island of Atlantis was similarly swallowed up by the sea and vanished; this is why the sea in that area is to this day impassable to navigation, which is hindered by mud just below the surface, the remains of the sunken island.

However incidental to the main theme of Plato's philosophy, this story of catastrophe reinforced a general belief in periodic upheavals that was to exert a pervasive influence on the earth sciences. The myth also justified the Greeks' reluctance to travel beyond the Pillars of Hercules (the Straits of Gibraltar) out into the Atlantic.

Aristotle's break with Plato came about at least in part because he wished to develop a philosophy that paid more attention to knowledge of the material world. Natural history played a major role in his research programme, and at one time it was fashionable to assume that it was the study of living things that inspired Aristotle to develop his essentially teleological approach to explaining natural processes. Scholars now believe that his works on natural history were written at a late stage in his career. They were an effort to put his philosophy of knowledge to work in practice, and for this reason we must be careful not to judge his natural history by modern standards.

Aristotle offered no cosmogonical theory because he held that the world was eternal. But he had clear views on the structure of the universe and his ideas remained popular through to the fifteenth century. He stressed the absolute distinction between the earth and the heavens. The earth was a sphere at the centre of the universe, with the sun, moon and planets carried around it by concentric spheres of a transparent ether. This fifth element was quite different to the four terrestrial elements: it moved naturally in circles, while earth, air, fire and water had a natural tendency to arrange themselves with respect to the centre of the world. Aristotle adopted the view that the earth could be divided into *klima* or zones rigidly defined by the amount of the sun's heat they received. Life was only possible in the temperate zones: the equatorial zone was too hot to support life, while the polar zones were too cold. The southern temperate zone was forever isolated from the northern, and many people believed that it was uninhabited.

Although the basic structure of the earth was eternally fixed, Aristotle realized that there are natural processes that are constantly changing its surface features. His *Meteorologica* describes these processes in some detail. He adopted the common view that earthquakes and volcanoes were due to the wind circulating in underground caverns (Aeolus, the god of winds, lived under the Aeolian Isles, which are volcanic). He knew that rivers were produced from the rain that fell as mountains condensed atmospheric moisture, and realized that the fertile land of Egypt was built up from deposits brought down by the Nile. Although fossils gave evidence that some parts of the earth had once been covered by water, there was no general process of drying up because elsewhere new land was being created. To explain this phenomenon, and also why the fertility of some areas has changed through time, Aristotle postulated a cyclic pattern in the rainfall, with 'great winters' in which the rainfall was higher than normal for long periods of time[4]:

> Men whose outlook is narrow suppose the cause of such events to be change in the universe, in the sense of a coming to be of the world as a whole. Hence they say that the sea is being dried up and is growing less, because this is observed to have happened in more places now than formerly. But this is only partially true. It is true that many places are now dry, that formerly were covered with water. But the opposite is true too: for if they look they will find that there are many places where the sea has invaded the land. But we must not suppose that the cause of this is that the world is in a process of becoming. For it is absurd to make the universe to be in process because of small and trifling changes, when the bulk and size of the earth are surely as nothing in comparison with the whole world. Rather we must take the cause of all these changes to be that, just as winter occurs in the seasons of the year, so in determined periods there comes a great winter of a great year and with it excess of rain.

For Aristotle, the universe itself could not be formed by a natural process, but he was anxious to explain all other changes in which structures, especially living structures, are formed. He drew a distinction between potentiality and actuality. The seed is not a tree, but it has the potential to grow into one, and the tendency for life to reproduce itself shows that Nature contains organizing powers that will impose form on matter. The form of a species is not guaranteed by a supernatural pattern existing in the world of ideas. The species continues because matter cannot exist in an unformed state, and the existing form of the species naturally perpetuates itself through the reproductive process.

Aristotle understood all constructive change in terms of four causes: the formal (the structure being created), the material (the matter upon which form is being imposed), the efficient (the actual force working on the matter) and the final cause (the purpose for which the new structure is created). Teleology (explanation in terms of purpose) was essential to his philosophy, but he did not neglect the natural limitations that must be imposed on how the purpose is fulfilled. Not surprisingly, he devoted a great deal of his biological research to the process of reproduction, making extensive studies in embryology. Here the final cause is the perpetuation of the species, while the formal cause is the structure of the parent's body that is being duplicated.

It is his pioneering role in the establishment of scientific natural history that concerns us here, though. Aristotle's *Historia Animalium* contains a vast fund of information about a wide range of species, much of it gained through original observation and dissection. Over 500 species are mentioned, including 120 fish and 60 insects. Much of his research into marine animals was conducted in the lagoon of Pyrrha on the island of Lesbos, where he spent two years (344–342 BC). Modern biologists are impressed by Aristotle's skills as an observer because he made some discoveries that were not confirmed until the nineteenth century. He observed that in one species of dogfish the young were attached to a placenta-like structure within the mother. This claim was rejected as nonsense by generations of naturalists until confirmed by Johannes Müller in 1842. Aristotle's description of mating in cephalopods (squid and octopus) also provided information not verified until much later[5]:

Some assert that the male has a kind of penis in one of his tentacles, the one in which are the two largest suckers; and they further assert that the organ is sinewy in character, growing attached right up to the middle of the tentacle, which is admitted into the nostril of the female.

In this case Aristotle did not make the observation himself; he is probably reporting information gleaned from the local fishermen.

But what was the purpose of all this research? Aristotle has often been regarded as the founder of biological taxonomy, and it was assumed that his observations were designed to reveal characteristics that would allow the naturalist to construct a comprehensive system in which all the species would be grouped according to their degrees of relationship. He writes of species being grouped into genera, and it is easy to imagine that his purpose in this was – as in modern biology – to construct a hierarchical classification of groups within groups. But modern scholars such as David Balme have shown that there is no comprehensive taxonomy in Aristotle's writings. The same species is sometimes described within different genera, and the reason for this is that Aristotle was using the term *genus* (pl. *genera*) to define any group that could be subdivided on the basis of contrasts between its members. He took this approach because his intention was not to classify, but to discover which characteristics of living things were the more basic.

Aristotle wanted to explain the observed structure of species by showing that some characters were possessed as necessary consequences of more fundamental characters. The natural history thus follows the methodology outlined in his *Posterior Analytics*; the grouping of differences is seen as a prelude to the study of causes. Aristotle was prepared to accept that some characters of each species were merely a material consequence of others. Only the most basic were defined by the form of the genus transmitted through the reproductive process. The characters distinguishing the species within the genus were fixed by the teleological requirement that the organism must function in a way that is adapted to its environment. For this reason, Aristotle's teleology did not require the supposition that the species was modelled on an ideal archetype, nor did it entail the fixity of species except as a consequence of the fact that individuals have a natural tendency to reproduce their own likeness.

Aristotle did not altogether succeed in eliminating false information derived from folk beliefs about animals, nor was he able to escape the largely anthropocentric viewpoint of his culture. Humankind was still the standard of comparison against which all other animals were measured. Humankind stood at the head of Nature, with all the lower creatures arranged in a scale beneath, leading down to the lowest, the most primitive forms of life. Aristotle also shared the popular fascination with species that seem to 'fall between two stools'. The seal bridges the gap between land and sea creatures, just as the bat links the land animals and those of the air. 'Zoophytes' (corals and other marine forms that grow into branches) are intermediates between plants and animals. Although Aristotle never openly supported the view that there is a single, linear scale of organization, later writers would combine various aspects of his thought to give an idea that was to shape the European view of Nature through into the eighteenth century. If living Nature forms a hierachy with humankind at the top, and if there are no 'gaps' that are not filled in by some intermediate form, then we can imagine a 'great chain of being' – a single, linear, unbroken sequence of species uniting the whole of creation.

At a more immediate level, Aristotle paved the way for further research into natural history by his pupils. His school, the Lyceum, seems to have offered both lectures and facilities for research. He was succeeded as head of the Lyceum by Theophrastus (*c.* 373–*c.* 275 BC), who extended the research programme to include plants and minerals. Theophrastus was more willing to admit that some aspects of Nature served no rational purpose, but he did not seek to replace the doctrine of final causes. His botanical works include descriptions of many species, along with information gleaned from farmers and gardeners about the habits and uses of plants. Book IX of his *Historia Plantarum* deals with the medical uses of plants, suggesting that there was still a strong link between botanical studies and the medical profession. His work *On Stones* describes and classifies many minerals, and makes a serious effort to free the subject from the superstitions surrounding it.

Late Antiquity

The naturalists and geographers of the Hellenistic era had to accommodate themselves to the vast range of additional information about the world revealed by Alexander the Great's conquests. The Roman empire also covered a vast territory, generating masses of information that would have to be incorporated into the programme that Aristotle had established. Eratosthenes of Alexandria (*d.* 192 BC) measured the circumference of the earth and gave a figure that may have been quite accurate (there is uncertainty over the modern value of the unit of length he used). He accepted the idea of an uninhabitable equatorial zone, although this was challenged by Posidonius. Posidonius also offered a significantly smaller estimate of the earth's circumference – a point that would resurface at the time of Columbus to influence debates over the feasibility of sailing westwards to the Indies.

The geographer Strabo (*c.* 64 BC–AD 24) wrote at a time when the emperor Augustus was encouraging trade throughout the empire. He described much of the known world, including detailed descriptions of ore deposits and mining techniques in Spain. He also gave accounts of earthquakes and volcanoes, although he accepted the view that they were caused by underground winds. Strabo's *Geography* represents the high point of ancient knowledge of the earth and its resources, but its inability to suggest new explanatory schemes suggests that the amount of information becoming available was almost too great for anyone to process effectively. In the second century AD the astronomer Ptolemy wrote a geography that described much of the known world. Ptolemy accepted that the torrid zone was too hot to support life, but postulated an unknown land mass to the south of the Indian Ocean, the 'terra australis incognita', that was still taken seriously in the seventeenth and eighteenth centuries.

Botany still flourished as a result of its link with medicine, as illustrated by the *De Materia Medica* of Dioscorides (*fl.* AD 60–77). As a physician to the Roman army under Nero, Dioscorides travelled widely in Asia and gathered a vast body of information on plants and their medical uses. He promoted the use of illustrations to help identify plants, founding a tradition that would continue through into the herbals of medieval times. Yet there was an overall stagnation in efforts to comprehend the natural world as a whole. Another writer of the first century AD, Pliny, collected his *Natural History* in an effort to preserve what was known from the past in an age that no longer made an effort to extend the boundaries of knowledge. He tried to investigate things personally, and met his death while studying the eruption of Vesuvius that buried Pompeii and Herculaneum. But he copied many passages almost wholesale from Theophrastus and other authorities. He also tended to be less critical of the folk beliefs that are mentioned by his sources, and he included many fabulous creatures in his descriptions of the animal world. His work shows how ancient science eventually became bogged down through an exaggerated respect for existing authorities and a sense of the enormous scope of the subject.

The theoretical developments of late antiquity explored ideas that we would now regard as harmful to the study of the natural world. Neoplatonists such as Plotinus and Macrobius were essentially philosophers and theologians, whose only interest in the world was as a symbol of divine perfection. The concept of the 'chain of being' was formulated to express the view that God must necessarily create all conceivable forms of life from humankind down to the lowest and most imperfect. The principle of plenitude was combined with the hierarchical view of Nature and the conviction that there are intermediate forms linking all natural groups. On this view of Nature, 'imperfection' was merely an expression of the need to complete the work of creation at all levels. The suffering caused by carnivorous animals was a necessary component of the universal pattern. Along with uncritical compilations of ancient knowledge, such ideas would eventually be transmitted into Christian Europe and would shape the attitudes of medieval thinkers towards Nature.

THE MEDIEVAL PERIOD

Scholars no longer dismiss the middle ages as a stagnant period in the development of western culture. True, the 'dark ages' following the collapse of the Roman empire saw the loss of much ancient knowledge, and the early Christian Church did not encourage the study of Nature for its own sake. The traditions of Greek science were explored more actively by Muslim scholars, whose work eventually helped to transmit the legacy of the ancient world to the new culture developing in medieval Europe. The first sign of a revival of learning in western Europe was the twelfth-century 'renaissance' sparked off by the translation of Aristotle and other classical thinkers into Latin. Universities replaced monasteries as great centres of learning. Some authorities see the major developments in later medieval thought as paving the way for the true Renaissance of the fifteenth and sixteenth centuries.

Throughout this period there was an intense awareness of

Nature on the part of artists, craftsmen and those who had to make their living in the fields and forests. The clearance of the forests to make room for the expansion of agriculture marked a major advance in humanity's ability to control the environment and to produce permanent effects on the earth itself. Christianity assumed that God had given humankind dominion over the earth, adopting the Stoic view that Nature has been created for our benefit. Our fall from grace in the garden of Eden may have precipitated a decline from the earth's original paradisiacal state, but through hard work we could still hope to fulfil some of the Creator's intentions. As yet, though, the Christian Church did not encourage a purely exploitative interpretation of humankind's presumed dominion over Nature. Nature could be tamed, but it should not be destroyed for personal gain.

Christendom was only part of the known world. Most medieval thinkers had direct experience of Europe, although this would have been supplemented by travellers' tales and by those ancient geographies that had survived. Maps of the world were often highly stylized affairs in which the inhabited earth was divided by a T shape into the three continents of Europe, Asia and Africa. Jerusalem was positioned more or less at the centre. It would be wrong, however, to dismiss this as a period of total stagnation in Europe's knowledge of the rest of the world. The travels of Marco Polo to China in the thirteenth century represent only the best-known example of medieval exploration. Some modern scholars believe that we consistently underestimate the efforts by medieval traders and sailors to explore the world. Perhaps some of the stories that America was reached before Columbus did so should be taken seriously. If this is so, the fifteenth-century expansion that culminated in the opening up of the New World may represent only a continuation of earlier trends.

The activities of the traders did not necessarily interact with the world of educated men, however – Marco Polo, for instance, was ignorant of prevailing geographical theories.

Medieval thinkers had an attitude towards Nature that was quite different to our own. It is easy to dismiss their work as unscientific because they included, for instance, fabulous beasts in their natural histories. But we are now dealing with a time in which the study of Nature was almost by definition subordinated to the worship of God. The early fathers of the Church were suspicious of ancient learning because it tended to deflect people's attention away from salvation. If Nature had anything to offer at all, it was as a way of illustrating the nature of God. In such a climate of opinion, it was difficult not to take seriously the ancient legends of beasts whose behaviour seemed to carry a moral lesson. Those lessons were as much a part of natural history as the description of the beasts themselves – indeed, from the moralist's viewpoint it did not matter if the beast was purely fabulous.

People who thought in this way could nevertheless observe Nature closely. Admittedly, observations that we now know to be false were repeated over and over again. Yet this may be due to the fact that writers on natural history, and the artists who illustrated their works, were often copying classical texts that dealt with animals and plants unknown in northern Europe. In the later middle ages, the mythical dimension was coupled with a delight in Nature and a willingness to observe those plants and animals which were actually available for study. We now pick out the true observations as evidence that certain thinkers were transcending the limitations of medieval culture. But this is the judgement of hindsight, and it scarcely does justice to the long process by which European culture changed its value system to admit the possibility of a scientific natural history.

The attitude of the early Church towards the study of Nature was influenced by the Neoplatonism favoured by St Augustine (AD 354–430). The *Timaeus*, in particular, was used to develop the theme that the material world is but a copy of an ideal archetype conceived by the Creator. Even here, the emphasis was on the moral lesson to be learned from the marriage of Platonism and the Genesis story – only later were the more detailed aspects of Plato's creation myth studied at length. The true purpose of philosophy was to draw the mind towards God, and the study of Nature had value only to the extent that it contributed to this end. Many ancient texts were lost in the barbarian invasions.

Bestiaries and Herbals

Those who retained an interest in Nature had to rely on compilations and encyclopedias preserving fragments of earlier learning. Pliny's *Natural History* was available, with all its fabulous stories that could be absorbed into the search for moral lessons. The fourth-century *Physiologus* provided another fund of stories about the behaviour of animals, each with its moral value explained. From this developed the 'bestiary' or book of beasts, large numbers of which were written throughout the early medieval period. These works mixed real and imaginary animals, and attributed imaginary behaviour to the real ones. Even the more realistic characteristics were given a moral lesson, thus[6]:

> The wild goat (Caprea) has the following peculiarities: that he moves higher and higher as he pastures, that he chooses good herbs from bad ones by the sharpness of his eye; that he ruminates these herbs, and that, if wounded, he runs to the plant Dittany, after reaching which he is cured.
>
> Thus good Preachers, feeding on the law of God and on good works, as if delighting in this sort of pasture, rise up from one virtue to another.

Some historians have dismissed the bestiaries as having contributed nothing at all to the development of natural history, but recent studies suggest that they should be taken more seriously. The *Physiologus* may have been compiled in North Africa, and would thus have included real animals that were unknown in Europe. The unicorn, for instance, may have been an antelope with a missing horn. When the writers and illustrators of the bestiaries were dealing with animals known to them directly, their work shows a much greater degree of realism. Later medieval writers were able to incorporate more accurate observations into the moralizing tradition established by the bestiaries. Alexander Neckham (1157–1217) adopted a slightly more critical attitude towards the tales derived from Pliny and other ancient sources in his cosmological treatise *De Naturis Rerum*.

Herbals were intended to give practical descriptions of plants useful to physicians. There was seldom any attempt at classification – the plants were usually described in alphabetical order. Illustrations were widely used, especially in the *Ex Herbis Feminis* incorrectly attributed to Dioscorides. The usual assessment of historians has been that – like the bestiaries – the illustrations show little familiarity with Nature, becoming gradually stylized and debased so that the plants are hardly recognizable. Some degradation may have been due to the fact that classical herbals described species from the Mediterranean region that were unknown to the scholars and artists of northern Europe. But recent assessments suggest that the illustrations are not as bad as has been claimed, since many of the plants can be identified from them. Things began to improve after the twelfth century, with increasing attention being paid to naturalistic representation. The thirteenth-century herbal of Rufinus is a good example of improved observation.

The Revival of Scholarship

The greatest stimulus to the study of Nature in the twelfth century was the translation of Aristotle and other ancient philosophers into Latin. At a time when these classics had been lost in the west, they had been translated into Arabic and studied by Muslim scholars. Latin editions were now made available (often from the Arabic, not the original

Greek), and the new learning began to spread from centres of translation in Spain and Sicily into northern Europe. Aristotle was at first greeted with suspicion, in part because his theory of the eternal universe was at variance with the Genesis story of creation. But gradually a number of thinkers were able to effect a reconciliation, accepting the Aristotelian cosmology with the proviso that the structure of the universe thus described was created at a fixed point in time by God.

Among the scholars who turned Aristotelianism into the dominant philosophy of the later middle ages were Albertus Magnus (*c.* 1200–80) and, more especially, St Thomas Aquinas (*c.* 1224–74). They were able to define a new relationship between theology and the study of Nature, accepting that the world was the work of divine providence although its day-to-day activity was governed by natural causes. This left human reason free to study Nature, always bearing in mind that the results must illustrate the divine origin of the material world. The teleological aspect of Aristotelianism fitted into this approach, and St Thomas was able to use the orderliness of Nature as one of the proofs of the existence of God in his *Summa Theologiae*.

Although St Thomas provided the most original adaptation of Aristotelian philosophy, it was Albertus Magnus (Albert the Great) who showed the greater interest in natural history. Albert joined the Dominican order and travelled widely throughout Europe, both as a scholar and on diplomatic missions. He wrote extensively on mineralogy, botany and zoology, drawing upon the existing medieval literature and trying to incorporate the knowledge of the time into an Aristotelian framework.

Albert's *De Mineralibus et Rebus Metallicis (On Minerals and Metals)* drew upon earlier lapidaries (books on stones) but created a more coherent science of minerals. Where many of his contemporaries saw fossils as the remains of Noah's flood, Albert followed the Arab scholar Avicenna in arguing that they were the mineralized remains of once-living animals[7]:

> It seems wonderful to everyone that sometimes stones
> are found that have figures of animals inside and
> outside. For outside they have an outline, and when
> they are broken open, the shape of the internal organs
> is found inside. And Avicenna says that the cause of
> this is that animals, just as they are, are sometimes
> changed into stones, and especially salty stones . . .
> And in places where a petrifying force is exhaling, they
> change into their elements and are attacked by the
> properties of the qualities which are present in those
> places, and the elements in the bodies of such animals
> are changed into the dominant elements, namely
> Earth mixed with Water; and then the mineralizing
> power converts the mixture into stone, and the parts
> of the body retain their shape, inside and outside, just
> as they were before.

Albert's account follows the traditional scheme of the four
elements and accepts that there is a mineralizing force
capable of altering a body's qualities. But this is not an
anticipation of the modern view, since he compares the
force to the reproductive power of living things and also
invokes an astrological influence[8]:

> . . . when dry material that has been acted upon by
> unctuous moisture, or moist material that has been
> acted upon by earthy dryness, is made suitable for
> stones, there is produced in this, too, by the power
> of the stars and the place . . . a power capable of
> forming stone – just like the productive power in
> the seed from the testicles, when it has been drawn
> into the seminal vessels; and each separate material
> has its own peculiar power, according to its specific
> form. And this is what Plato said – that the heavenly
> powers which act upon things in nature are poured
> into matter according to its merits.

Although fossils are thus presented as the remains of living things, they are not seen as clues about the formation of the earth's crust. The invocation of reproductive powers in this context created confusion by implying that fossils grew within the rocks from seeds.

Albert's book is a treatise on mineralogy, not geology. His primary concern was the properties of minerals and metals, with their formation being dealt with only as a subordinate issue. Apart from studying the physical properties of minerals, he described at length the figures that can be carved on precious stones, and their magical significance. In some cases, he believed, human figures could appear naturally within precious stones. On more general matters concerning the structure of the earth, he followed Aristotle's *Meteorologia* in ascribing volcanoes to underground winds. He accepted that land could emerge from the sea, although he did not see fossils as evidence of this. Where Aristotle realized that streams and rivers were produced by rain, Albert followed Anaxagoras and Plato in postulating a vast underground reservoir that fed the streams.

Albert's *De Vegetabilibus et Plantis (On Vegetables and Plants)* offered a commentary on a pseudo-Aristotelian *De Plantis*. It included detailed descriptions of the structure of many plants, showing that Albert had undertaken extensive observations of his own. While allowing for 'mutations' from one species to another, as when mistletoe is formed from a rotting tree, Albert was anxious to preserve the stability of the natural kinds. He outlined a classification scheme for the plant kingdom following Theophrastus, with a scale running from fungi at the bottom to flowering plants at the top.

The bulk of Albert's *De Animalibus (On Animals)* followed translations of Aristotle's *Historia Animalium* and other zoological treatises. He dismissed many of the legends reported in the bestiaries, offering instead some careful descriptions of animals from the north that had been unknown to Aristotle. As in Aristotle's own writings, there is no classification scheme in the modern sense, but the scale of

Nature is again brought in to rank the classes in terms of their means of reproduction. Viviparous animals are, of course, the highest, with humankind at the top of the scale (unlike Aristotle, however, Albert classified whales and dolphins with the fish). Below them are ranked the oviparous animals, in descending order: birds, reptiles, fish, molluscs, crustacea, insects, etc. The 'lower' sea creatures including zoophytes come at the bottom of the scale because they are supposed to be produced by spontaneous generation from slime. Albert accepted the principle of continuity according to which there were intermediate forms between the various classes. Thus the image of a continuous linear chain of being emerged, all grades of existence being produced as part of the divine plan.

The writings of Albert show that the study of Nature was once again becoming respectable for a philosopher, although it was still influenced by many ancient preconceptions. But there was an increasing tendency for those with practical knowledge of animals and plants to put this at the disposal of others through the publication of books.

An important contribution to zoology came from the Emperor Frederick II (1194–1250) in his *De Arte Venandi cum Avibus* (*The Art of Falconry*). Falconry was a favourite sport of the nobility, and Frederick paid great attention to the birds in his collection. His book contained details of avian anatomy and behaviour, and was well illustrated. He was willing to criticize Aristotle when he thought that the Greek philosopher had departed from the truth. A century or more later Gaston de Foix's *Le Miroir de Phoebus* offered the same kind of practical information about hunting. Other books discussed angling, forestry and similar subjects of practical concern. If schoolmen such as Albert still took Aristotle as their starting point, there were many other sources of knowledge available to medieval thinkers concerned with the management and exploitation of the natural world. Medieval writers knew that humankind had the ability to control Nature and believed that such improvements were part of God's plan for the world.

There is also evidence of great delight in the natural world. The margins of medieval manuscripts were frequently decorated with naturalistic representations of plants and animals. The carvings that may still be seen throughout the cathedrals and churches built at this time show that the sculptors had a keen eye for Nature. Famous artists such as Giotto included naturalistic details in their paintings. The stories of St Francis of Assisi indicate a concern for the animals as fellow creatures within God's creation. If the early Church fathers had looked down on the study of Nature, we cannot attribute this attitude to the whole medieval period. Philosophers, artists and practical men were increasingly willing to observe Nature and to dismiss the fabulous stories of the past. If they had not yet developed a new world view to replace that of Aristotle, they were at least prepared to evaluate ancient knowledge with their own eyes.

Renaissance and Revolution

Historians used to dismiss the medieval period as a collapse into superstition. They assumed that the renewed attention given to ancient texts by the humanist scholars of the fifteenth- and sixteenth-century Renaissance led to a revival of Greek science and thus paved the way for the Scientific Revolution of Galileo, Kepler and Newton. Science picked up where it had left off when suppressed by the emergence of Christianity. The natural philosophers of the Renaissance could now appreciate the need for observation and mathematical analysis, and soon went on to transcend the theoretical achievements of the ancients.

Medieval science cannot be dismissed in quite so summary a fashion (see chapter 2). Modern studies of the Renaissance also confirm that the new scholarly initiatives of that period did not necessarily stimulate the growth of a scientific attitude. The humanists were at first so excited by their discovery of ancient texts that they had less interest than their medieval predecessors in the direct observation of Nature. Renaissance scholars were strongly attracted to a world view in which magic and science were closely intertwined. Far from escaping medieval superstition, they expanded the role played by spiritual and non-physical forces in Nature. Plants, animals and even minerals were still thought to have symbolic properties at least as significant as their physical structures. If the emergence of science became possible only when this magical view of Nature was overthrown, then we must look to the late seventeenth century before the revolution really got under way.

In those sciences where mathematics could be used to analyse natural regularities, a renewed interest in the philosophy of Plato did help to create a new attitude. But even in astronomy and physics it was necessary to combine the mathematical spirit with an increased accuracy of observation. In natural history the rise of the observational or 'empirical' method seems all-important. Humanist scholars eventually realized that the ancient texts did not provide a complete description of the natural world. They were forced to look for themselves to identify the limitations of the knowledge they had inherited. The expansion of geographical knowledge made possible by the great voyages of discovery was an important stimulus to this more critical attitude. Along with the artists, who were also developing a new respect for accurate representation of Nature, naturalists began to create a programme of research that would eventually revolutionize the study of the earth and its inhabitants. Historians of biology have often picked out those naturalists who reported accurate observations as precursors of the truly scientific approach to the subject. But Renaissance naturalists would not have separated their empirical from their humanistic interests, and it was only in the late seventeenth century that the old symbolism was eliminated.

Was the emphasis on observation really the foundation for a scientific revolution in natural history? Some historians have dismissed this possibility on the grounds that the study of the earth and its inhabitants did not experience the dramatic improvement in explanatory power we associate with physics and astronomy. There was no equivalent of Galileo or Newton in the biological sciences. The new empiricism led only to more accurate descriptions – it did not allow naturalists to *explain* why living things exist in the form we observe. Yet the naturalists of the late seventeenth century were quite certain that they *were* participating in a revolution. They threw off the old tradition in which each species was supposed to have a range of symbolic and magical properties, and described Nature only in terms of observable properties. This was the conceptual equivalent of the mathematical theories that were transforming physics. It was also a technique that led to significant disagreements on how best to classify natural objects. To classify may not be the same thing as to explain, but late-seventeenth-century naturalists thought that it was just as revolutionary an activity.

The significance of the process by which Nature was stripped of its symbolic value is immense. But the new science was not thereby freed from religious connotations. Newton firmly believed that the world was created by God and saw his science as part of a general programme to understand the divine purpose of Nature. If living species were still thought to have been supernaturally created, the naturalists who studied them were hardly less 'scientific' than the physicists who saw the overall structure of the cosmos as a divine contrivance.

If there was a 'scientific revolution', natural history participated as fully as the physical sciences. The revolution was itself a very complex affair, all the more difficult for us to understand because modern historical studies have shown that the new science had significant ideological overtones. Newton's synthesis of physics and astronomy may have been a triumph of the scientific method, but his world view was used quite deliberately to support the social hierarchy that had emerged in late-seventeenth and early-eighteenth-century Britain. The world might now be pictured as a mechanical system – but it was a machine designed and sustained by a God who expected His creatures to obey their betters. The magical view of Nature was increasingly identified with political radicalism, becoming, somewhat paradoxically, the foundation for the 'materialism' of later generations. Few historians believe that one can explain the scientific achievements of Newton in terms of his political stance, but it is no longer possible to treat the Scientific Revolution as a purely intellectual development.

The transition to a natural history based on the observation of material qualities also had ideological overtones. European societies were increasingly governed by an elite that drew its resources from trade rather than from land. The age of discovery was driven by a desire not just to explore, but to exploit an ever-increasing proportion of the earth's surface. It can be argued that such an attitude (which still called upon Christianity for its justification) demanded a more impersonal view of Nature, an image of living things as mere artefacts to be exploited. The sixteenth century saw a resurgence of the belief that everything in Nature is intended for humankind's benefit. By the end of the seventeenth century, naturalists were more willing to believe that species existed for other reasons than for our convenience, but they nevertheless reduced living things to material systems that could be studied and, by implication, manipulated without moral consequences. Magic was eliminated because Nature had to be despiritualized if people were to feel comfortable when they used the earth for their own selfish ends. The mechanistic view of Nature may have been created to legitimize the ruthless attitude of an age in which profit was the only motive that mattered.

HUMANISM AND THE NATURAL WORLD

The late fifteenth and sixteenth centuries were a period in which the horizons of western culture widened, although its foundations were as yet unchanged. As the French physician Jean Fernel wrote in his *De Abditis Rerum Causis* (*On the Causes of Secret Things*) of 1548[1]:

> The world sailed round, the largest of Earth's continents discovered, the compass invented, the printing-press sowing knowledge, gunpowder revolutionising

the art of war, ancient manuscripts rescued and the restoration of scholarship, all witness to the triumphs of our New Age.

For all the new discoveries, it was the revival of ancient learning that provided, at least for the time being, the framework through which they would be interpreted.

The most characteristic manifestation of the Renaissance was the enthusiasm of the humanist scholars who rediscovered the true character of ancient literature. Medieval scholars had used poor editions of classical texts, often translated from Arabic rather than from the original Greek. They wrote in a debased form of Latin and had argued endlessly over rival interpretations of a few ancient authors, especially Aristotle. The humanists of the fifteenth century wanted a return to the purity of classical Latin and Greek. They searched diligently for original texts, improving the quality of what was already available and discovering a host of ancient writers unknown to medieval scholars. Among the new discoveries was the text of Ptolemy's *Geography*, which created an interest in the problems of cartography. In 1417 the full text of Lucretius was discovered by the humanist scholar Poggio Bracciolini in a remote monastery, although it would be some time before the full impact of Lucretius' materialism would be felt.

Secret Powers

More influential at first was the revival of interest in Plato and the Neoplatonists. Plato's vision of an ideal world underlying the confusions of everyday life was to have important consequences for the physical sciences because it encouraged the search for mathematical laws. But Neoplatonism was to play a more general role in Renaissance thought by promoting the view that visible Nature was permeated by spiritual forces that could be controlled by the human mind. The same approach was fostered by the Hermetic tradition, based on writings attributed to the Egyptian magus or 'magician' (i.e. a wise man) Hermes Trismegistus. Here was knowledge that seemed to come from before the time of Moses himself, a powerful alternative to traditional Christianity. While the printing-press was disseminating orthodox knowledge to an ever-widening audience, small groups of scholars pored over ancient texts that they hoped would uncover the secrets of Nature.

Owing to these influences, Renaissance scholars took what we now dismiss as 'magic' very seriously indeed. For them, knowledge of the world could be gained both through the senses and through the study of the underlying spiritual and symbolic relationships that bind everything in the world together. The Hermetic tradition in particular fostered a secretive approach to knowledge in which only the initiated could understand the true significance of Nature. Humans were the centrepiece of the universe; everything revolved around them and everything could be understood by them if only the secret keys could be found to unlock the mysteries of creation. Every aspect of the 'microcosm' (humankind) was reflected somehow in the 'macrocosm' (the universe at large). Nothing in Nature was without spiritual significance – animals, plants and even minerals all had purposes that could be discovered through their 'signatures' or symbolic resemblances to parts of the human body. The world itself was alive, and was thus capable of growth and decay. Such an attitude did not prevent the study of the natural world or its accurate depiction by writers and artists, but it did ensure that the world revealed by the senses would be clothed in a symbolism that we can no longer appreciate.

The effect of this attitude on what we today call 'science' can hardly be underestimated. A Renaissance scholar studied the world through a conceptual framework that seems to combine objective research with what a modern scientist would dismiss as a fog of misconception. The Elizabethan mathematician John Dee (1527–1608) was deeply involved in the attempt to provide sailors with new navigational techniques – yet he was also a magician and astrologer who saw number symbolism as the secret key to the understanding of the universe. For him, the stars had both a practical and a symbolic role to play. To appreciate the developments that took place in the fifteenth and sixteenth centuries, we must be willing to suspend the judgement of hindsight. Any attempt to extract what was 'really' scientific from the background of magic can produce only an artificially distorted account of Renaissance achievements. Instead, we must be prepared to follow the progress of this world view as a whole, looking for those factors which led eventually to the separation of what would become science from the framework of Renaissance thought. In the seventeenth century the magician would become the natural philosopher, as the practical component of natural magic was extracted to become the key to a new way of conquering Nature.

Historians seeking to evaluate Renaissance natural history have concentrated on the growth of naturalistic representation. The mid sixteenth century saw the publication of numerous books containing accurate descriptions and beautiful illustrations of animals, plants and minerals. The visual side of this enterprise was linked to the triumphs of artists such as Leonardo da Vinci (1452–1519) and Albrecht Dürer (1471–1528). The artist and the naturalist both saw their task as the study of Nature through the light of human reason, and both were aware of the symbolism of natural forms. The elimination of symbolism from the study of Nature was the work of the late seventeenth century, and this was accompanied by a switch from visual depiction to verbal description of natural forms. Renaissance naturalism certainly played a role in the growth of scientific natural history, but we must be more flexible in our efforts to identify the factors responsible for the transition.

The Riches of Nature

The initial outburst of humanist enthusiasm did not have a beneficial effect on the study of Nature. Humanists were concerned with the actual texts of the ancient writings, not with how well their authors had described what they saw. If anything, there was a decline in the standard of illustrations in early-fifteenth-century herbals – the pictures were meant to illustrate the text, not the natural objects under discussion. Important work by medieval scholars such as Albert the Great was now dismissed as outdated scholasticism.

By the sixteenth century, however, a distinct tradition of literature describing Nature had begun to emerge. Its authors certainly did not separate the study of the natural world from its humanist and magical connotations, but they did go beyond what was described by the ancients. Renaissance scholars were forced to look for themselves as the limitations of ancient knowledge became more apparent. Artists were being hired and even trained by the naturalists to specialize in the accurate depiction of species not found in any classical text. Botanical gardens and museums were being set up in the universities to promote the study of Nature from real life, or from preserved specimens. Padua had a botanical garden in 1542, and was soon followed by other universities. Luca Ghini (*d.* 1556), a medical teacher at Bologna, developed the technique of preserving plants by pressing, thus pioneering the herbarium as a tool for botanical reseach. In a host of ways beyond mere accurate depiction, Renaissance scholars were establishing the study of Nature as a recognizably distinct enterprise.

A communications network was being established so that information and specimens could be transmitted between scholars in various parts of Europe. The Swiss naturalist Conrad Gesner (1516–65) has been described as a one-man

Royal Society, so extensive were his efforts to gather and disseminate information. Gesner ranged over the whole panorama of Renaissance learning. He was professor of Greek at Lausanne before being appointed town physician of Zurich. He wrote extensively on classical literature in addition to the works on natural history discussed below. Gesner was no armchair commentator on Nature; he was one of the first Europeans to indulge in mountain-climbing to enhance his appreciation of alpine beauty. But his main purpose in life was to summarize the whole range of human knowledge and to transmit it to his fellow scholars around Europe.

One of the main reasons why naturalists were forced to make more active use of direct observation was the increasingly obvious limitations of ancient botanical and zoological works. As humanists, they were anxious to develop a full understanding of the ancient texts, and this required them to be sure that they had correctly identified the species being discussed. But to the naturalists of northern Europe it was obvious that writers such as Aristotle and Theophrastus had not had access to the animals and plants with which they themselves were familiar. Having identified the Mediterranean species in the ancient texts, it was necessary to provide some account of the northern species that were not mentioned. In the absence of a technical vocabulary for describing the animals and plants, good-quality illustrations were vital to prevent confusion. Renaissance naturalists were thus drawn into the project of providing a complete description of the European flora and fauna.

The significance of this project was highlighted by the rapid expansion of knowledge of the world outside Europe. The fifteenth and sixteenth centuries were also the age of discovery, when Portuguese, Spanish and British sailors discovered the New World and opened up the sea route around Africa to Asia. There was a continuous tradition of Portuguese exploration around Africa, which culminated in Vasco da Gama's voyage around the Cape of Good Hope to

India in 1498. The possibility that Asia might be reached by going west across the Atlantic was suggested by the Florentine geographer Paolo Toscanelli some time before Columbus decided to put it to the test. Contrary to popular belief, no one thought that Columbus would fall off the edge of a flat earth – everyone knew that it was round, but there was a serious dispute over just how large the circumference was. Toscanelli and Columbus dramatically underestimated the distance that would have to be sailed to reach the Indies by crossing the Atlantic. They were also encouraged by the discovery of islands such as the Azores, which suggested that other Atlantic islands might be found that would serve as staging posts on the way westwards. In fact, their critics were quite right; had America not existed, Columbus would have perished long before he reached the coast of Asia.

The explorers demanded better navigational techniques, and this helped to stimulate the map-makers and the mathematicians who hoped to devise new techniques for working out the position of a ship at sea. But in the long run it was the discoveries that were made in America and Asia that were to have the most profound effect on European science. Scholars did not respond immediately to the reports of new animals and plants brought back by the sailors, but they could not ignore the introduction of tobacco, the potato and other exotic species. The discovery of the New World drove home the point that the ancients had not known everything, and naturalists would eventually have to buckle down and provide descriptions of the ever-widening array of new species. The *History of the Indies* written by Oviedo y Valdes (1478–1557) included much information on the new species, while the translation of Nicholas Monardes' *Joyfull Newes out of the Newefound World* (1577) performed the same function in English.

The men who set off to colonize America and the Indies were not scholars – they were traders out to make a profit. In the end, their more practical approach to Nature was to have an even greater effect on the growth of science than the actual discoveries they made. Even at home, there were practical men writing books that may have been scorned by the humanists, but were welcomed by farmers, miners and others who were concerned to make a living from the natural world. The German mining engineer Georg Agricola (1494–1555) wrote his *De Re Metallica* of 1556 to explain new techniques, but also discussed the nature and location of minerals. The Scientific Revolution of the seventeenth century came about because scholars began to realize that practical developments were generating material results far outstripping the promises made by the Renaissance exponents of natural magic. The humanists themselves could extend natural history to include careful descriptions of the newly discovered species, but they were unable to strip Nature of the magical and symbolic overtones with which it had been clothed. All the sciences, including natural history, would be revolutionized when the new commercial empires began to demand an ideology that presented Nature only as a material system to be exploited.

Animal, Vegetable and Mineral

Since zoology lacked botany's connection with medicine, it had lagged behind during the medieval period. The incentive to study animals still derived from the example of ancient encyclopedias such as Pliny's *Natural History*. The indefatigable Conrad Gesner produced his *Historia Animalium* (1551–8) as an attempt to summarize the knowledge gained since classical times. His work was based on much first-hand experience, including his own observations and those reported by his many correspondents. But Gesner was a Renaissance scholar who studied Nature so that he could comment upon and extend the knowledge of the ancients. His work included detailed discussions of the names given to the various species, their use and significance for humankind, and an evaluation of previous commentaries upon them. The full title of Edward Topsell's

1608 translation of extracts from Gesner's work illustrates the flavour of the whole:

> *The Historie of Foure-Footed beastes. Describing true and lively figures of every Beast, with a discourse of their severall Names, Conditions, Kindes, Vertues (both naturall and medicinall), Counties of their breed, their love and hate for Mankinde, and the wonderfull worke of God in their Creation, Preservation, and Destruction. Necessary for all Divines and Students because the Story of every Beaste is amplified with Narratives out of Scriptures, Fathers, Phylosophers, Physitians, and Poets; wherein are declared divers Hyeroglyphicks, Emblems, Epigrams, and other good Histories.*

Gesner continued to report at least some fabulous beasts, but he was honest enough to admit that he had no evidence to confirm their existence. He produced no classification of animals in the modern sense, adopting an alphabetical arrangement that allowed him to group together only those species that could be included under the same basic name.

Compilations such as Gesner's have sometimes been dismissed as unscientific mixtures of fact and fantasy. Contemporary studies of particular groups of animals, including works by Pierre Belon (1517–64) and Guillaume Rondelet (1507–66) on fish, have been pointed out as closer to the spirit of modern science. Belon also studied birds, and his description includes a famous comparison of the skeletons of a bird and a human, which can be seen as an anticipation of modern comparative anatomy. Yet these authors could be just as credulous as Gesner when reporting facts outside their own experience, and in general the Renaissance zoologies seem to offer little foundation for later developments. There was as yet no serious effort to study the structure of the various types of animals with a view to arranging them into a natural system of classification expressing degrees of anatomical relationship.

Even the concept of a biological species was unclear.

Fifteenth-century studies of monstrous births attributed them to the combination of characters from two distinct species as the result of unnatural sexual unions. The *characters* were fixed by Nature, but the particular combination defining a species was maintained only by the normal tendency for like to mate with like. There was also a widespread belief that animals could be formed by 'spontaneous generation' from putrefying matter – the chemist van Helmont gave a recipe for producing mice from old underclothes. Within such a world view there could be no prospect of a classification based on degrees of resemblance between species, because the species themselves were not rigidly determined by the process of heredity.

Judged by the standards of modern science, the addition of accurate descriptions and illustrations did little to undermine the basically traditional framework within which animals were studied. But such a negative evaluation ignores the fact that these naturalists were in no position to make a direct leap into the world of modern science. They saw themselves as reformers, not as revolutionaries: their task was to evaluate existing knowledge, not to erect a new conceptual framework. They could hardly be expected to throw out all those aspects of traditional lore that we now recognize as unscientific, especially as they were still trying to create a system that would allow specialists to communicate information without ambiguity. Gesner realized that zoology ought to move beyond mere description towards recognizing the natural affinities of animals, but he had to concentrate on gathering information in the hope that others could build upon the foundation he had established.

In one respect the new literature created a framework within which further changes would become possible. The use of accurate illustrations made natural history a public rather than a secretive science. The old magical symbolism was still there, but it was no longer part of an arcane body of information accessible only to the initiated. Ordinary people could read the descriptions of animals and feel that the study of Nature was something that they could participate in. Eventually the demand for knowledge of practical value would encourage the transition from Renaissance magic to the natural philosophy of the early modern era.

Botany already had a practical basis in medicine, but was slowly acquiring an independent character of its own. The humanists provided new editions of Dioscorides and set out to identify as many as possible of the plants mentioned by the classical writers. Because of the demands made by physicians, there was an even greater requirement for accurate identification of both old and newly discovered plants. By the early fifteenth century, herbals were being produced with extremely accurate woodcuts. Particularly important was Otto Brunfels' *Herbarium Vivae Icones* (*Living Portraits of Plants*) of 1530, where the quality of the illustrations far surpassed that of the text. Leonhard Fuchs (1501–66) carefully supervised the artists preparing the illustrations for his *De Historia Stirpium* (*History of Plants*) of 1542. Although the herbal was, by definition, a medical handbook, the emergence of botanical gardens and of specialized lecturers in botany created a growing awareness that the field could be studied in its own right. In 1592 the Bohemian botanist Adam Zaluzniansky wrote[2]:

> It is customary to connect medicine with botany; yet scientific treatment demands that in every art, theory must be disconnected and separated from practice, and the two must be dealt with singly and individually in their proper order before they are united. And for that reason, in order that botany, which is (as it were) a special branch of natural philosophy, may form a unit by itself before it can be brought into connection with other sciences, it must be divided and unyoked from medicine.

It would take another century of effort before Zaluzniansky's demands would be fulfilled. In the meantime, botanists concentrated on describing an ever-widening array of plants –

including many from the New World – within the traditional approach. Medical and other practical uses were
included in the description, and the arrangement was
often purely alphabetical. The significance of each plant
for humankind was often linked to the doctrine of 'signatures', a typical manifestation of natural magic in which
any physical resemblance between the plant and a part of
the human body was thought to indicate the appropriate
medical application. The Creator had, in effect, built notices
announcing the value of each species into its very structure.
Physical characters should not be merely described by the
botanist; they should be analysed for their symbolism
because this offered a short-cut to knowledge of the
species' practical application. For all their emphasis on
accurate description, the Renaissance herbalists were still
a long way from adopting a materialistic interpretation of
their science.

The vast mass of new information was, however, exposing
the limitations of the classical authorities and creating
a framework within which it would eventually become
possible to develop new initiatives. The growth of specialist
knowledge also created a body of natural philosophers
increasingly aware of their status as the possessors of
accurate information. At first, the search for new plants
had depended upon the local knowledge handed down for
generations among country people. But the botanists soon
discovered that they needed to create a new nomenclature
to avoid the proliferation of local names that had been
applied to species. Gaspard Bauhin's *Pinax* of 1623 supplied
Latin names for over six thousand species. Naturalists also
discovered that many of the 'old wives' tales' about animals
and plants had no foundation. The specialists thus became
increasingly contemptuous of folk beliefs. Their own ideas
might sometimes seem equally strange to a modern reader,
but the trend towards specialization was an indication of the
path that would be followed in the future development of
science. Sir Thomas Browne's *Pseudodoxia Epidemica* (often

known under its English title of *Vulgar Errors*) of 1646 was the product of an age in which scholars had become acutely aware of the gulf separating their knowledge of Nature from that of the common man.

As yet, though, there was little appreciation of the fact that the new knowledge might form the basis for a system of classifying species entirely independently of their significance for humankind. A pioneering effort to create a system of plant classification based solely on biological characters was made by the Italian physician and botanist Andrea Cesalpino (1519–1603) in his *De Plantis* of 1583. Cesalpino had little influence on his own time because his starting point was the Aristotelian philosophy that had now been discredited because of its association with medieval scholasticism. Nor did his arrangement of the species bear much resemblance to what we now recognize as natural groupings. But the naturalists who began to create the modern system of biological taxonomy in the late seventeenth century recognized that their concerns had to a significant extent been anticipated by Cesalpino.

It was Cesalpino's respect for Aristotle that led him to challenge the assumptions of the herbalists and to seek a classification of plants that rested on the characters of the species rather than upon their value to humankind. He sought knowledge of Nature based on the essential structures of the species themselves, and went beyond Aristotle in recognizing that, to make this feasible, it was necessary to determine which characteristics are the more important for identifying the 'essence' of each species. Trivial characteristics such as colour and smell must be ignored in favour of those which were related to the basic functioning of the plant, its 'vegetative soul'. Cesalpino identified the two most important functions as nutrition and reproduction, and hence concentrated on the structure of the root and the flowers. He thus pioneered the principle of the subordination of characters, which made it possible to divide Nature up into genera or groups of species related in their most fundamental characters. The object was to create a 'natural' system of classification, and, although later naturalists rejected the groupings that Cesalpino created, they acknowledged his influence by adopting the reproductive parts as the basis for their own systems. Far from holding back the development of science, in this case the influence of Aristotelian philosophy stimulated the breakthrough that established a new approach to Nature.

If Cesalpino's work pinpoints issues that would come to the fore as the Renaissance world view began to crumble, the study of minerals illustrates just how far removed that world view was from our own. We now take it for granted that the earth's surface contains many structures that are relics of the processes that have created its present form. We make a sharp distinction between minerals that have been created by physical processes and the fossils that preserve the structures of once-living things. But Renaissance thinkers were unable to make the sharp distinction that allows us to distinguish a fossil from a mineral – to them a 'fossil' was anything dug up from the earth, and even if the resemblance to a living thing was too obvious to ignore, their world view encompassed processes that provided ready alternatives to our modern explanation of how that structure became embedded in the rock. To imagine that large sections of the earth's surface had been formed from sediment laid down under water was to postulate changes of a magnitude that seemed inconceivable. Noah's flood could be used to reconcile such a possibility with the Bible – a favourite technique used by later naturalists – but in the sixteenth century even this seemed an unnecessary magnification of the changes to which the earth had been subjected.

Conrad Gesner published a book *On Fossil Objects* in 1565, although this was only a sketch and did not include the full descriptions that characterized his work on living animals. The book included excellent illustrations, an innovation of particular significance in a field where the true nature of the objects under discussion was unclear. Gesner saw the obvious resemblance between some fossils and living sea creatures such as shellfish; in his book on animals he also noted the similarity of the so-called *glossopetrae* or 'tongue-stones' to sharks' teeth. But Gesner was not thereby led to adopt the modern view that these fossils are indeed the remains of once-living things entombed in sediment that has been petrified. There were many other structures where the resemblance to a living body was less clear, and Gesner was thus forced to classify them by the closest analogy he could recognize. Ammonites, which we now accept as extinct molluscs, were closest to rams' horns, while belemnites looked like arrowheads. There was no sharp dividing line between these structures and purely mineral bodies such as crystals. The earth was full of mysterious shapes, and Gesner would have seen little point in trying to make a distinction based on the degree of apparent relationship to living things.

Had Gesner produced a complete study of fossils, he would almost certainly have included a discussion of their significance for humankind. Gems had always been endowed with mystical significance, and Renaissance astrology merely extended the supposed analogies that existed between the earth, the heavens and humankind. The similarities (real and imagined) between fossils and living things merely confirmed the fact that Nature was pervaded by mysterious analogies and symbols. The resemblance of fossils to living things was not accidental, but nor was it evidence that fossils had once been alive. The earth itself was capable of growth; crystals grow just like a living thing, while many miners believed that the minerals they extracted were perpetually renewed. Nature was full of petrifying influences, as illustrated by pearls, corals and human gallstones. Some Renaissance thinkers even supposed that the 'seeds' of living things could grow by absorbing mineral nourishment, thus producing a stony counterpart to the parent form. Within such a world view the significance of fossils and minerals was bound to be quite different to that which emerged after the foundation of the modern earth sciences. There was no need to postulate great revolutions in the earth's surface, because the organized structures it contained were interpreted as the products of modern processes, not as relics of the past.

There was some practical interest in useful minerals. In addition to his great work on mining, *De Re Metallica*, Georg Agricola published his *On the Nature of Fossils* in 1546. This provided a classification of minerals based on their material properties that served as the basis for later developments in mineralogy. The French potter Bernard Palissy (*c.*1510–90) wrote his *Admirable Discourses* of 1580 to ridicule the beliefs of the learned in the name of practical experience. This utilitarian side to natural history would grow in significance during the following century, providing the inspiration for an even greater emphasis on the direct study of natural objects, living and non-living. The framework of natural magic that sustained the Renaissance scholars' fascination with analogy and symbolism would crumble in the face of this greater emphasis on practical information. The demand for realism and personal experience would be uncoupled from its foundation in Renaissance humanism and used as the basis for a new philosophy in which the study of Nature would take on a more materialistic character, thereby opening up entirely new questions about the character of the earth and its inhabitants.

THE GREAT INSTAURATION

The philosophers and naturalists who created the framework for this new science were quite deliberate in their desire to transcend Renaissance values. Francis Bacon emphasized the need to observe Nature in order to control its operations by purely physical (as opposed to magical) means. Descartes stressed the need for clarity of thought when Nature was analysed as a mechanical system. Observation, experiment and the search for mathematical regularities or laws were to be the methods of the new natural philosophy. The result was a form of science that made a significant break with both the Aristotelianism of medieval thought and the Hermetic symbolism of the Renaissance.

Yet historians now recognize that the break with tradition was not as abrupt as its heirs – the modern scientists – might imagine. For Bacon and others this was to be the 'great instauration' or renewal: humankind would seek to control Nature by new means and would thereby regain the level of understanding and power once enjoyed in the garden of Eden. The magicians' search for secret knowledge would become a public research project that would benefit the commercial enterprises that were changing the structure of European society. The new materialism did not reject Christianity; on the contrary, it hoped to redesign the traditional message in a way that was acceptable in the modern world. Nature lost its human symbolism, but it was not yet reduced to a random dance of atoms in the void. To maintain the impression that we live in an orderly cosmos created for us to exploit, the new science retained the concept of a natural hierarchy of causes imposed by divine providence. Materialism was qualified in ways that conveyed a definite ideological message.

Natural History and Revolution

The most obvious triumphs of the Scientific Revolution of the seventeenth century were the developments in mechanics and cosmology by Galileo, Kepler, Descartes and Newton. The sun-centred astronomy of Copernicus was reinforced by a theoretical foundation in which the whole structure of the universe was maintained by the laws of motion and gravity. Newton's theory was seen as a triumph of the human mind, a clear indication that the application of rational methods would lead to the uncovering of all Nature's secrets. In fact, though, the development of a scientific account of the origins of the earth and its inhabitants was delayed until the nineteenth century. Natural history seems to have been left behind in an age of description and classification.

This image of the natural sciences trailing behind their physical counterparts runs through many older accounts of the Scientific Revolution, which imply that the search for accurate descriptions begun in the Renaissance was merely extended by new tools such as the microscope. There was no conceptual transformation equivalent to that taking place in astronomy and physics. It is now clear that there are major limitations to the validity of this interpretation. The search for a mechanistic biology made only limited advances in the seventeenth and eighteenth centuries, but the naturalists who set out to describe and classify the world according to the new principles certainly thought that they were engaged in a revolutionary activity. As the English naturalist John Ray (1627–1705) wrote[3]:

> I am full of gratitude to God that it was His will for me to be born in this last age when the empty sophistry that usurped the title of philosophy and within my memory dominated the schools, has fallen into contempt, and in its place has arisen a philosophy solidly built upon a foundation of experiment . . . It is an age of daily progress in all the sciences, especially in the history of plants.

Ray was fully conscious of all the other achievements of the new science, but he expected his own botanical work to rank alongside the more visible triumphs of the physical scientists.

The naturalists of the seventeenth century made only limited moves towards explaining the origin of the order they observed. There was as yet no conceptual space within which the idea of biological evolution could have been constructed, and it would be a century or more before the development of geology would create that space. In the meantime, Ray was content to assume that the order he observed in material Nature was imposed by its Creator. In so doing he was not betraying the spirit of the new science, since even Newton himself had gone out of his way to insist that the mechanical philosophy endorsed rather than undermined belief in a God who superintends the world in which we live.

The claim that natural history was revolutionized along the same lines as the physical sciences is reinforced by the work of historians seeking to reinterpret what has traditionally been seen as the mainstream of the new science. A new world view was being created to sustain the whole scientific enterprise, and we need to assess both its conceptual foundations and its broader implications within the context of seventeenth-century thought. We no longer accept that the era of Newton saw a sudden transition from ancient superstition to the objectivity and rationalism of modern science. Without denying that the physical and astronomical sciences saw a dramatic improvement in their ability to describe certain phenomena, historians have become aware of a wider story lying behind the decision to apply the new methods. The natural philosophy of the late seventeenth century was the foundation upon which modern science was built, but it was not a straightforward anticipation of modern ideas. Newton and his contemporaries were deeply concerned about the philosophical consequences of what they were doing. For the most part they were determined to link their science with their religion to form a conceptual system that could be used to uphold their social and political values. The significance of 'Nature' for human affairs was thereby transformed in ways that continue to influence modern attitudes towards the environment.

It has long been suspected that the emergence of the new science was linked both to the Reformation and to the growth of a commercial economy. Where the Catholic Church had maintained the social hierarchy based on the ownership of land, the Protestant 'work ethic' allowed each individual to make his or her own way in the world. Success in business was the visible symbol of spiritual worth, since God intended us to use our energy and initiative to exploit the gifts of His creation. Such a view of humankind's relationship to God and Nature encouraged the view that the search for practical knowledge was in itself a religious activity.

In seventeenth-century Britain, the most obvious manifestation of this new attitude was to be found among the Puritan sects, which led the challenge to the old order that came to a head with the civil war of the 1640s. Many of the early converts to the new science came from a Puritan background. John Ray sacrificed his comfortable position at Cambridge University because he refused to sign Charles II's Act of Uniformity designed to purge Puritans from the Anglican Church. In his later life he was supported by a wealthy patron, Francis Willoughby, who shared his passion for natural history. Men such as Ray and Willoughby saw the study of Nature as a duty laid on them by God; it gave them a better appreciation of His power and wisdom, but also provided information that would be of practical value to those who were engaged in the creation of wealth through industry. Such an attitude encouraged the pursuit of natural philosophy, but it also tended to favour the view that Nature was a resource to be exploited. Christianity, which had long nurtured a relatively harmonious view of humankind's position in the world, now became a foundation for the belief that the material universe was created for our benefit.

Much of the groundwork for this new philosophy was laid by Francis Bacon (1561–1626) in works such as his *New Atlantis* of 1627. Where some Protestants stressed the sinfulness of humankind, Bacon boldly proclaimed our ability to regain the state of knowledge and power that Adam enjoyed before his fall from grace. By careful study we uncover the secrets of Nature, which can then be used to control the material world[4]:

> The end of our foundation is the knowledge of causes,
> and secret motions of things; and the enlarging of the
> bounds of human empire, to the effecting of all things
> possible.

Nature would be 'put to the question' (tortured) to reveal
these hidden secrets. To some extent it is clear that Bacon
was manipulating rather than destroying the old idea of
natural magic, but he was also transforming it to become
the philosophy of a commercial age. Where the Renaissance
magus sought to decipher the symbolism of Nature by
secret rituals, Bacon's followers would co-operate in a
public enterprise to discover facts that became visible only
through careful observation and experiment.

For the exponents of such a world view, Nature could
not be a quasi-divine system to be venerated and cherished.
The world may have been created by God, but it was not
in itself divine and it could be changed according to our
will. In the words of the chemist and natural philosopher
Robert Boyle[5]:

> The veneration wherewith men are imbued for what
> they call nature has been a discouraging impediment
> to the empire of man over the inferior creatures of
> God: for many have not only looked upon it, as an
> impossible thing to compass, but as something impious
> to attempt . . .

Reducing Nature to an essentially material system gave us
both the means and the right to exploit that system as we
wish, on the assumption that God had created it solely for
that purpose.

There were new tools available to help the process of
investigation. The introduction of the microscope had a dramatic effect upon naturalists' perception of the world. The
compound microscope (which has two lenses, the objective
and the eyepiece) became common in the late seventeenth
century. In the hands of Robert Hooke (1635–1703) and

others, it revealed the detailed structure of hitherto despised animals such as the louse and the flea. Anatomists such as Marcello Malpighi (1628–94) and Jan Swammerdam (1637–80) used the microscope in their dissections and showed that even the humble insects had complex internal organs. The early instruments were, however, of very limited use, and the famous Dutch microscopist Anton van Leuwenhoek (1632–1723) obtained better results using a tiny, but carefully made, magnifying glass with a single lens. Leuwenhoek revealed the profusion of micro-organisms that swarm among the more familiar things accessible to the naked eye, thus emphasizing the immense variety of the natural world.

As yet, though, there was little hope of understanding the significance of these revelations. It was now less easy to believe that insects were produced by spontaneous generation from rotting matter, a point confirmed by Francesco Redi's famous experiment of 1668 in which flies were prevented from reaching meat by muslin, with the result that no maggots appeared on the meat as it decayed. Redi insisted that the maggots came only from flies' eggs, and were not spontaneously generated as the meat decomposed. This fitted in neatly with the observations confirming that insects were, in fact, very complex organisms, not merely animated blobs of slime. But the microscope could reveal only the structure, not the function of minute organs. It expanded the range of natural species to be classified, but offered little help in understanding the intimate processes of life.

The Baconian methodology was more significant in encouraging the extension and formalization of the network of local studies that had already begun to emerge in the previous century. The natural philosophers who founded the Royal Society of London in the early 1660s regarded themselves as Bacon's heirs. Local societies had existed in Italy for some time, but the Royal Society (and its French equivalent, the Académie des Sciences of Paris) expressed a more positive desire to promote science on a national and even an international scale. Journals such as the *Philosophical Transactions* of the Royal Society allowed new ideas and information to be transmitted far more effectively. In addition to formal scientific papers, these journals included detailed accounts of local industries, mineral resources and natural history. The desire to study the environment for practical ends was thus incorporated into the network of science. Natural history was firmly included within the new world view along with the latest triumphs of mechanics. The practical benefits resulting from this accumulation of information were not always apparent, but in principle, at least, the gathering of information about the natural world was part of the new ideology of utility.

The Mechanical Philosophy

This increasingly well organized search for knowledge was bolstered by the creation of a new image of Nature. Feminist and other historians have pointed out the strongly masculine terminology used by what was certainly a male-dominated research community. Science would 'strip the veil' from Nature so that she revealed her innermost self – a metaphor that meshes all too well with that of industry 'raping' the natural environment. But the Baconians certainly did not want to endow Nature with the traditional female attributes of wisdom and fecundity. In order to justify their possessive attitude, they portrayed the material world as a passive, mechanical system that had no moral dimension. Where Nature had once been seen as an organic whole, a source of mysterious constructive powers, it now became nothing more than a gigantic piece of clockwork with which humankind could tinker at will.

Various aspects of this new world view have been singled out by the critics of modern science to illustrate the link between the study and the exploitation of Nature. Since Nature was no longer governed by self-integrating powers, it did not have to be treated as an organic whole, but could be broken up into isolated elements for the purpose of study. The whole was no more than the sum of the parts, so scientists (to use the modern term) could specialize in the study of a single component without worrying about the broader implications of what they were doing. The study of Nature could thus become increasingly fragmented, with no one having the responsibility to assess how manipulation of one part of the environment might affect all the others. From this point onwards, science became increasingly specialized. Ordinary people might provide evidence of interesting phenomena, but the interpretation of those phenomena was the province of specialists who had spent a great deal of time mastering the technical literature. The growth of a systematic approach to the classification of species by John Ray and others is symptomatic of this trend.

The 'mechanical philosophy' was also quite explicit in its desire to treat the world as a system of material particles in motion. Robert Boyle (1627–91) played a leading role in promoting a revival of the atomic or corpuscular view of Nature. His *Origin of Forms and Qualities* of 1666 argued that material objects could be understood as collections of rigid particles governed by the laws of motion. Matter and motion were 'the two grand and most catholick principles of bodies'. The only real properties (the primary qualities) were those of shape, size, solidity and (after Newton) mass. All other qualities such as smell and colour were merely secondary – they had no existence in Nature but were produced in the human mind as our sense organs interacted with the particles of the material world. Heat, for instance, was a manifestation of the particles' vibration. The scientific model of Nature was thus mechanical in the most literal sense of the term. All the more 'human' qualities had been stripped away to leave only the bare bones. In such a world view, Nature could not be the object of worship or reverence. It contained no special powers, and the symbolic relationships of natural magic were illusory.

The mechanical philosophy was certainly extended to include living things. The French philosopher René Descartes (1596–1650) insisted that animals were nothing more than complicated machines. Even the human body was nothing more than a mechanical system, although Descartes and many others admitted that there must be a soul somehow integrated into the system. But the 'animal machine' doctrine had limited applicability in an age when the best possible model was, literally, a piece of clockwork. Even with the help of the microscope, the mechanical philosophy could make little headway in the understanding of the more complex processes of life, especially reproduction. In natural history, the most characteristic manifestation of what French historians call the 'classical' era of European thought was not the attempt to reduce life to mechanics, but the search for a comprehensive pattern linking the various kinds of structures to be found in the living world. If the symbols that had once given each species a context in relationship to humankind were to be dismissed, Nature became a vast and potentially confusing mass of different material forms. If human reason was to take hold of this mass, it must reduce it to order by fitting each form or species into an intelligible system of relationships.

It is thus no accident that Ray and others established the origins of the modern system of biological taxonomy (classification) at this point in time. Their work offered the equivalent in natural history of the order being imposed on the physical world by the discovery of Newton's laws. Ray openly repudiated Gesner's fascination with the classical world and the whole network of symbolic relationships that had once bound animals to humankind. The new science would classify by physical appearance alone, and, instead of relying on accurate pictures, it would develop a technical language that would itself be part of the process by which order was introduced. The limited impact of the microscope is illustrated by the fact that classical natural history took only external appearances into account; there was as yet no systematic effort to develop a comparative anatomy based on dissection and the study of the (still mysterious) internal processes of the organism. The great triumph of eighteenth-century natural history would be, not the direct exploitation of the animal machine doctrine, but the erection of a comprehensive system by which the variety of living structures could be understood, and thus in a sense dominated, by the human mind.

The exponents of the new natural philosophy of the seventeenth century still saw the search for order as the uncovering of a pattern that existed in the world as part of the Creator's plan, intelligible to humankind because of our God-given power of reason. The alternative was a Lucretian picture of the world as an unplanned result of the random combination of atoms. This ancient alternative to the idea of a divinely planned universe was certainly available and was promoted by a few radical materialists including the social philosopher Thomas Hobbes (1588–1679). But the majority of the natural philosophers found this alternative abhorrent and took active steps to qualify the mechanical philosophy so that the vast machine of the universe could be seen as the product of divine workmanship.

The new science and the new religion thus went hand in hand. Protestants wanted to believe in a God who had given all individuals the freedom to better themselves spiritually and materially through the fruits of their industry, and it made sense to argue that the same God had designed the world so that human reason could both understand and control it. The link with commercialism is obvious, and in Britain the new science had an even more specific ideological dimension. The Puritans had worked actively to challenge royal authority, but the chaos of the English civil war allowed even more radical groups to bid for power. As the old social hierarchy crumbled, the advocates of what we should today call anarchism and socialism called for the abolition of private property altogether. This threatened the very basis of the Protestant work ethic, and the Puritans soon realized that their new commercial society would flourish best if the social order was changed rather than destroyed.

The radicals used a non-hierarchical image of Nature to bolster their challenge to the social hierarchy. This derived not from Lucretian atomism (Hobbes saw the imposition of royal power as the only alternative to chaos) but from the traditional image of Nature as an active, self-organizing system maintained by its own spiritual powers. Materialism was, in effect, pantheism: the belief that the whole world was suffused with divine power, so that no one part could proclaim superiority over another. The mechanical philosophy became a device for resisting this ancient, now dangerous, world view. It denied the self-organizing power of Nature by reducing the world to matter in motion, but argued that the organization we observe was imposed upon matter by a higher authority, the Creator.

Following the Restoration of Charles II, Boyle and his followers used their synthesis of science and religion to uphold the ideal of a constitutional monarchy that allowed freedom for commercial activity. The new physics proposed by Isaac Newton (1642–1727) was soon incorporated into their programme. Newtonian science upheld a view of the universe that provided the model for an orderly society. The social hierarchy ensured the stability necessary for commercial progress, just as God's power maintained the order and activity of Nature. These ideals were enshrined in the 'glorious Revolution' of 1688, in which William of Orange was given the throne to prevent a slide back into Stuart absolutism.

The success of the Revolution necessarily brought the new elite into conflict with Louis XIV of France, whose own very powerful form of absolutism exploited the old social hierarchy in a way that resisted the demands of the commercial classes. Louis expected his Académie des Sciences to bolster his own regime by providing both useful information and a model of Nature that emphasized its static, hierarchical character. As a moderate French reformer, Voltaire (1694–1778) saw both Newtonian science and the British system of government as ideals to be followed, but he had little success in promoting this alternative against the power of the French monarchy. A more radical republicanism seemed the only way out, and, for this reason, the materialist/pantheist tradition flourished as an underground movement among French intellectuals throughout the eighteenth century until the explosion of the French Revolution in 1789.

Boyle and the Newtonians created a version of the mechanical philosophy that avoided complete materialism. Descartes had pictured animals as machines, but even he accepted the reality of the human soul. Some seventeenth-century thinkers saw the animal machine doctrine as potentially dangerous. They argued instead that the dead machine of Nature was maintained in its orderly structure by powers delegated for that purpose by the Creator. The Cambridge Platonist Henry More invoked the concept of 'plastick virtues' as powers designed to impose order on brute matter. John Ray followed this lead in biology: Descartes' extreme mechanism was quite incapable of explaining complex functions such as reproduction, and therefore it was necessary to postulate forces that actively arranged the matter into the structures we observe and classify. Newton himself was deeply concerned about the sources of natural order, and was willing to see the divinely originated pattern of the universe maintained by subordinate powers to which the Creator delegated some of His authority. The fact that such powers seemed reminiscent of the old magical world view was carefully concealed; Newton read deeply in alchemical literature, but cloaked his public pronouncements in the language of the mechanical philosophy.

Boyle and Newton were anxious that the link between science and religion be proclaimed as widely as possible. Shortly before his death in 1691, Boyle established a series of lectures to be named after him and devoted to the defence of the Christian religion. Most of the Boyle lecturers were Newtonians, including William Derham (1657–1735), whose *Physico-Theology* gave its name to the whole programme by which natural philosophers sought to demonstrate the wisdom and goodness of God through the study of His creation. Newton himself thought that God's power would best be demonstrated by showing how the cosmos was held together by physical laws, but Derham's study (originating from his Boyle lectures for 1711 and 1712) drew heavily on the 'argument from design' in living Nature expressed in John Ray's *Wisdom of God Manifested in the Works of the Creation* of 1691 (see chapter 5). The mechanical philosophers were only too glad to point out the complexity of the living world, since the only conceivable explanation of that complexity was the designing hand of the Creator.

The world may have become a machine, but it was a machine with many interlocking parts. The desire to study and classify the various species of living organisms represented the more analytical side of the new philosophy of Nature, but natural theology ensured that there would still be respect for the system as a whole. Ray and Derham believed that humankind should both use and exploit the natural world, but they did not see every species as created for our benefit, at least not directly. Each served a function in the overall pattern of creation, serving as food for another, or helping to keep others in check. In response to the objection that God seemed to have created many species harmful to humankind, Derham answered[6] that 'the fierce, poisonous and noxious Creatures serve as Rods and Scourges to chastise us, as means to excite our Wisdom, Care and Industry, with more to the same purpose'. God had more than one way of ensuring that humankind was kept up to the mark.

The useful species of animals and plants were, of course, to be exploited, but as yet there was no recognition of the possibility that God's bounty might have a natural limit.

In taking control of Nature, the human race was merely regaining the place originally intended for it by the Creator. True, shipbuilding and the charcoal industry had wreaked havoc with Britain's remaining forests, providing a clear example of our ability to deplete the environment. But as John Evelyn noted in his *Sylva: A Discourse of Forest Trees and the Propagation of Timber in His Majesty's Dominions* of 1662, the obvious solution was to adopt a policy of conservation and replanting, thus maintaining the balance established by the Creator. Efforts were, in fact, made to maintain the supply of local wood, although Europeans soon began to exploit the timber resources of the rest of the world. Natural theology encouraged the idea of humankind's stewardship of God's creation, and the resources of the whole earth seemed as yet inexhaustible.

4

Theories of the Earth

Historians often treat the late seventeenth and eighteenth centuries as a unified period in the development of western culture. The triumphs of Newtonian science played a major role in convincing many Europeans that new ways of looking at the world were preferable to the old. This was to be the 'Age of Enlightenment' in which the power of human reason would sweep away ancient superstitions and allow a new social framework to emerge. Science would extend our knowledge of material Nature and reinforce the commercial and industrial power that was allowing Europe to dominate the globe. The application of reason to the study of the human mind itself would allow the creation of a new social order.

The Scientific Revolution of the seventeenth century helped to usher in the age of reason, and has received a great deal of attention from historians. Yet the scientific developments of the eighteenth century remain comparatively neglected. The period is something of an enigma, lacking the interest of the truly revolutionary period that preceded it, but not modern enough to allow the scientist-turned-historian to feel at home in it. It is a period in which the expansion of knowledge was already leading to specialization, yet where the boundaries between the modern disciplines were only just beginning to be recognized. The confidence made possible by the triumphs of the physical sciences led those who studied the earth and its inhabitants to look for all-embracing systems that would impose a conceptual structure on the world, but there was no consensus on the basic character of the system that should be employed. Apparently modern ideas rubbed shoulders with thought-patterns that are difficult for us to take seriously today.

Nowhere is this fluid character more apparent than in the environmental sciences. The eighteenth century was the period in which modern disciplines such as geology began to emerge from the welter of traditional natural history. To begin with, however, the boundaries between the disciplines were not precisely drawn, and it was possible for topics that we regard as quite distinct to be treated as integral aspects of the same basic problem. Natural phenomena such as fossils were acquiring a theoretical meaning for the first time, but as yet there were disagreements over what that meaning should be. Old-fashioned ideas, including the attribution of organic powers to the mineral kingdom, lingered on despite the prevailing support for the mechanical philosophy. By the end of the century the sciences were beginning to take on their modern aspect, but the process by which this came about was complex and has proved difficult for historians to interpret.

When historians familiar with modern science first began to look back at the eighteenth century, they focused on those topics which were most easily interpreted in familiar terms. Geography and the great voyages of exploration were left to specialists and were seldom discussed in general histories of science. Geography has a problematic status within the modern academic world: its physical dimension seems to link it with sciences such as geology, yet its human and economic dimensions give it a character more in line with the social sciences (which many regard as not truly scientific at all). It was easier for historians to exclude eighteenth-century geography from the sciences altogether and concentrate on those areas which seemed to anticipate the disciplines with which the modern scientist feels comfortable. Meteorology and oceanography are also problematic disciplines that have been ignored except by specialists. Yet eighteenth-century students of the natural world did not make these distinctions; they gathered information on a host of different phenomena and began with the expectation that all would be included in the same theoretical framework. By excluding those problematic areas which do not fit our modern preconceptions, historians have artificially fragmented the conceptual framework of Enlightenment science.

Even within those areas which could be identified with modern disciplines, hindsight has been used to pick out themes that seem to anticipate later developments. The technique of looking for the 'precursors' of modern concepts such as geological gradualism led to a highly selective study of an unrepresentative sample of eighteenth-century naturalists. Passages were taken out of context in an effort to show how one writer or another had almost (but not quite) put together the conceptual framework that we accept today. To facilitate the search for precursors, certain aspects of eighteenth-century theories were artificially highlighted to emphasize the extent to which they contributed to the development of what we now take to be the correct position. Those naturalists and geologists who favoured the 'forward-looking' theories were hailed as heroes of scientific discovery, even if this meant ignoring some very unmodern aspects of their thought. It was assumed that anyone who favoured the 'modern' alternative did so because he or she was committed to the methodology of observation and experiment. Rival theories were dismissed as obstacles to the development of science, supported only by conservatives who wished to retain traditional values. Respect for the Genesis story of the creation was identified as the most important conservative influence, on the assumption that science and religion must automatically come into conflict.

The result was a restricted and highly distorted picture of what happened in the eighteenth century. The period was fitted into the history of science by pretending that it was an epoch of initial chaos during which the burgeoning scientific method was only gradually beginning to make sense out of a vast range of new information. In the last couple of decades historians have begun to realize how much harm this overenthusiastic application of hindsight has done to our understanding of the period. It would be useless to pretend that we are as yet in a position to outline all the main currents in the development of eighteenth-century science. Many unfashionable areas still await serious investigation. But we are beginning to recognize that here is a period that must be studied in its own right, without being forced into a Procrustean bed defined by our modern interests. In some areas, at least, we have been able to see that the model of a neat, step-by-step replacement of ancient superstition by modern theory gives a hopelessly distorted impression of how science developed. If we are to appreciate what eighteenth-century naturalists were actually trying to do, we must be prepared to put our current preconceptions to one side.

Classification and Explanation

Some historians have concentrated on trying to identify the underlying principles that shaped the Enlightenment view of Nature. The French historian and philosopher Michel Foucault has argued that the eighteenth-century view of the world was profoundly different to our own.[1] The goal of the natural sciences in what is called the 'classical' era was to impose a rational order on the variety of Nature. Taxonomy (classification) consisted of working out a system of mental pigeon-holes in which every conceivable species would have an appropriate place. Fitting the species actually observed in Nature into this conceptual system would then be a more or less automatic process. Gaps in the arrangement would merely be empty pigeon-holes waiting to be filled in by later discoveries. There could be no unexpected discoveries because the conceptual network was a closed system imposed by the laws of thought itself. Foucault argues for an abrupt breakdown of this self-contained image of Nature at the end of the eighteenth century, ushering in the modern view in which Nature is essentially an open-ended and unpredictable system.

No one doubts that classification was an important preoccupation of eighteenth-century naturalists, but some historians are suspicious of Foucault's attempt to depict it as such a rigid procedure. A more conventional way of representing the key development in eighteenth-century ideas about the environment is to emphasize the extent to which naturalists began to explore the concept of a changing universe. Recognizing that the earth and its inhabitants have changed through time was certainly a major challenge to the traditional view of divine creation – hence the assumption that science and religion have always tended to come into conflict. But the image of Nature as an ever-changing system seems equally incompatible with the desire to impose a rigid order on the observable world. Foucault counters this point by insisting that early attempts to depict the universe as a changing or developing system were quite different to the evolutionary theories of later times. In the eighteenth century, change involved merely the sequential filling in of pigeon-holes in the universal scheme of things. What happened in the course of time was preordained from the beginning. Change was merely the entirely predictable unfolding of a predetermined pattern.

Some historians have been impressed by Foucault's characterization of eighteenth-century ideas. They accept his claim that the classical era mounted a holding action to defend the idea of an essentially fixed universe against the onslaught of the evidence for change. Development was admitted, but only so long as it did not introduce an element of instability or unpredictability into the system. Other historians find it difficult to accept that the appearance of a time dimension in the natural science can be dismissed so easily. If the period was so devoted to the idea of stability, why did scientists spend so much time investigating the geological evidence for change? In the end we may be forced to conclude that there was an underlying tension in the Enlightenment view of the natural world. The drive to understand the forces that shape the world came into conflict with the desire to impose order upon it. The result was a complex mixture of theoretical systems that admitted some degree of change, but often tried to limit the impact of the new discoveries in ways that seem quite strange today.

In addition to their efforts to identify the characteristics of Enlightenment thought, historians have also begun to wonder about its social dimension. Exploring the concept of a changing universe certainly had ideological implications. The concept of a static creation was taken to uphold traditional religious views, and hence the social hierarchy. Ideas based on predictable development encouraged the hope of steady reform. The radical materialists who postulated a chaotic world in which humankind played only a trivial role were striking at the roots of the existing social order. We are beginning to see the outlines of the ideological tensions within eighteenth-century ideas about Nature, but much more work needs to be done before the lessons learnt in the study of the scientific revolution can be effectively applied in this later period.

At a less political level, it is clear that significant changes were at work in Europeans' attitudes towards nature. The new science encouraged the view that the world was a purely material system that could be exploited for human gain. But there was also a growing willingness to accept that wild Nature had a beauty of its own. Mountains came to be seen as sublime expressions of Nature's power, where once they had been feared as ugly and dangerous. These changing sensibilities may be linked only indirectly to the growth of science, but we need to understand how the prevailing attitude to Nature may have influenced those who studied it in detail.

Even the gathering and publishing of information had a social purpose, and here again there is a great deal of work for historians still to do. Governments financed the voyages of discovery and established mining colleges and other forms of technical education, presumably with a view to reaping commercial profit. Some scientists became government functionaries – but what effect did this have on their work? Was science still (as is often claimed) truly international, or did the rivalries of the European powers impose restrictions on what was done? Wealthy amateurs also spent vast amounts of money on collecting and displaying natural objects, while publishers sold well illustrated scientific books in great numbers. In such cases the social purpose of the activity is less clear, leaving historians with many questions still to answer.

The Earth Described

The eighteenth century saw a continuation of the voyages of discovery that were bringing Europe face to face with the rest of the world. Greatly improved mapping techniques were introduced. The search for better means of navigating a ship across trackless oceans was continued, and the problems largely solved. The Académie des Sciences of Paris undertook a programme of research designed to allow the determination of longitude by detailed observations of the moon, and by the late eighteenth century improved lunar tables had dramatically increased the accuracy of this technique. In 1714 the British government offered a prize of £20 000 for a method that would allow longitude to be determined to an accuracy of within half a degree. This prize would eventually be won by John Harrison in 1765 for a chronometer accurate enough to allow local time to be compared with time on the Greenwich meridian. At the same time, the French clockmaker Pierre Leroy improved the theoretical principles upon which chronometers were based. By the end of the century, Europeans could at last navigate their ships around the globe with confidence.

At the same time, governments were increasingly willing to send expeditions to explore those vast areas of the globe that were still unknown. Emphasis was concentrated on the Pacific and on the polar seas. Captain James Cook's first voyage (1768–71) was undertaken to observe the transit of Venus across the face of the sun in the South Seas, but revealed valuable information about the coastline of New Zealand and Australia. Cook was an inspired navigator, exploiting the best that the new techniques had to offer. He was also concerned to preserve the health of his seamen, protecting them from the scurvy that arose when fresh food ran out on a long voyage. Trained naturalists, including the wealthy amateur Sir Joseph Banks, accompanied the ship to collect specimens from the countries visited. Cook's second voyage (1772–5) was explicitly undertaken to determine the extent of the southern land masses and demonstrated that there was no great southern continent to 'balance' those of the northern hemisphere. Although ostensibly in search of pure knowledge, the Admiralty sent Cook and other explorers out in part for nationalistic reasons, and the naturalists on board sometimes complained that they could not collect properly while the ship's main priority was surveying. There were also important consequences for economic botany (see chapter 6).

By the end of the century, those parts of the world accessible by sea had been largely explored. Only the polar regions remained, and the search for the North West Passage from the Atlantic to the Pacific to the north of Canada was to cost explorers' lives well into the nineteenth century. The interiors of the continents were not so well known. The western parts of North America remained largely unexplored by Europeans until the early nineteenth century. The first crossing of the continent was made by the Scottish explorer Alexander Mackenzie in 1793, following his expedition to the great river of the

Canadian northland that now bears his name. The threat of disease ensured that the interior of Africa would remain the 'dark continent' as far as Europeans were concerned until the late nineteenth century. But the geographers of the Enlightenment could rest assured that some ancient myths had been decisively refuted. No part of the globe was uninhabitable by humans, although tropical regions were certainly unhealthy for Europeans. The problem now became one of explaining why the various regions of the globe experienced their very different conditions. If there was as yet no basis for a grand geographical synthesis, the natural philosophers were determined to study the forces that shape the environment.

These studies were undertaken in the spirit of the Baconian programme that inspired the members of the early Royal Society. The programme involved detailed investigation of the air, the water and the land itself, but was often linked to theoretical questions arising from the physical sciences. The revolution in physics led to a debate on the shape of the earth, which was only resolved in the 1730s when the Académie des Sciences sponsored expeditions to Peru and to the Arctic to measure the length of a three-degree meridian arc. The results confirmed that the Newtonians were correct in their prediction that the earth was flattened slightly at the poles. Blaise Pascal's work on air pressure in 1648 led to the invention of the barometer and the demonstration that air pressure diminishes with altitude. Benjamin Franklin's well known demonstration that lightning is a natural electric discharge (1749) was the prelude to a detailed investigation of the nature of electricity itself.

Another area of interaction between theory and observation was in the study of tides. Newton had explained the tides in terms of the moon's influence on the earth, and detailed studies of tides were undertaken by Edmond Halley (1656–1742) and others to determine the extent to which winds and coastlines affected the local tidal pattern.

Henry Oldenburg, the first secretary of the Royal Society, encouraged the investigation of the salinity, temperature and pressure of the ocean. Robert Boyle worked on methods to determine the salt content of sea water, while Robert Hooke devised mechanisms to bring up water for testing from various depths. Halley even suggested that the age of the oceans might be estimated by measuring the rate at which their salt content was increasing through time. In the early decades of the eighteenth century, Count Luigi Marsigli (1658–1730) conducted extensive studies of the Mediterranean Sea and noted that the continents are surrounded by shallow areas of ocean that seem to be related to the land itself (the modern idea of a continental shelf). In the 1770s the eminent French chemist Antoine Lavoisier (1743–94) worked on the chemical composition of sea water.

A closely related issue centred on the source of rivers and springs, which many still believed to issue from immense underground caverns. In 1674 Pierre Perrault (1611–80) estimated the rainfall over the Paris basin to see if this would be adequate to explain the outflow through the River Seine without the need for an underground source. He concluded that the rainfall was more than adequate, rejecting the claim that his methods were too inaccurate to be convincing[2]:

> I well know that this deduction has no accuracy; but who could give one which could be precise? I believe, however, that it ought to be more satisfactory than a simple negative like Aristotle's or the premise of those who hold, without knowing why, that it does not rain enough to furnish the flow of rivers. At any rate, until someone makes more exact computations, by which he proves the contrary of what I have advanced, I shall remain convinced . . .

Perrault's results were, in fact, confirmed by more careful studies made a decade later by Edmé Mariotte. The same point was made by Antonio Vallisnieri in 1715, and from this time onwards the concept of a vast underground water source began to fall into the background. In 1752 the French geographer Phillippe Bruache (1700–73) used the concept of river basins defined by mountain ranges in an attempt to divide the earth into natural regions.

Bruache's *Essai de Géographie* was intended to provide an overarching synthesis of knowledge about the earth, but historians of geography treat the early eighteenth century as a largely stagnant period. A textbook such as Bernhardus Varenius' *Geographia generalis* of 1650 was reprinted for a century or more in the absence of anything to replace it. Only in Germany was a serious effort made to produce a general theory of geography, with the philosopher Immanuel Kant (1724–1804) lecturing on the subject from 1756 onwards. In part, the difficulty of producing such a synthesis was created by the fact that the human dimension to geography was seen as an aspect of history rather than of science. Study of the physical character of the earth was broken up into a host of topics such as those described above. The Baconian method encouraged concentration on individual problems rather than the search for unifying factors on a global scale. It has been argued that the search for a global synthesis revived only when Kant's philosophical innovations began to undermine the foundations of Baconian empiricism.[3] This paved the way for the work of Alexander von Humboldt in the 1790s – an innovation that would profoundly affect the development of the environmental sciences in the first half of the next century (see chapter 6).

In the meantime, if there were efforts to provide a comprehensive 'theory of the earth', they came from the introduction of a historical dimension. The natural philosophers of the late seventeenth century devoted a great deal of effort to the study of the earth's surface. They climbed mountains and studied the rocks of which they were composed. They puzzled over the fossils found in the rocks, many of them apparently resembling marine creatures. They also investigated the natural processes that affected the earth: volcanoes (the eruption of Mount Etna in 1669 attracted much attention), earthquakes and the effects of the various processes of erosion.

These studies generated renewed interest in the question of the origin of mountains. It was possible to maintain, as had Varenius, that the present structure of the earth's surface was essentially the same as that originally created by God. But other scholars were inclined to believe that the mountains might actually have been formed by natural processes at the time of Noah's flood. This would account for the fossils contained in stratified rocks, which could be interpreted as deposits laid down when the earth had been covered with water. But was it possible that the shape of the mountains was changed by natural processes *after* they had been formed? The effects of volcanoes and of denudation by rivers and streams suggested that processes were still at work that could profoundly alter the surface of the earth.

The ultimate product of these investigations was a rash of theories attempting to explain the existing structure of the earth in terms of natural processes operating through time. But this period also witnessed a profound change in people's attitudes towards the earth. Those who depicted the mountains as products of the flood were able to draw upon a common feeling that such massive irregularities of the surface were an indication that the earth was subject to a process of decay. Many Protestants believed that the earth itself suffered as a result of humankind's sins; mountains were ugly and dangerous and thus formed a permanent reminder of our fallen state. As one traveller noted[4] in 1693, after passing through the Alps, mountains 'are not only vast, but horrid, hideous, ghastly Ruins'. The sheer irregularity of the hills and valleys was an affront to the tidy minds of the seventeenth-century rationalists. They would have been happier if the mountains and continents were formed in regular geometrical shapes.

It is difficult for us to realize that mountains have not

always been seen as places of beauty and wild grandeur. But this more positive attitude is a product of education, and did not begin to enter deeply into European consciousness until the eighteenth century. The great formal gardens of the seventeenth century are certainly expressions of human power over Nature, but they are also characteristic of an age that felt very uncomfortable when faced with the wild irregularity of Nature. The botanical gardens of the time were laid out along geometrically regular lines, not as an expression of human rationality, but because it was thought that the garden of Eden had been created to an artificial pattern. Humankind was merely trying to reimpose divine order upon a world that had degenerated into chaos.

In the eighteenth century, attitudes gradually began to change in a way that allowed the irregularity of Nature to be admitted as a source of beauty. Europeans were becoming more tolerant of natural forms, more willing to see beauty in diversity. The ordered gardens of the seventeenth century gave way to the more natural appearance favoured by Lancelot 'Capability' Brown, who designed parks and gardens to enhance the aesthetic appeal of the existing landscape. This created a pastoral ideal, allowing the wealthy to escape the harsher transformations being forced upon the landscape by the new industries. Only with the emergence of the Romantic movement at the very end of the century did mountains come to be worshipped as expressions of the sublime in Nature.

It may seem paradoxical that an age that was becoming increasingly conscious of the earth as a source of minerals to be exploited for industrial development should develop an enhanced awareness of natural beauty. The paradox may be resolved by recognizing that there was an underlying movement towards treating the earth as a natural system rather than the direct product of the Creator's hand. Even those who believed that the universe was governed by divinely instituted laws were prepared to accept that the earth itself was a product of natural processes. This made it easier for the commercial classes to justify humankind's interference with the system, but it also seems to have encouraged others to believe that Nature is something worthy of study in its own right. The poet and the painter could worship the power of Nature, seeing the wildness and irregularity of mountains as a symbol of that power, while the scientist could contemplate the underlying unity imposed by the regularity of the physical laws themselves.

THE ORIGIN OF THE EARTH

Historians have invoked a number of factors to explain the sudden growth of interest in cosmogonies (theories of the origin and development of the earth) in the late seventeenth century. The new Baconian philosophy encouraged the study of rocks and fossils, thereby highlighting the evidence that the earth's surface has not always been as we see it today. The mechanical philosophy could be applied to the earth itself in the hope of uncovering the physical nature of the processes at work. This attitude was also encouraged by the astronomical revolution initiated by Copernicus. In the Aristotelian world view the earth was a unique object at the centre of the universe, but now it was merely the third planet orbiting the sun. Presumably, then, it was a material body like any other, and could have an origin along with the rest of the solar system in a natural process. There was also a theological tension between those who believed that the earth was designed by God and those who saw it as a ruined planet fit for sinners. Newton and some of his followers thought that the social upheavals of their time heralded the imminent end of the world according to biblical prophecy. This millenarianism ensured that speculations about the origin of the earth would be coupled with predictions of its end.

The early theories thus had a strong link to the Bible. A literal interpretation of Genesis leads one to assume that the universe and the human race were created more or less together. Some Christians felt that God would have seen no purpose in creating a world that was not inhabited by rational beings. James Ussher, Archbishop of Armagh, published a chronology based on biblical scholarship (1650–4) that estimated the date of creation to be 4004 BC. Ussher's figure is treated with scorn by modern geologists, but he was an eminent scholar and his estimate was printed in many Bibles. It would take the best part of a century before even the geologists would feel comfortable with a timespan much in excess of Ussher's. Their willingness to attempt even a limited modification of Genesis was, however, encouraged by discoveries that seemed to cast doubt on the story of Adam and Eve as the founders of the human race. Travellers reported that the Chinese claimed to have been civilized for many thousands of years, thus undermining the claim that Genesis offered a universally valid account of human origins. As faith in the literal accuracy of the Genesis story of human origins diminished, so it became possible to speculate about the origin of the earth itself.

The resulting theories consist of a curious mixture of scientific speculation and biblical literalism. Many historians of geology have dismissed them as of little real importance on the grounds that science so bound up with religion must necessarily be unproductive. More recent assessments have adopted a less critical attitude, in part because we now recognize that most late-seventeenth-century science included similar components. These apparently clumsy speculations provided a foundation for trying to organize the ever-expanding amount of information about the earth. Building upon this foundation, the naturalists of the eighteenth century would put together an impressive array of observations throwing light on both the structure of the earth's crust and the processes still affecting it. The creation of geology as a recognizable discipline was a product of this steady accumulation of information. The resulting science was often significantly different in different countries – so new a field was inevitably subject to local influences, both social and environmental. But the various schools of geological thought could nevertheless interact with one another because they were all trying to address the same problem.

The Meaning of Fossils

The observations that generated this mine of information included surveys of local landforms, revealing phenomena that seemed to indicate the power of erosion to destroy the existing features. There were also the strange objects that could be dug out of the earth. The urge to collect objects of natural interest expanded dramatically during the seventeenth century among those able to afford a 'cabinet of curiosities'. Such cabinets would certainly include unusual and visually attractive minerals, along with those structures with a resemblance to living things that we would now call 'fossils'. In the eighteenth century this passion continued and led to the setting up of a network of dealers who specialized in obtaining rare specimens for their wealthy clients. Some of these wealthy naturalists played an important role both in communicating information and in encouraging expeditions to newly opened-up parts of the world. Many universities already possessed museums, and in some cases these became centres of important scientific collections. The motivations that lay behind the building up of these collections are not always easy for us to understand, but historians are increasingly inclined to see the whole movement as an effort to incorporate natural knowledge into the world of property. Those who possessed important specimens were, in a sense, the owners of the knowledge that could be derived from the objects themselves.

For many of the naturalists who studied the earth's crust, the most important phenomenon was the appearance of fossils within those rocks which were stratified (laid down in layers). Today we take it for granted that fossils are the remains of once-living things trapped in sediment that was laid down under water and then solidified to form rock before being elevated to its present position. Fossils are thus a clue to the earth's past, and the nature of the sedimentary rocks provides clear evidence that much of the earth's surface has been created by natural processes. But not every naturalist interested in the earth shared this view. Many Renaissance thinkers preferred to see fossils as the products of mysterious forces operating within the rocks. This view was still shared by some seventeenth-century fossil collectors, including the noted Welsh naturalist and traveller Edward Lhwyd (1660–1709). To him, fossils were merely natural objects to be classified along with all the other products of the earth. They carried no message about the past. This does not mean that Lhwyd was blind to the implications of his observations, since he admitted that vast periods of time would have been required to explain the effects of denudation to be seen in the Welsh valleys. But fossils were not part of his effort to understand the processes that might have shaped the earth's surface.

Lhwyd's view was shared by Martin Lister, who pointed out that ammonites, although superficially resembling a seashell, bear no resemblance to any living mollusc. But other naturalists looked at the very close resemblance that could be seen between fossil sharks' teeth and those of the living sharks, and accepted this as clear evidence that the fossils were derived from once-living creatures. A pioneer in such studies was the Scandinavian anatomist Niels Stensen, better known under the anglicized version of his name, Nicholas Steno (1638–86). Steno spent much of his life in Italy, at first as physician to the Grand Duke of Tuscany, and it was here that the opportunity to dissect a shark made him aware of the true nature of the fossil teeth. He then went on to explore the stratified rocks of Tuscany with the intention of seeking to explain how they had been formed. His theory was published in his *Prodromus to a Dissertation concerning a Solid body Enclosed by Process of Nature within a Solid* of 1669. Steno stripped the fossils of their old symbolic meaning and treated them solely as a clue to the origin of the rocks within which they were contained. The layers of rock had once been deposited as sediment on the bottom of the ocean, with the remains of living things becoming trapped and ultimately petrified within the sediment.

Steno also went on to discuss the geological history of Tuscany, postulating two epochs of sedimentation, each followed by a period in which great caverns were hollowed out beneath the surface by flowing water. The subsequent collapse of the overlying strata into the caverns accounted for the broken and irregular shape of the once-level deposits. Although his views on the nature of fossils were remarkably modern, Steno had no intention of challenging the biblical story of the earth's history. He accepted that the first period of deposition occurred immediately after the creation, while the collapse of the overlying strata constituted the great flood. He did, however, admit that the later period of collapse represented an episode in the formation of the modern crust that was not mentioned in the Scriptures.

Steno's theory attracted wide attention, especially in England. Henry Oldenburg (the Secretary to the Royal Society) published an English translation in 1671. As a result, there were numerous efforts to explore the implications of the fossil-bearing rocks in more detail. One of the leading British fossil collectors was John Woodward (1665–1728), who lectured at Gresham College in London. Woodward was an avid collector of all antiquities, drawing no clear distinction between what we would now call archaeology and paleontology. He dug up Roman remains to throw light on the early history of Britain, just as he dug up fossils to study the early history of the earth. For his pains he was ridiculed on the London stage as someone more interested in the dusty past than in the amorous adventures of his own wife. Yet Woodward published very accurate depictions of fossils, and his *Essay toward a Natural History of the Earth* of 1695 offered an extension of Steno's explanation of how they were formed. Woodward supposed that the whole of the earth's original surface was destroyed at the time of the deluge, the material subsequently settling out of the water to form the layers of stratified rocks. He noted that particular fossils are found only in certain types of rock, suggesting that this could be explained by the heavier organic remains settling to the bottom more quickly.

Woodward's suggestion is occasionally resorted to by modern creationists hoping to restore faith in the literal truth of Genesis, but even at the time it was pointed out by other fossil collectors that it was not borne out by observation. One such critic was the eminent naturalist John Ray, who postulated both a deluge and earth movements caused by underground explosions to account for the present state of the earth. Like Lhwyd, Ray was aware of numerous phenomena that made it difficult to stay within the biblical chronology. But his most active theological concern was the threat that the fossils seemed to offer to the notion of divine providence. Ray believed that all species were designed by a wise and benevolent Creator, implying that the world is static and unchangeable. Yet if ammonites were once living shellfish, they must have become extinct, thus leaving a gap in the plan of creation. Writing to Lhwyd, he confessed that, if one adopted this view[5]:

> . . . there follows such a train of consequences, as seem to shock the Scripture-History of the novity of the World; at least they overthrow the opinion generally received, & not without good reason, among Divines and Philosophers, that since the first Creation there have been no species of Animals or Vegetables lost, no new ones produced.

For all his experience as a naturalist, Ray could thus see points in favour of Lhwyd's claim that fossils were not organic remains. It was by no means easy to accept that God's creation was subject to change.

Ray and Woodward rejected the claim that the current state of the earth is one of ugliness and decay. For them, the mountains were useful to humankind, serving as the sources of streams and playing a vital part in the world's economy. They were also willing to treat the earth as an object of beauty, something to be studied as evidence of God's bounty to humankind. The old idea that the earth had fallen into decay was gradually being replaced by a more positive view.

It was Robert Hooke (1635–1703) who made the boldest effort to explore the consequences of the new discoveries in his *Discourse of Earthquakes* (completed in 1668 but not published until 1705). Hooke was convinced of the organic origin of fossils, having used the microscope to compare the structure of fossil and living wood. He thus shared Steno's view that the fossils were the remains of living things trapped in sediment, offering plain evidence of the past state of the earth[6]:

> ... if the finding of Coines, Medals, Urnes, and other Monuments of famous persons, or Towns, or Utensils, be admitted for unquestionable Proofs, that such Persons or things have, in former Times, had a being, certainly those Petrifactions may be allowed to be of equal Validity and Evidence, that there have been formerly such Vegetables or Animals. These are truly Authentick Antiquity, not to be counterfeited, the Stamps and Impressions, and Characters of Nature that are beyond the Reach and Powers of Humane Wit and Invention, and are true universal Characters legible to all rational Men.

Hooke had little patience with those who saw all the sedimentary rocks as the remains of Noah's flood. He noted the evidence that earth movements might alter the shape of the surface and suggested that violent upheavals in the past had elevated the stratified rocks from the sea bed to dry land. He was prepared to accept that some species might have been made extinct in the catastrophes. Hooke thus moved towards the concept of a changing earth, but even he could not visualize a greatly extended timescale, hence his reliance on the violence of prehistoric earthquakes. He suggested that the various legends of catastrophic floods and upheavals were evidence that our earliest ancestors had actually witnessed such events.

New Cosmogonies

The drive to study the earth's crust generated information that encouraged speculation about changes taking place on the earth. But not all of the 'theories of the earth' were based on detailed observation. This is evident from one of the most famous and most controversial examples, the *Sacred Theory of the Earth* of Thomas Burnet (*c.* 1635–1715), published originally in 1681–9. As his title indicates, Burnet believed that the Bible should be taken seriously as a guide to the earth's history, although he warned against following the Genesis story too closely. His natural inclination was to believe that the earth as created by God must have been a beautiful place, a fit habitat for humankind in our original state of grace. For Burnet, 'beauty' meant order and regularity, whereas the earth was now a 'dirty little planet', overgrown with ugly and dangerous mountains.

He believed that it would be possible to explain the emergence of this state of affairs as a consequence of the great flood. As well as killing off all except Noah and his family, the flood had dramatically altered the original state of the planet to produce its present degraded state. Burnet's intention was to explain these changes as the inevitable consequence of physical processes, thereby showing that natural philosophy and the Bible were in harmony. He also wished to account for what he believed would be the imminent end of the world, again by physical causes.

The philosopher Descartes had suggested that the earth might have originated as the ball of ash left by a burnt-out star. Interpreting sunspots as an indication that the cooling and solidification of a star began from the outside, Descartes argued that the earth had originally been covered by a shell of solid matter with the oceans lying beneath the crust. Burnet borrowed this idea and suggested that the smooth original shell represented the paradise upon which humankind had originally been created. At that time there had been no seasons, with their inconvenient alternation of hot and cold weather. Eventually, however, the crust had dried out and cracked apart, collapsing in fragments into the waters beneath and precipitating the great flood. The irregular fragments of the original shell constitute the ugly face of the land masses we know today. The igniting of underground fires would eventually lead to the whole earth being consumed in a great conflagration.

The sequence of events postulated in this theory bore little relationship to the observations made by naturalists such as Ray and Woodward. Yet Burnet was not unaware of the evidence for a changing earth. He noted that erosion would eventually destroy the mountains – clear evidence that the event that had produced them had occurred only a few thousand years ago. Burnet refused to accept the possibility of new mountains being raised by earthquakes, as Hooke and Ray supposed.

Burnet's book aroused both interest and antagonism among his contemporaries. It was soon followed by an alternative, more in line with the Newtonian theory, the *New Theory of the Earth* of William Whiston (1667–1752), which appeared in 1696. Whiston argued that the deluge was caused by a comet striking the earth. There was much opposition from conservative thinkers, who objected to the liberties Burnet and Whiston had taken with Genesis. Their arguments seemed to undermine the traditional belief that the flood was a direct (i.e. miraculous) consequence of divine anger. Burnet had justified his provision of a purely natural explanation by arguing that the Creator could foretell the need for sinful humankind to be punished. But however hard the early theorists might try to link the sequence of events to the Bible, their ideas reduced the earth itself to a material system governed only by natural law.

The philosophers of the eighteenth-century Enlightenment were increasingly willing to challenge the authority of established churches, and were thus less tolerant of the biblical story of creation. The outburst of theoretical speculation that marked the last decade of the seventeenth century left a legacy of interest in the basic problem of explaining how the earth's surface acquired its present form, but Noah's flood was no longer assumed to be the principal mechanism of change. As yet, however, few theorists were willing to go along with Hooke's suggestion that the surface of the earth itself has been unstable. The most popular alternative to the deluge as an explanation of the sedimentary rocks was the assumption that the earth had been completely covered with water immediately after its creation. The sedimentary rocks would thus have been laid down when the oceans were much deeper than they are today, and gradually exposed as dry land as the oceans retreated. This retreating-ocean theory was often known as 'Neptunism', after the Roman god of the sea.

The theory was expounded by the German philosopher G. W. Leibniz (1646–1716) in his *Protogaea*, written in the late seventeenth century although not published in full until 1749. Leibniz made no attempt to challenge the orthodox chronology, but a far more radical application of the idea was advanced by the French writer Benoît de Maillet (1656–1738) in a book written and circulated clandestinely in the early eighteenth century, although not published until 1748. In an effort to evade censure by the Church, de Maillet (who had been French consul in Egypt) presented the theory as the work of an oriental philosopher whose name formed the title of the book: *Telliamed* (de Maillet spelled backwards!). The theory assumed that the earth was immensely ancient and made no reference to a deluge in recent times. The Egyptians had kept detailed records of the

flooding of the Nile, and de Maillet argued that this provided evidence of a decline in the sea level during historic
times. His theory extended this trend back into the past
via the supposition that the whole earth had originally
been covered with water. When the decline in the sea
level exposed the first dry land, erosion began and sediment built up on the flanks of the submarine mountains, only to be exposed as the retreat of the ocean
continued.

This model was followed by the French naturalist G. L.
Leclerc, comte de Buffon (1707–88), when he proposed a
theory of the earth to introduce the survey of the animal
kingdom contained in his monumental *Histoire naturelle*,
which began publication in 1749. Buffon was the keeper
of the royal zoological garden in Paris and was determined
to produce the most comprehensive account of the earth
and its inhabitants then available. He realized that the geographical dimension of natural history – the distribution of
the different families of animals around the globe – was the
product of a historical process. A complete natural history
must therefore include an account of how the present state
of the planet's surface has been formed. Buffon set out to
provide such an account in his introductory volumes of
1749, and – although censured by the Church for challenging the biblical chronology – he extended his ideas in
a supplementary volume of 1778, which was given the
separate title *Les époques de la nature*. There were seven
epochs, providing a superficial parallel to the seven days
of creation.

As a Newtonian, Buffon sought an explanation of our
planet's origin that was compatible with the new physics. He
proposed that the planets were struck off from the surface
of the sun by a colliding comet. Globules of incandescent
matter were thrown into space, where they gradually cooled
and solidified. Each planet would cool at a different rate
depending on its size – Buffon estimated that the earth
would have taken at least 70 000 years to reach its present

state. The granite mountains represent the only parts of the originally solidified crust still visible today – all the rest has been covered by overlying material deposited in later epochs. As soon as the crust was cool enough, a rain of nearly boiling water began and covered all, or almost all, the surface with a vast ocean. The original surface was then eroded and the resulting debris laid down as beds of sedimentary rocks. The fossils indicated that the ancient ocean was already inhabited, although Buffon knew that the fossils often represented creatures no longer alive today. He invoked the gradual cooling of the earth to explain extinction by supposing that the earth's earliest inhabitants were adapted to life in oceans of almost boiling water, and had died as the water cooled!

Buffon believed that volcanoes were the result of chemical activity that began only in modern times. He attributed their explosive power to steam produced by water penetrating into the earth, for which reason volcanoes are only found near the seashore.

When the dry land at last became inhabited, the surface temperature was still much higher than it is today. Buffon appealed to the discovery of the remains of elephant-like creatures – the mammoth and the mastodon – in Siberia and North America to prove that northern regions had once enjoyed a climate as warm as the modern tropics. Only as the planet cooled were these heat-loving species able to migrate to their present abodes[7]:

> One cannot doubt ... that the earliest and greatest formation of animated beings occurred in the high, elevated regions of the north, from whence they have successively passed into the equatorial regions under the same form, without having lost anything but their great size; our elephants and hippopotamuses, which appear large to us, had much larger ancestors during the time in which they inhabited the northern regions where they have left their remains.

This vast southwards migration had been interrupted by the collapse of some parts of the original surface into the oceans, dividing the Old World from the New. The elephants that had moved into America had evidently disappeared in modern times. Buffon was inclined to believe that the Americas provided a generally less favourable climate and were inhabited by creatures of lesser stature than their Old World equivalents.

Buffon's account pushed the speculative power of the old-style theories of the earth to its limits. He sought to link his vision of the earth's origin and physical development to the growing evidence suggesting that life itself had a history. He made a pioneering effort to discern a historical dimension underlying the geographical distribution of species. Yet in the end his search for a comprehensive explanatory system overreached itself and brought this whole approach into disrepute. There was too much information to fit comfortably into so simple a scheme. The evidence for a southwards migration, for instance, was suspect almost from the start. The remains of the mammoth had been brought to the attention of naturalists owing to an expedition to Siberia led by Peter Simon Pallas in the years 1768–74 – but Pallas himself noted that the giant northern creatures may have had long hair, adapting them to a colder climate.

Although he was alert to evidence of past changes, Buffon had not tried to reconstruct the history of life on the basis of the evidence alone. Instead, he had imposed upon the evidence a predetermined historical sequence defined by his theory of the earth's origin. The underlying physical trends – the decline in the sea level and in the temperature – constrained his explanations within rigid limits. The later decades of the century saw few efforts to carry on the search for an all-embracing theory of the earth. Geologists abandoned the attempt to link their science with cosmology via an explanation of planetary origins. As their work became more specialized, they concentrated on the need to provide a more detailed sequence of events that would account for the distribution of minerals on the surface.

FIRE AND WATER

The conventional image of late-eighteenth-century geology depicts the science as a battleground for progressive and conservative forces. The progressive wing is represented by the 'Plutonists', who believed that forces deep in the earth's interior were responsible both for volcanic activity and for the elevation of new land masses. This position can be identified as the source of progress because it anticipates the modern viewpoint. Especially important is the form of Plutonism advocated by the Scottish geologist James Hutton, who insisted that all of the changes affecting the earth's surface have taken place slowly and gradually. Opposing this essentially modern view of geological change stands the Neptunist theory advocated by the German mineralogist A. G. Werner. Since it was based on the retreating-ocean model, Werner's theory was obviously wrong, and must have been supported mainly by those geologists who wanted to use the possibility of a resurgence of the waters to defend the biblical story of the flood.

Modern historians have exposed this interpretation as an artifice of hindsight, an attempt to make the period conform to the conventional notion of scientific progress. The Neptunists were by no means the backward-looking caricatures presented in the older histories. Werner had no interest in the Genesis story, nor did most of his continental followers. Only in Britain did a few scientists from an older generation seek to link the theory to the great flood. Furthermore, the Wernerians did valuable work in establishing the principles upon which a historical explanation of the earth's crust would be based, even though their interpretation of the underlying causal trend would have to be abandoned. Hutton had little influence outside his native Scotland, and his theory was not a direct anticipation of later ideas. If Werner's supporters gradually abandoned the retreating ocean in favour of earth movements, this was not because of Hutton's arguments.

Neptunism

Werner's revival of Neptunism drew its inspiration from a mineralogical tradition quite distinct from the cosmological theorizing of de Maillet and Buffon. The German states derived much of their income from the mining of metals, and their rulers made sure that the local industry would be supervised and encouraged. Government officials regulated what was going on, and provision was made for the education of future generations of mining engineers. Suspicious of the old universities, many states, including Saxony and Prussia, set up mining schools and employed eminent mineralogists to teach in them. The French government followed suit in 1783 by setting up the Ecole des Mines and encouraging the production of a map displaying the country's mineral resources.

The German mining schools fostered a quite different theoretical tradition that traced its origins back to the writings of Agricola in the sixteenth century. From this tradition came an interest in the formation of mineral veins as a chemical process, following the chemical theories of J. J. Becher and others. The formation of the earth's crust was seen as a process in which precipitation and crystallization – usually assumed to be from solution in water – played the major roles. The laying down of the stratified rocks by mechanical sedimentation was of much less interest. The German mineralogists resisted the claim that minerals could be classified along the same lines as the species of animals and plants. They wanted to identify the basic types of minerals with their formative processes and then to use this information to locate the most likely sources of mineral veins in the earth's crust.

It was assumed that any rock with a crystalline structure must have been deposited from water. The existence of granite mountains thus seemed to provide evidence that the ocean had once covered the whole earth. It was discovered that the Baltic Sea was slowly getting shallower, which seemed to support the idea of a steady retreat of the waters. (Geologists now account for this as a belated consequence of the ice ages: the surface of northern Europe is still rising after having been relieved of the weight of the ice.) Following the retreating-ocean model, mineralogists such as J. G. Lehmann (*d.* 1767) developed a division of the earth's crust into classes of rocks with different mineral properties. In 1756 Lehmann distinguished three classes of rocks. The primary rocks were supposed to have been formed by crystallization from the great ancient ocean. They formed the core or heart of the great mountain ranges and contained most of the useful mineral veins. The flanks of the mountains were covered with secondary rocks, formed by sedimentation during a later inundation that Lehmann still identified with the biblical deluge. More recent tertiary rocks were formed by ongoing processes of erosion and deposition of sediment in the post-diluvial period.

It was Abraham Gottlob Werner (1749–1817) who built most effectively upon the foundation laid by Lehmann and the mineralogical school. He began teaching at the mining school in Freiburg, Saxony, in 1775. Although he wrote comparatively little, his fame became so great that students flocked to Freiburg from all over Europe, returning home to spread the word about their master's teachings.

Because the theory depended upon the retreating-ocean concept, it has often been caricatured as the basis for an oversimplified 'onion-skin' model of the earth's crust as a series of regular, concentric layers. Werner believed that the various formations of rocks have been laid down in sequence, one upon another. But he was aware of the complexity of the formations and explained this as a consequence of the irregularity of the earth's original shape. The primary rocks were crystallized directly onto the surface, following the original contours. The Wernerians taught that a layer of granite, the oldest rock of all, covered the whole surface of the earth. As the great ocean retreated, some of the primary rocks were at last exposed to the air, whereupon they became subject to erosion. The secondary and tertiary rocks were formed from the erosion products, laid down in successive layers on the flanks of the granite mountains. In this theory the sea level at the time of formation defined the highest elevation of the rocks, so the younger rocks do not reach as far up the mountainside as the older ones upon which they have been deposited. Werner postulated some settling of the strata after deposition to account for the folding and faulting observed in many regions.

The power of the Wernerian theory lay in its ability to organize all the mineralogical data by identifying each formation with a particular period in the formation of the earth's crust. The stratigraphical principle of superposition – that younger rocks are deposited on top of older ones – was firmly entrenched in this approach. The logic of the retreating-ocean theory imposed a rigid framework upon the sequence of events. Primary rocks such as granite and basalt were formed by crystallization from chemicals dissolved in the ancient ocean, and there was no possibility that such rocks could be formed at any later stage in the earth's history. In principle, each mineral type belonged to a particular epoch of deposition – although in practice Werner accepted that different kinds of rocks were associated together in certain formations. His theory thus began to depart from the strict framework imposed by the retreating-ocean model and took on the characteristic of a genuinely historical geology.

Werner was even prepared to use the fossils as a means of dating the rocks, although the full development of paleontology as a stratigraphical tool came later. The fossilized organisms of the oldest rocks were primitive creatures adapted to life in an ocean that contained many dissolved chemicals. As the waters became purified, more advanced sea creatures were introduced, along with terrestrial forms to inhabit the ever-expanding land surface.

Some of Werner's own followers accepted the growing evidence that basalt, at least, was formed from an originally molten state. Werner had minimized the extent of vulcanism by supposing that localized melting of rocks occurred as the result of chemical reactions in recent periods. He did, however, make the valid point that the crystalline structure of granite was quite unlike the lava produced by a modern volcano. His theory was also criticized for offering no explanation of where the water of the great ocean came from, or where it went to in the course of the earth's history. On this point Werner adopted the stance of a specialist: he was a geologist, not a cosmologist, and it was not his job to speculate on such matters. The evidence for a decline in the sea level was contained in the rocks themselves, and he merely interpreted that evidence as best he could. Werner's conviction was based on the assumption that the earth's crust is absolutely rigid. Only the growth of evidence suggesting that substantial elevation is possible would undermine this conviction and expose the weakness of the retreating-ocean theory's basic postulate. Werner's students were to play an important role in providing this evidence, although many of them saw their work as an extension, not a contradiction, of his basic stratigraphical programme.

Werner's theory played an important role by setting the scene for the 'heroic age' of geology in the early nineteenth century. The Wernerians' 'formations' were the basis for the geological periods we recognize today. Earlier historians failed to recognize this because their attention had been focused on a handful of older Neptunists who used the theory to revive interest in the story of

Noah's flood. Following the outbreak of the French Revolution in 1789, conservative forces in Britain saw any challenge to the authority of scripture as a threat to the social order. Scientists such as J. A. Deluc (1727–1817) and Richard Kirwan (1733–1812) took it upon themselves to resist the anti-biblical trend in geology by adapting Neptunism to provide an explanation of the flood story. Their writings created a sense of crisis within the British scientific community that was lacking elsewhere, and has led English-speaking historians to overestimate the extent to which Wernerianism had been associated with scriptural geology.

A far more characteristic Wernerian was Robert Jameson (1774–1854), professor of natural history in the University of Edinburgh. In its teaching of the natural and earth sciences, Edinburgh was far ahead of the universities south of the border. Its chair of natural history was founded in 1767, and the second professor, John Walker, had done much to promote the teaching of the subject. By the time Jameson took over in 1804 he had already studied under Werner at Freiburg and was determined to spread the influence of the theory in the British Isles. He was an active teacher and did much to build up the university's museum as a teaching aid. Geology was seen as an important part of natural history, and – since the university was non-sectarian – it was taught for its practical implications, not for its bearing on Genesis. Jameson also founded the Wernerian Natural History Society to encourage research under the umbrella of the Neptunist paradigm.

The third volume of Jameson's *System of Mineralogy*, published under the title *Elements of Geognosy* in 1808, is one of the most complete accounts of the Wernerian theory in English. It shows how the theory could be defended on purely scientific grounds, without reference to the story of Noah's flood. In Jameson's eyes, it was Werner who had shown that the mountains themselves provided the best evidence for the retreat of the oceans[8]:

It was reserved for WERNER to give this theory stability. With his usual acuteness, he soon discovered that the important documents for the illustration of this great phenomenon, were not to be sought for in the formations that have taken place within the limits of human history, but in the mountains themselves, those mighty aquatic formations.

The proof was the fact that the older rocks always stretched further up the side of a mountain than the younger ones overlying them. Jameson was convinced that the granite tops of the mountains were but the exposed peaks of a vast layer of this primitive rock stretching uninterruptedly over the whole surface of the earth. In the early years of the new century such ideas were still acceptable, but they were soon overtaken (chapter 7).

Plutonism

James Hutton (1726–97) has often been portrayed as a forward-looking geologist who saw the evidence for vulcanism and earth movement and led the campaign to discredit the retreating-ocean theory. Since he described the changes affecting the earth's surface in the past as being no different from those observable today, Hutton has been hailed as the founder of modern uniformitarianism (the belief that the earth is subject only to uniform, i.e. gradual, modification). By eliminating all links with the biblical flood, Hutton's vulcanism is seen as the foundation upon which later developments in geology would be built.

This neat image of science triumphing over superstition is false on almost every count. Many Neptunists were not devoted to the reality of Noah's flood. They could advance plausible arguments in favour of many aspects of their theory, and played a valuable role in establishing the foundations of stratigraphy. Hutton himself was not an exponent of the modern scientific method. He made no

FIG: 4.1 *Alternative explanations of mountain structure*

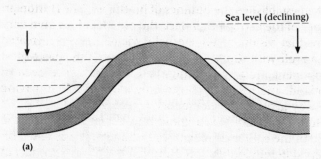

Sea level (declining)

(a)

The Neptunists and Plutonists offered different explanations of the structure of granite mountains. In the Neptunist theory (a), the mountain is built on an original elevation in the earth's surface. The granite core (shaded) is part of a world-wide layer of primary rock crystallized from the ocean when the water covered the whole surface. The overlying strata of sedimentary rocks were deposited as the sea level fell; their maximum elevation on the mountainside is thus determined by the level at the time of deposition, and the older strata are exposed at higher altitudes.

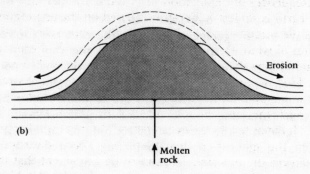

Erosion

(b)

Molten rock

In the Plutonist theory (b), the mountain is younger than the sedimentary rocks. The granite (shaded) was injected into the strata as molten rock rising from deep in the earth. It solidified slowly owing to the overlying blanket of sedimentary rock. The granite has subsequently been exposed at the top of the mountain because erosion is more active in elevated regions and has worn away the sedimentary rocks. For the same reason, the older sedimentary rocks are exposed part-way up the mountainside. Both theories thus explain the same *visible* structure, although they disagree on what lies below the surface.

detailed study of volcanic activity and discovered some of the best evidence for significant earth movements only after he had established the outlines of his theory. The Huttonian theory of the earth was, in fact, the product of his own very individual vision of an eternally viable universe created by an all-wise God. Owing in part to this theological component, Hutton's writings had little influence even in Scotland, and his ideas played only a minor role in paving the way for later developments in geology.

Hutton's theory was a grand intellectual scheme on the scale of the earlier cosmogonies – although it was based on a very different foundation. Hutton himself was a product of the 'Scottish Enlightenment', the brilliant circle of intellectuals who flourished in late-eighteenth-century Edinburgh. He knew Adam Smith, the economist, and James Watt, the inventor of the steam engine. Hutton set out to devise a theory of the earth because he had spent some time as a gentleman farmer and had become fascinated by the consequences of denudation or erosion. The old idea that the earth is subject to decay was based on the assumption that the mountains are being gradually worn away by the action of wind, frost, streams and other agents of erosion. Hutton knew that the soil upon which we depend to grow our food is produced by erosion – and is ultimately swept out to sea by the rivers. He believed that there is a balance in which the production of new soil exactly makes up for the silt deposited on the ocean floors. But what will happen when the mountains themselves have been destroyed? Whatever the timescale, Hutton could not accept that the Creator would allow the earth to become uninhabitable at some time in the future. Therefore, he reasoned, there must be a corresponding elevation of the land surface to renew the mountains and maintain the cycle of erosion.

Hutton thus postulated an eternal balance between uplift and erosion, turning the earth into a perpetual-motion machine built by the perfect workmanship of God. The sediment laid down on the ocean floor is ultimately consolidated to form the strata of sedimentary rock, and then elevated to form new areas of dry land. Volcanoes are an indication that the centre of the earth is molten, but Hutton argued that rising lava often fails to reach the surface. Instead, it intrudes beneath the sedimentary rocks, lifting them to form mountains and gradually cooling to form masses of granite or basalt. Meanwhile the mountains are constantly being eroded as streams carve out the valleys to establish the drainage systems we observe today. In Hutton's system there is no such thing as a primary rock dating back to the earth's origin. Granite can be formed at any time whenever molten matter intrudes into the crust, and the sedimentary rocks are the erosion products of earlier rocks – which in their turn may have been laid down following the erosion of yet earlier continents. The whole surface of the earth has been worked and reworked over and over again in an eternal system designed to keep the planet perpetually habitable. As Hutton wrote in his most often-quoted sentence[9]: 'The result, therefore, of this physical inquiry is, that we find no vestige of a beginning, – no prospect of an end.'

Only after conceiving the outlines of this system did Hutton begin to search for evidence that significant uplift does occur. In 1785 he observed a classic instance of granite veins intruding into sedimentary rocks at Glen Tilt, Perthshire. Two years later in the Tweed basin he noticed horizontal layers of sedimentary rock deposited on top of steeply inclined beds of a much earlier age, proving that extensive earth movement and erosion must have occurred between the two episodes of deposition. His theory was thus provided with extensive 'proofs and illustrations' – yet when his friend, the chemist James Hall, offered to test the claim that granite was solidified from a molten state by experimenting with a blast-furnace, Hutton objected and the experiment was only performed after his death.

After many years of development, the Huttonian theory was published in the *Transactions* of the Royal Society of Edinburgh in 1788, and as a two-volume book in 1795. It was the first of these publications that sparked off the attacks by Deluc, Kirwan and other scientists who found Hutton's total repudiation of the Genesis story unacceptable. They claimed that, by denying that the rocks provided evidence of the creation, and ignoring the story of a catastrophic flood, Hutton was encouraging atheism. But to geologists less concerned about Genesis, the problem with Hutton's exposition was that it relied too heavily on the notion of an earth that has been designed to perfection by an all-wise God. Hutton's own book was seldom read, and his arguments found their way into the geological literature through adaptations by friends determined to stress the science at the expense of the theology. James Hall was particularly active in promoting the Plutonist theory that the centre of the earth is intensely hot. But even Hall found it difficult to accept the idea that all past changes have been slow and gradual.

The uniformitarian aspect of Hutton's ideas was preserved by John Playfair (1748–1815) in his *Illustrations of the Huttonian Theory of the Earth* of 1802. Playfair set out to describe the theory in plain language, and to free it from the theological foundation built into Hutton's own writings. Although he insisted that Hutton had no intention of denying the creation of the earth, he acknowledged that in the Huttonian system there could be no trace of such an event. All the rocks we actually observe are composed of debris from yet earlier rocks[10]:

> Thus we conclude, that the strata both primary and secondary, both those of ancient and those of more recent origin, have had their materials furnished from the ruins of former continents, from the dissolution of rocks, or the destruction of animal and vegetable bodies, similar, at least in some respects, to those that now occupy the surface of the earth.

Here, once again, was the image of an earth maintained in a more or less steady state throughout an indefinite period of time.

Hutton is remembered because he pioneered the modern view that the earth has been subjected to intensive uplift by plutonism, and that erosion is a slow and gradual process brought about by the kind of forces we still observe in operation today. But his work had limited impact because it linked these claims to the highly controversial image of an eternally self-sustaining earth. It was this creed, rather than a purely scientific empiricism, which sustained Hutton's commitment to the use of only observable causes to explain the earth's past transformations. Hutton's theory of the earth's central heat certainly provided a comprehensive alternative to Neptunism. Linked to Buffon's notion that the planet is cooling down from an originally molten state, Plutonism would form the basis for most nineteenth-century theories of the earth's history. But in the steady-state version presented by Hutton the idea was unacceptable even to those geologists who had little respect for the Genesis story of creation.

Hutton played only a limited role in the process by which Werner's theory was transformed to become the foundation stone of modern stratigraphy. In his published works, Hutton expressed no interest in the problem of establishing the relative dates of the various episodes in the formation of the earth's crust. He was more concerned to identify the factors that interacted to sustain the eternal cycle of creation and destruction. It was the Wernerians who set out to define a sequence of geological formations that could be linked to distinct episodes in earth history. Their conception of the underlying trend was gradually abandoned, but their emphasis on establishing the sequence in which the rocks were laid down was to become the basis for the 'heroic age' of geology in the early nineteenth century (chapter 6).

Evidence that the earth's crust was unstable was pointed out by continental geologists who had no interest in the Huttonian steady-state theory. The earliest challenges to Neptunism came from the school of 'Vulcanism', which emphasized the role of volcanic activity in shaping the existing surface of the earth. The Neptunists assumed that volcanoes are a very recent and very localized phenomenon. But there was evidence that volcanic activity had not only occurred in the distant past, but had occurred on a scale far beyond anything observable today. In 1752 J. E. Guettard (1715–86) reported to the Académie des Sciences that many of the mountains in central France seemed to be extinct volcanoes. Nicholas Desmarest (1725–1815) extended this argument to include other examples of what he took to be igneous rock, including the Giant's Causeway in Ireland. Werner insisted that rocks such as basalt and granite had crystallized from solution in water, but by the end of the century the majority of his supporters had come to admit that these rocks were indeed of igneous origin.

By itself, such evidence could not undermine the retreating-ocean theory. Only phenomena indicating massive uplift of sedimentary rocks could establish an alternative based on earth movement rather than a decline in the sea level. Long before Hutton and his followers began to stress this evidence, some Neptunists were already modifying the theory to include at least a limited role for elevation. Peter Simon Pallas (1741–1811) – whose reputation was based on his extensive explorations of Siberia – argued that subterranean explosions had thrown up many secondary rocks to form the contorted strata of the mountains. He was prepared to accept that the whole structure of the Alps might have been formed in this way, so that the ocean from which the secondary rocks were deposited might have been only a few hundred feet deeper than the seas of today.

It was a better knowledge of the structure of mountains that did most to undermine Neptunism. Attitudes towards mountains had changed significantly, and many travellers now explored them for their scenic grandeur. One of the most enthusiastic exponents of mountaineering was Horace-Bénédict de Saussure, whose *Voyages dans les Alpes* (1779–96) contained much information on the structure of mountains and the forces now shaping them. De Saussure accepted that valleys were formed by the flow of rivers and streams, and was quoted by Hutton on this point. Although reluctant to admit that uplift could be the result of volcanic forces, de Saussure nevertheless described many structures that indicated that sedimentary rocks had been folded and eroded after their original deposition.

In the decades around 1800 many of Werner's followers began to accept this kind of evidence and turn to Plutonism – but without abandoning the underlying programme that sought to establish the sequence of events that had built up the present structure of the earth. Unlike Hutton and Playfair, they continued to insist that the earth had an origin and a history of cumulative change. However superficially modern, Hutton's gradualism was wedded to a steady-state view of history that denied any room for development through time. Few geologists could take this seriously, and the rival developmental viewpoint continued to dominate the science even as Plutonism triumphed over its Neptunian rival.

Nature and the Enlightenment

The late seventeenth and eighteenth centuries saw a growing recognition of the possibility that the earth did not provide a uniformly stable environment. The first serious efforts were made to challenge the Genesis story of the creation of life. The modern system of classifying species by grouping together those which display an obvious 'relationship' to one another was introduced. But the relationship was seldom taken to imply descent from a common ancestor, even by those naturalists who expressed an interest in the mutability of species. Some earlier histories of biology were written with the intention of locating the 'precursors' of Darwinism in this period, but this approach is distrusted by modern historians of science. The naturalists of the age of Enlightenment found it more difficult than we might imagine to assimilate the idea of change into their thinking.

As in the earth sciences, eighteenth-century natural history combined apparently modern concepts with ideas that now seem quite bizarre. Historians need to understand the cultural environment that made such non-modern patterns of thought seem reasonable at the time. Only then can we make a realistic effort to understand the developments that led towards the modern pattern of knowledge. By making a sympathetic effort to appreciate the period on its own terms, rather than imposing modern categories upon it, we may be able to see how ideas of permanent value could be produced within a conceptual framework that was eventually abandoned. We can also hope to see *why* the conceptual framework changed, without falling back on the simple view that it was all due to the accumulation of more factual knowledge. The eighteenth century acquired a vast amount of additional knowledge of the world of life, but we need to understand why naturalists were eventually forced to interpret that information through different conceptual schemes.

By the end of the seventeenth century people had already begun to view Nature without fear. Just as mountains came to be seen as objects of beauty, so living things could be observed with pleasure. Natural history became fashionable, with aristocratic patrons supplying a social framework that rivalled and in some cases eclipsed the moribund universities. The fact that wealth sought to display itself through the collection of natural-history specimens tells us something about the morality of the time, but we must beware of assuming that an interest in observing Nature necessarily boosted scientific study. The connection between popular natural history and science has often been indirect, since the study of Nature can be undertaken for artistic, recreational, or social purposes rather than in an effort to understand what is observed. A rich enthusiast might hire gardeners, huntsmen and naturalists to satisfy his or her tastes, each assigned to a particular rank in the local social hierarchy. But the indirect connections may have been important in creating the opportunity for science to flourish. At a time when professional openings for scientists were few and far between, the patronage of the wealthy often provided an opportunity for serious students to get on with their work.

THE DIVERSITY OF LIFE

The Classical Era

Collections of animals and plants – both local and exotic – had to be described and classified if they were to be of any scientific value. The late seventeenth and eighteenth centuries saw the creation of the modern system in which species are classified according to their physical resemblances. The old interest in the medical and symbolic importance of plants and animals was replaced by a reliance on visible properties as the only measure of relationship. The French historian and philosopher Michel Foucault characterized this as the 'classical' era in the development of western thought, a period in which the urge to classify natural objects symbolized humankind's determination to impose a rational order upon the world (see chapter 4). According to Foucault, the mechanical philosophy did not lead directly to a search for the physical causes that maintained the structure of Nature. Life was reduced to a mechanical process, but, since that process could not be understood, science would impose order on the world by creating mental pigeon-holes into which all the various kinds of physical structure – including those complex structures we call living things – can be arranged in a rational pattern. This pattern might be explained as the divine plan of creation, but there was an increasing tendency for classification to be seen as the imposition of order upon Nature by the human mind.

Enlightenment naturalists were obsessed with the desire to create a system in which every species could be assigned a unique position in a comprehensible pattern. The modern technique of biological classification was created at this time to allow the vast flood of new species from abroad to be marshalled into a pattern determined by observable relationships. Foucault rightly emphasized that the whole project depended upon the assumption that the species are eternally fixed, the pattern of relationships unchanging. One could not impose order upon a constantly changing world. Where eighteenth-century naturalists were forced to confront the possibility of change through time, they frequently minimized its unsettling implications by seeing the appearance of new species as merely the filling in of hitherto empty pigeon-holes in a system whose overall structure had been fixed from the beginning. Foucault claimed that the conceptual scheme of the classical era was incompatible with the modern idea of evolution as an open-ended process driven by the demands of local adaptation, and this is an important antidote to the old-fashioned search for the precursors of Darwinism.

And yet there are many aspects of Foucault's attempt to reconstruct the classical way of thought that seem less plausible to historians of science. Traditionally, the advent of materialism was supposed to have paved the way for the first theories in which the existing structure of the earth and its inhabitants had been produced by natural processes rather than by divine creation[1]. Theories of the earth's origin are characteristic of this movement, and it would thus be no surprise to find the more radical Enlightenment thinkers searching for natural explanations of organic origins. Contrary to Foucault's interpretation, the materialists of the eighteenth century can thus be seen as the first evolutionists.

Both sides in this dispute accept that the mechanical philosophy tended to reduce living things to mere machines – the philosopher Descartes openly proclaimed that animals were no more than complex mechanical systems (self-duplicating robots, we might call them today). If what we call 'life' was merely a function of complex mechanisms, there could be no such thing as 'biology' since there would be no unique set of functions for such a science to study. Animals and plants could be studied and classified just like any other physical object. The crucial question confronting historians is this: Did the mechanical philosophy encourage a search for natural processes of change, or did its advent merely coincide with the determination to treat Nature as an absolutely stable system that could be reduced to a rational order by the human mind?

The situation is complicated by the fact that Foucault and his followers seem reluctant to admit that religion played a significant role in the thinking of many naturalists. Lip-service was paid to the claim that the order of Nature was imposed by the Creator, but this was only to escape the censure of the Church. English-speaking historians are more willing to see religion as a powerful force in the early part of the classical era, a force that came under assault from a rising tide of materialism as the Enlightenment's confidence in the power of human reason increased. British and American historians have been conditioned to see the rise of evolutionism as a materialist challenge to the concept of divine creation. Foucault and his followers represent a French school of historiography that has been less influenced by the concept of a 'war' between science and religion. They see rationalism as a key force in the history of ideas, and are less interested in the empirical approach that seems so important to English-speaking historians of science.

There is an important lesson to be learned here. One's approach to the history of science is conditioned by one's own cultural environment, and even two neighbouring European countries can have intellectual traditions so distinct that their historians take very different views on the origins of modern science. Both science and the history of science are cultural artefacts and can only be understood if their intellectual backgrounds are taken into account. The fact that many British and American historians have found something of interest in Foucault's position offers us the hope that a fruitful synthesis may emerge from the interaction between the two traditions. Each is of value in illuminating different aspects of what was actually going on in the late seventeenth and eighteenth centuries.

Foucault's approach is most fruitful when dealing with the passion for order and stability that sustained the early efforts to develop a system of classifying the objects making up the natural world. But the belief that the order of Nature was imposed by its Creator cannot be dismissed so lightly. Nor was classification the only passion of these early students of the plant and animal kingdoms. Much effort was expended on the simple description of species and their behaviour, with the complex habits of insects and other lower animals being seen as evidence of the divine workmanship by which the Creator adapted His living structures to their environment. The founder of modern biological taxonomy – the Swedish naturalist Linnaeus – also saw the behaviour of animals as part of a very different conception of natural order, an 'economy of Nature' in which each species depended on others for its food, and in turn was depended upon by its own natural predators. Since the whole system worked to impose natural limits on the fecundity of each species, Nature could be seen as a vast, carefully structured machine based on the laws of economics. This idea, anticipating some aspects of modern ecological thinking, was a powerful alternative to the rather abstract ordering of species for which Linnaeus is best known.

Religion still played a role in shaping emotional attitudes to Nature. Linnaeus' vision of Nature's economy can be seen as a manifestation of the tendency to view Nature as a system designed for humankind to exploit. The naturalist marvels at the diversity of life, but nevertheless interprets it in essentially practical terms. However complex, it is our property, something we can comprehend and ultimately modify to our own ends. But respect for God's creation could also generate a more reverential attitude, a feeling that Nature is a system to be cherished for its own sake. Even where we have taken control, as in any well farmed landscape, we should delight in studying the behaviour of our fellow species. In its most extreme form, this delight in Nature could sustain an almost romantic, Arcadian view of Nature as a source of peace and beauty in our lives. As the effects of industrialization began to make themselves apparent, more and more people would begin to long for a past world in which humankind was supposed to have been more in tune with Nature.

Both the passion to classify species and the concept of a

natural economy depended upon the assumption that the world has been created to a stable pattern. The species must be fixed, and there was much concern about how this fixity might be maintained. The concept of a non-physical force imposing the character of the species on successive generations was no longer tenable. How, then, did reproduction (or 'generation' as it was then called) ensure that offspring were essentially identical to their parents? It was difficult to believe that animals, as mere machines, could duplicate their own structure by natural processes. The only way of guaranteeing the fixity of species was to assume that there was no real 'generation' in Nature. The offspring grew from a miniature structure already contained within the ovum or egg supplied by the mother – and since she could not have formed the miniature by herself, it must have already been contained within her ovaries when she was born. In the most extreme version of this 'preformation theory' (more properly known as the theory of pre-existing germs or *emboîtement*, encapsulement), the whole human race was created directly by God as a series of miniatures, stored one within the other like Russian dolls, within the ovaries of the first woman, Eve. The same was true for the first created female of every animal species.

The preformation theory is often depicted as a bizarre manifestation of a period that had not yet learned the truths of Mendelian genetics. But in an age when the concept of divine creation was still taken seriously, it seemed a natural way of reconciling the mechanical philosophy of life with the presumed stability of Nature. The theory also alerts us to the fact that we cannot draw a sharp distinction between the study of species in the world at large, and the study of the internal processes that maintain life. Whether species are stable or change in the course of time, this must be explained in terms of how each generation is formed. The rise of evolutionism cannot be understood merely as an event within natural history: from Linnaeus through to

Darwin and the modern genetic theory of natural selection, concepts of fixity or transformism have required a synthesis in which the study of reproduction interacts with the environmental sciences.

The materialists of the later eighteenth century repudiated the preformation theory because it provided no real explanation of generation. They sought alternative theories in which there was nothing to provide an absolute guarantee that the offspring would mirror the structure of their parents. Within this alternative conceptual scheme, radical thinkers began to suggest that species might not remain the same for ever. Before we leap to the conclusion that such theories represent the origins of Darwinism, however, we must recognize that they functioned within a very different intellectual framework. Some opponents of the preformation theory did not become transformists (the term 'evolution' was not used in the modern sense until the post-Darwinian era), and those who postulated change did so in a way that evades what we now see as crucial aspects of Darwinism.

Buffon's theory of the earth's history was extended to include the history of living species, but he postulated only a limited modification of fixed specific types. Buffon is a classic example of a materialist who could not shake off the conviction that species are cast in eternally fixed moulds. When J. B. Lamarck formulated his pre-Darwinian theory of adaptive evolution, he made no effort to link it to the study of the history of life through the fossil record. Nor did he study the geographical distribution of species, although Buffon had already argued that this could only be understood as the outcome of a historical process of migration and adaptation. In the eighteenth century, the history of life could still be investigated without the idea of organic evolution, and a theory of evolution could be constructed purely from ideas concerning the existing processes that govern living things. It was Darwin's great triumph in the mid nineteenth century to show that a successful theory of evolution should relate the processes of change observable in the modern world to our knowledge of the history of life and its distribution around the globe. Eighteenth-century naturalists studied all three areas, but never appreciated the possibility that they could be combined in the way we take for granted today.

The Social Environment

The newly founded learned societies of the late seventeenth century encouraged the collection of animals and plants from Europe and the world at large. The passion for natural history was also reflected in the creation both of local societies and of groups dedicated to the study of particular aspects of the natural world. As early as 1689 there was an informal botanical club based on the Temple Coffee House in London, with a membership that included a number of reputable collectors and naturalists. The butterfly net came into use in the early eighteenth century, and there was a short-lived society of lepidopterists (butterfly collectors) in London during the 1740s. Books dealing with the natural history of local regions, or with particular groups of organisms, were already selling widely. There was a thriving market in natural-history specimens, including rare flowers, shells, insects and birds. Travellers setting off for remote parts of the world were pestered by dealers asking them to collect specimens and willing to provide information on what was of most value.

The scientific value of this activity was variable. Many wealthy people collected natural-history specimens and books merely because it was fashionable – and interest in particular areas waxed and waned along with fashions elsewhere. The 1720s saw a surge of interest in botany coinciding with a fashion for floral decorations. Ladies developed a passion for painting flowers and insects. Nature had become something to be played with, and would eventually become an object of reverence. This represents an important shift in European attitudes, which helped to create a framework within which more serious study was encouraged. For every collector who prized his or her specimens for the social status they conferred, there was a serious naturalist who travelled the country gathering, describing and classifying. The social interest provided economic support for the specialists. The serious student of animals or plants could write books for the wider market and might even find employment cataloguing a wealthy person's collection. In Britain, the Duchess of Bedford and Lady Margaret Bentinck acquired vast collections and employed specialists both to collect for them and to work on existing specimens. When specialist societies were localized and often short-lived, wealthy amateurs provided a social framework within which serious study could proceed.

This was especially important in England, where formal scientific organization became moribund in the eighteenth century. Even the Royal Society degenerated into a rich man's social club. The universities had chairs (professorships) of botany, but they were occupied by nonentities who did no research and no teaching. Botanic gardens were neglected and fell into disuse. Things were better north of the border, where the University of Edinburgh had a chair of natural history from 1767 onwards. In France, the state provided an alternative structure for the scientific community by financing the Académie des Sciences and the Jardin du Roi (the royal zoological and botanical garden). It was from his position as superintendent of the Jardin du Roi that Buffon produced his mammoth – and immensely popular – survey of the animal kingdom, the *Histoire naturelle*. In some countries, the universities updated themselves and remained centres of research. Linnaeus was professor of botany at Uppsala in his native Sweden; he was still officially part of the medical faculty, but he had enough freedom to maintain his international reputation as the founder of a new system of classification.

Individual patrons could lay the foundations for more permanent scientific institutions. George III of England and his family were keen on botany, and the Queen's sister, Princess Augusta, established a botanical garden at Kew that eventually became a national institution. Another wealthy patron, Sir Joseph Banks (1743–1820), was for a time the unofficial director of Kew while the gardens were still in royal hands. Banks illustrates how private wealth, when under the control of a dedicated naturalist, could become the basis for a substantial scientific enterprise. As an undergraduate at Oxford he had been forced to hire a private tutor for lessons in botany, but he became determined to devote his fortune to science. He explored in Newfoundland, and then accompanied Captain Cook on his first voyage to the South Seas (1768–71). At his own expense, Banks ensured that the expedition had an expert botanist, Daniel Solander (1736–82), and a staff of naturalists and artists. The expedition made Banks' reputation; he went on to become not only the unofficial director at Kew, but President of the Royal Society, which he was able to revive from its lethargy at least in the field of natural history. His house in Soho Square, London, became the centre from which many expeditions were organized or supported.

Some time after the death of Linnaeus, Banks was offered the Swedish naturalist's plant collection and library. Although he had once hoped to purchase the collection, he now passed on the opportunity to a wealthy medical student, James Smith (1759–1828), who in 1788 used the material as the basis for founding the Linnean Society of London. Smith himself soon left the capital, but the Society survived and eventually purchased the collection. The Linnean Society remains Britain's oldest natural history society and went on to play a major role in fostering the scientific study of animals and plants. Although he endorsed

Smith's initiative, Banks used his position as President of the Royal Society to oppose the creation of other special-interest societies, fearing that increased specialization would undermine his authority.

Banks was also associated with the Admiralty's plan to transplant breadfruit trees from Tahiti to the West Indies, which led to the ill-fated voyage of HMS *Bounty* under Captain Bligh in 1788. This expedition was part of a burgeoning programme by the European nations to exploit newly discovered species in the services of their global empires. As Bligh himself wrote[2]:

> The Object of all former voyages to the South Seas, undertaken by command of his present majesty, has been the advancement of science and the increase of Knowledge. This voyage may be reckoned the first, the intention of which has been to derive benefits from these distant discoveries.

It is often forgotten that Bligh led a second expedition (1791–3), which successfully transferred breadfruit to the West Indies, as well as bringing some of the plants back to Kew.

The fact that the Admiralty financed voyages of exploration and research reminds us that natural history had a practical side that was now being recognized. In the next century both the state and private enterprise would play an ever-increasing role in the exploitation of the world's natural resources. But natural history also had broader ideological implications. The passion for describing the structure and habits of exotic species exploited the public interest in Nature in order to convey a social message. To begin with, at least, this message was linked to traditional religious beliefs and their function in maintaining social order. Natural theology used the complexity of natural adaptations to proclaim the existence of a God who had gone to great trouble to create a stable world for humankind to inhabit and exploit. The implication was that anyone who rocked the boat was threatening a divinely ordained social hierarchy. The mutiny on the *Bounty* neatly encapsulates the radical challenge to this tradition of authority, just as the purpose of the voyage itself demonstrates the expansionist capacities of the existing power structure.

The Argument from Design

The 'argument from design' thus had a social message (chapter 3). Natural history offered one of the most fruitful areas in which the argument could be developed, since the vast numbers of animal and plant species offered an endless source of examples to illustrate the complexity of the material universe and the care with which its component parts had been designed by God. One of the most active naturalists in the early Royal Society was John Ray, who made his name in the collection and description of British plants. Ray was a deeply religious man, and in the later part of his career he brought together his views on how the study of Nature illustrated the scope of divine creation in a book, *The Wisdom of God Manifested in the Works of Creation*. This first appeared in 1691 and went through numerous editions in the following century. It was used as a source of ideas and information by later apologists seeking to defend the claim that science demonstrated the existence and power of the Creator.

Ray was determined to show the falsity of the claim that the universe could have been brought to its present state of complex organization merely by the continued operation of natural laws. Although he compared God to a skilful workman who designs structures to fulfil a purpose, Ray was not a dogmatic follower of the mechanical philosophy. He regarded it as absurd to suggest that animals were only machines: the fact that they could reproduce themselves showed that there must be a non-physical 'plastic virtue' responsible for maintaining the form of each species. God was the only conceivable source of the original structure of the species and of the force that guaranteed the preservation of that structure through time. Ray's was thus an essentially static view of Nature. As he said in the preface to the *Wisdom of God*[3]:

> Note, That by the Works of the Creation in the Title, I mean the Works created by God at the first, and by him conserv'd to this day in the same state and condition in which they were at first made; for *Conservation* (according to the *Judgment both of Philosophers and Divines) is a continu'd Creation.*

This created problems for Ray when he confronted the evidence for geological change and extinction (chapter 4), but it was an essential component of the argument that Nature alone could create nothing. Given the timescale available to him, the concept of gradual evolution was unthinkable to Ray. God's power and wisdom was the only explanation of how the world came to contain complex structures.

The best way of demonstrating this was to describe a large number of species, showing how the parts of their bodies interacted to give a harmoniously functioning organism. Since in every case the structure was carefully adapted to the species' way of life, the argument also proved the benevolence of the Creator by showing that living things were meant to enjoy their lives. Ray described the anatomy of the human body, emphasizing how the eye and the hand were designed to make an active life possible. But he was prepared to accept that the animals and plants reflected a similar degree of care and attention on the part of their Creator. Many were, of course, useful to humankind, but Ray was convinced that they also had their own independent lives to live as part of a great universal scheme. He included careful descriptions of many species, linking their physical structure to their way of life to show that the adaptation of means to ends was universal. Each species was also endowed with instincts designed to control its behaviour, ensuring that it would follow its intended lifestyle and would reproduce itself in perpetuity.

Ray and his followers updated the concept of design by God to make it consistent with the mechanistic viewpoint of the new science. For well over a century, natural history was accepted as a suitable pastime for a rural Anglican clergyman, since it combined the study of Nature with the worship of God. Ray's style of argument was revived in William Paley's *Natural Theology* of 1802, ensuring that the argument from design would continue to be taken seriously by a new generation of naturalists. Paley stressed even more strongly the comparison between the Creator and the engineer who builds complex machines – a fitting metaphor in an age of increasing industrialization.

The naturalists of other European countries lost interest in natural theology as the age of Enlightenment progressed. But in the early decades of the eighteenth century the tradition flourished even in rationalistic France, as witnessed by the highly successful *Spectacle de la nature* of l'abbé Pluche. At the Académie des Sciences, René–Antoine de Réaumur (1682–1757) made the extensive studies of insects reported in his *Mémoires pour servir à l'histoire naturelle des insectes* of 1734–42. Réaumur continued the work of the early microscopists by exploring the internal structure of insects. But he also conducted extensive observations of the behaviour of many species, noting in each case how the organs were adapted to its lifestyle. Réaumur was convinced that the organs of the butterfly are hidden within the caterpillar, and was thus led to endorse the preformation theory of reproduction. His protégé Charles Bonnet (1720–93) made his name by discovering that female aphids could reproduce for many generations without access to the males. Here was further evidence that reproduction stemmed from minute germs or miniatures created by God and stored up within the females of each species. Bonnet devoted the rest of his life to exploring this image of Nature as a divinely created system.

There were some discoveries, however, that seemed to undermine the logic of the divinely created world machine. In 1741 Abraham Trembley (1700–84) noted that the freshwater hydra (or polyp, as it was then called) possessed the curious property of being able to regenerate itself. If a polyp was cut in half, both portions regenerated the missing organs and became whole organisms once again. This simple discovery was greeted with consternation. The polyp's abilities seemed to disprove the preformation theory, suggesting instead that living matter had the ability to create new structures by itself. A generation of new and more radical materialists seized upon Trembley's discovery as a means of disproving the claim that all complex structures in the world owe their origin to God. They proclaimed that Nature itself was the source of all creativity, thereby reviving the pantheistic tradition that had been driven underground in the late seventeenth century. This new materialism attributed organic properties to matter itself. The universe was portrayed not as a static system, but as a scene of constant and unpredictable change. This philosophy was picked up by radicals hostile to the existing social order. In France, writers such as Denis Diderot and the baron d'Holbach combined the materialistic view of Nature with calls for political change.

The observations of the eighteenth-century naturalists were thus drawn into the great debates of the time. The behaviour of simple organisms such as the aphid and the polyp served as examples for conservatives and radicals arguing over the desirability of maintaining the existing social order. If natural theology provided a framework in which the adaptation of structure to function served to illustrate divine benevolence, materialism seized upon any natural function that seemed to indicate that Nature had creative powers of its own. The static world view of the late seventeenth and early eighteenth centuries thus came increasingly under fire as the age of Enlightenment progressed.

THE SYSTEM OF NATURE

Réaumur and Trembley described individual species but made no effort to produce a comprehensive system of biological classification. Other naturalists saw the search for an orderly pattern of natural relationships as the chief task of natural history. Ray made his original reputation through his efforts to classify plants. Bonnet became deeply involved in the search for a system of Nature after failing eyesight forced him to give up his experimental work. Linnaeus gained an international reputation by producing a workable system for naming and classifying species. Whatever the tensions introduced by the rise of materialism, the original search for order characteristic of the Scientific Revolution manifested itself in the desire to find a rational pattern underlying the apparently bewildering variety of species. Where natural theology studied individual species to show how their structure was adapted to their way of life, the taxonomist sought a pattern that would reveal how all the various species fitted together into a rationally ordered system.

The search for order created problems of its own, however, and these problems can all too easily be glossed over in the rush to see Ray and Linnaeus as the fathers of modern biological taxonomy (taxonomy is the science of classification). Conservative thinkers certainly assumed that the universe was an orderly system, and that the order represented the divine plan of creation. But the system of representation we use must be created by the human mind. The old idea that God had incorporated 'signatures' into the things He had created to indicate their purpose had now been abandoned. The classifier had to work with the visible structure of the various species, trying to see resemblances that would allow them to be arranged into a coherent pattern. But in a world of bewildering variety, could one really be sure that the patterns we discern correspond to the real pattern imposed by God? Some naturalists created deliberately artificial systems designed merely for human convenience. They might hope that the system approximated to the divine plan, but there was always the chance that – even with the most carefully thought out system – the human mind had been deceived into seeing a pattern that was not really there. There were some who eventually began to argue that any humanly made system is bound to be false, since Nature is too complex for our feeble minds ever to comprehend properly. It was a short step from this to the claim that Nature had no order at all, because the ceaseless activity of natural forces is constantly dissolving the structures we see and replacing them with new ones in an unpredictable way.

There was also the problem of the units to be used in classification. Traditionally naturalists had assumed that living things can be divided up into recognizably distinct species. There are cats and dogs in the world, and we do not see a range of cat–dog intermediates that would be difficult for us to fit into either category. But the ancient concept of the chain of being had been associated with the view that God had created all conceivable forms of life – in which case the intermediates that we can conceive in our minds must have been granted physical existence. Classification into distinct species would then become a purely arbitrary procedure, perhaps depending upon the comparative rarity of some intermediate forms. In the end, sheer practical necessity forced most naturalists to accept the reality of distinct species, but there were a few who insisted that this procedure was unjustified. In their view, Nature presents an absolute continuity of forms and any system based on the classification of distinct units is bound to be artificial. This principle of continuity has sometimes been mistaken for an anticipation of the modern concept of evolution. But the naturalists concerned wanted to break down the barriers between the living species, not between species and their ancestors. Modern evolutionism accepts the continuity of change *through time*, but uses the idea of branching or divergent change to justify the belief that species are distinct from one another.

The Chain of Being

The principle of continuity was associated with the ancient philosophy of the chain of being (see chapter 2) and implied that there was no sharp dividing line between minerals and the simplest living things. This assumption was now endorsed by the mechanical philosophy, which treated animals and plants as merely complex assemblies of material parts. There was no such thing as 'life', and animals and plants could thus be classified in the same way as minerals.

The chain of being, with its associated principles of continuity and plenitude, also made it difficult to accept the possibility of extinction. If there were no gaps in Nature because God had created every conceivable form of structure, then the extinction of any form would destroy the coherence of the divine pattern. As the poet Alexander Pope wrote in his *Essay on Man*[4]:

> Vast Chain of Being! which from God began,
> Natures aetherial, human, angel, man,
> Beast, bird, fish, insect, what no eye can see,
> No glass can reach; from Infinite to thee,
> From thee to nothing. – On superior pow'rs
> Were we to press, inferior might on ours:
> Or in the full creation leave a void,
> Where, one step broken, the great scale's destroy'd:
> From Nature's chain whatever link you strike,
> Tenth or ten thousandth, breaks the chain alike.

John Ray found it difficult to accept extinction despite the evidence of the fossil record, and this way of thinking held back acceptance of the evidence for change throughout the eighteenth century.

The chain of being was the simplest possible pattern that could be imagined to link the species into a rational order. Each species would have but two close neighbours, one above and one below, and subdivisions of the chain into families of similar species would – because of the principle of continuity – be entirely arbitrary. By the early eighteenth century, the sheer number of new species being discovered was forcing naturalists to consider more complex patterns of relationship. But the underlying philosophy of the chain was still important to those conservative naturalists who used it to support the image of a static hierachy in Nature and society.

Charles Bonnet was one of the most eloquent exponents of the chain concept in the mid eighteenth century, and his ideas help to illustrate what Foucault regards as the preoccupations of the classical era. Bonnet was convinced that God had created the world to an orderly pattern, but he appreciated that our human models of that pattern are likely to be imperfect. The image of the chain represented one possible model for the pattern of Nature, but it was not the only conceivable pattern, and in the end Bonnet was prepared to admit that there might have to be 'branches' in the chain. This would destroy the principle of continuity because there would be gaps between the branches. Bonnet also came to admit that there might be a gulf between the mineral kingdom and the world of living things, thus paving the way for the emergence of a distinct science of 'biology'. He favoured the chain because it epitomized his vision of a static arrangement of species, created by God and maintained by the pre-existing germs responsible for reproduction. The pattern of Nature was eternal – even though Bonnet eventually admitted that some species had achieved physical manifestation later in the earth's history than others.

The chain of being represented a pigeon-hole system in

which the boxes are arranged in a single vertical line stretching down from humankind to the most unstructured mineral substance. Its great weakness was its artificiality; in order to sustain the principle of continuity, one had to have 'bridges' between all the natural groups. Flying fish were called upon to serve as intermediates between birds and fish, while corals (or 'zoophytes' as they were then called) bridged the gap between plants and animals. Such ideas began to seem ever more implausible as naturalists became more sophisticated, while the vast number of species now being discovered in foreign parts made it seem increasingly less likely that the plan of creation was quite as simple as the chain concept supposed. The challenge faced by eighteenth-century naturalists was that of balancing the human passion for imposing order upon the world against growing evidence that the world was so complex that its true order would forever remain unknowable.

System and Method

There were two ways of dealing with this problem. One was to create a system of classification based on variations in a single important characteristic. Most of these systems were created by naturalists who admitted that the order thus imposed was artificial or arbitrary, although they hoped that further refinement would bring them ever closer to the goal of a truly natural arrangement, i.e. one that corresponded to the pattern of creation imposed by God. The alternative approach was to work according to a method in which all characters would be taken into account so that the resulting system was natural from the beginning. Species that naturalists intuitively recognized as being closely related would automatically be ranked closely together by such a method. Artificial systems sometimes broke up such natural groups because the overall similarity was not apparent in the one character upon which the system was based.

The tension between the supporters of the system and the

method reflected a deep philosophical problem created by the new mechanistic world view. The origins of the systematic approach lay in the old philosophy of Aristotelianism, which had ostensibly been rejected during the seventeenth century. Aristotle had declared that true knowledge rested upon knowing the essence of things, the inner character that expressed the true nature of what each natural thing really is. If the naturalist knew the true essence of each species, he or she would be able to classify them by comparing and contrasting their essential characters. But Aristotle himself had proposed no such classification (chapter 2) and seemed to have thrown doubts upon the ability of naturalists to know what the true essence of a species is. In the Renaissance, Cesalpino had solved this problem for plants by boldly declaring that the essential functions of all plants are nutrition and reproduction (chapter 3). Knowledge of two characteristics alone was enough to allow the naturalist to classify species according to their true essences. In particular, he focused on the flowers and fruit, declaring that variations in the reproductive system must be the true basis for a system of classification.

Cesalpino's suggested groupings were not very successful, and his ideas were ignored until the late seventeenth century, when the new philosophy began to focus everyone's attention onto the conceptual problems of taxonomy. Many naturalists now proposed artificial systems, and in botany these often assumed that the reproductive parts of plants are the only basis for determining relationships. The demonstration of plant sexuality by Camerarius in 1694 only highlighted the importance of these parts. A leading exponent of the system was the French botanist Joseph de Tournefort (1650–1708), whose *Elémens de botanique*, also of 1694, proposed a simple and easily applicable breakdown of plants based solely on the structure of the flowering parts. When questioned on the justification for using only this structure, Tournefort responded by insisting that the reproductive system was the most important part of the plant, and by declaring that the Creator must have provided us with a simple way of recognizing the essence of species so that we could perceive the rational order He had imposed on the world.

Because it created groups defined solely by similarity in a single character, Tournefort's system was easy to use. Unfortunately, being artificial, it sometimes grouped together species that were dissimilar in most other characters, while it broke up some groups that were widely recognized as 'natural'. The Renaissance naturalists had already begun to group together those species which appeared to be 'related' in most of their characters, each group forming a 'genus' (pl. 'genera'). Such groupings are based on an intuitive recognition of overall similarity, as when we associate the lion, tiger and leopard together as 'big cats'. In the late seventeenth century, John Ray emerged as the champion of the search for a method that would allow *all* characters of the species to be taken into account when working out the pattern of relationships. In his *Methodus Plantarum Nova* of 1682, Ray acknowledged the significance of Cesalpino's ideas, but insisted that other characters besides the flowers and fruits would have to be taken into account. The same policy was extended to the animal kingdom later in Ray's career.

To support this policy, Ray argued that the new mechanical philosophy did not permit one to know the essence of any species by a simple test. According to the empiricist philosophy of John Locke (1632–1704), our senses reveal only the external characters of things to us, not their inner nature. All natural objects are material structures, and to assume that there is one character that reveals their essence is to fall back into the old way of thinking in which God has written a mysterious 'signature' into each of His creations. The inner nature of each species can only be studied indirectly by gathering as much factual information as possible, and hence classification must be based on similarity or difference in a wide range of characters. Other naturalists took up this call in the eighteenth century, including the French botanist Michel Adanson (1727–1806).

Using this technique, one could be sure that the groupings were as natural as possible. In fact, almost everyone agreed that genera should be established by overall similarity, and Tournefort recognized more 'natural' genera (as judged by modern standards) than Ray. But Tournefort's artificial system was based on the assumption that the genera themselves should be grouped into orders and classes by a single character. The basic pattern was created by applying a rational analysis based on first principles and then slotting the genera into it. Ray insisted that the broadly based method of searching for overall similarities should be used at all levels of classification – genera should be grouped into orders by *their* overall similarities. He wanted to discover the pattern of Nature by working upwards from observation, not downwards from first principles. In theory, his method would give a truly natural arrangement because it took every character into account. But, in practice, it was very difficult to use, because so much information had to be processed that only a very experienced naturalist could judge where a species ought to be fitted into the pattern. The great advantage of the artificial system was that, in an age of ever-extending horizons, it could be used by anyone who could read a textbook to work out at least the basic category into which any newly discovered species should be fitted.

Linnaeus

It was this practical necessity that ensured the triumph of the artificial system, and the naturalist who symbolized that triumph was the Swedish botanist Carolus Linnaeus, also known as Karl von Linné (1707–78). After an expedition to Lappland, Linnaeus studied medicine in Holland and then returned to his native Sweden as professor of botany at Uppsala. His mission in life was to revolutionize natural history by deciphering the divine order of creation. He was convinced that God had made this possible by providing clues that human reason could understand. The first step in the process was to reduce chaos to a semblance of order by imposing an artificial system, although Linnaeus held that the ultimate goal of his enterprise was the recognition of the true natural order. The first edition of his classic *Systema Naturae* appeared in 1735, and subsequent editions were gradually extended to provide a complete classification for the plant and animal kingdoms.

Linnaeus was primarily a botanist, and here his artificial system was based on the prevailing view that the sexual or flowering parts were the only characteristic that should be taken into account. The plant kingdom was divided into twenty-four classes, of which the first eleven were defined solely by the number of stamens in the flower, while the others depended upon structural variations in the stamens. Each class was then divided into a series of subsidiary orders depending upon the number or structure of the pistils. In those classes defined solely by the number of stamens, Linnaeus set up a closed network of relationships; his system provided a rigid set of pigeon-holes that predetermined the number of classes and orders that could exist. Nature would have to fit into the system imposed by the human mind, and it would be possible to determine if there were any gaps in our knowledge, i.e. pigeon-holes created by the system but unoccupied by any known species. In practice, of course, the system was not as rigid as this, but in principle Linnaeus' reliance on numbering created a system that left no room for the unexpected because all possible structures available to the Creator were defined by the grid or network of the system itself.

Each order would contain a number of genera, each in turn consisting of a number of closely related species. Linnaeus defined the genera in terms of the overall similarity of the species and introduced the modern technique of giving each species two Latin names. This binominal nomenclature requires that each species be assigned to a genus; its first name is that of the genus, while the second identifies the individual species. Generic relationship is thus indicated in the actual name of the species: the dog is *Canis familiaris*, the wolf (obviously a close relative) is *Canis lupus*. The modern system of botanical names is traced back to Linnaeus' *Species Plantarum* of 1753, while zoological names start with the tenth edition of his *Systema Naturae* of 1758. The success of this system ensured that those naturalists who controlled it had great intellectual power. It became accepted that each species was founded upon a single 'type specimen' to which its name was first applied – hence the importance of Linnaeus' own collection to later naturalists such as James Smith, the founder of the Linnean Society. New species could be named after the discoverer, or a wealthy patron, offering an effective system of rewarding those upon whom the naturalist depended for information or support. Enemies could be ridiculed, as when Linnaeus named an obnoxiously smelling plant after his great rival, Buffon.

The Linnaean system of classification gained widespread popularity. It was especially valued in newly emerging countries such as America, where its simplicity recommended it to naturalists who did not have access to the great collections of Europe. In such a situation, a natural system depending upon close comparison with many related species was impossible to apply in practice, while Linnaeus' artificial system could be learned from his books. But the highly artificial character of Linnaeus' own botanical system gradually began to give way – as he himself had hoped – to a more natural one in which the grouping of genera into orders, and orders into classes, was based on the recognition of overall similarities. The rigid grid of the Linnaean classes and orders was replaced by a more open-ended system that did not impose a preconceived order upon Nature. Most newly discovered species can be fitted into an existing genus, but occasionally something entirely different turns up, requiring the creation of a new genus. The process of classification does not predefine what can exist,

Fig: 5.1 *Natural groupings and their modern explanation*

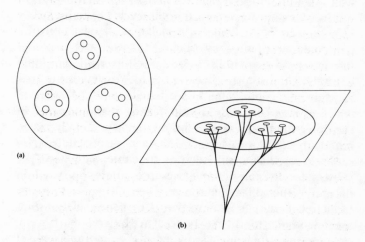

(a)

(b)

The method of grouping species by overall resemblance gives relationships that can also be represented two-dimensionally, but there is no predetermined pattern and new components can be added at any level in the taxonomic hierarchy. Species (represented by the smallest circles in (a)) are grouped first into genera; the genera are then grouped into orders by more fundamental similarities (larger circles). Modern biologists use an additional rank, the family, between the genus and the order. The orders are then grouped into classes (not shown here). Since there is 'open space' between the circles representing the groupings, new species, new genera, or even new orders can be inserted into the system when hitherto unknown forms are discovered. The system is thus 'open-ended'.

To give a single example, the dog, *Canis familiaris*, is a member of the genus *Canis* of the family Canidae (the dog family) of the order Carnivora (flesh-eaters) of the class Mammalia (mammals).

For most eighteenth- and early-nineteenth-century naturalists the relationships depicted here in (a) were still purely formal, representing similarities inherent in the divine plan of creation. Modern biologists see these relationships as a cross-section through a branching tree of evolution, as shown in (b). The degrees of similarity used to classify species depend upon how recently they shared a common ancestor; the species in a single genus began to diverge from one another quite recently, while the genera making up a family share a common ancestor at a more distant point in time.

and the naturalist can always hope to discover something entirely new.

The rigidity of the original Linnaean system gradually gave way to a situation in which this kind of open-endedness or unpredictability was accepted at all levels of the hierarchy of classification. In France especially, the application of a more flexible method by Michel Adanson and Antoine-Laurent de Jussieu (1748–1836) led to the creation of more natural relationships. Following Ray, de Jussieu divided the vegetable kingdom into classes based on the number of seed-leaves or cotyledons found in the embryonic plant. In the flowering plants he established the distinction between monocotyledons (single seed-leaf) and dicotyledons (two seed-leaves) that is still seen as fundamental today. The subsequent division of the classes into orders was based on natural relationships, and the very flexibility of this process facilitated a willingness to recognize that such relationships are too complex for them to be predicted by a rational system. New species were being discovered all the time, and some of them at least represented hitherto unknown genera or even unknown orders.

The growing recognition that Nature often does the unexpected or unpredictable helped to undermine the passion for order so typical of the classical era. Instead of hoping to see an artificially regular pattern linking the species, taxonomists became prepared to accept that their search for relationships must be based on the assumption that Nature is bewilderingly complex. In the end, the relationships would come to be seen not as patterns in the mind of the Creator, but as the necessary consequence of a process of divergent evolution.

THE ECONOMY OF NATURE

The development of an efficient system for classifying species was essential if naturalists were to have any hope of gaining an overview of the immense variety of living

Nature now available to them. But for all the efforts to create a natural system, this was still an essentially abstract way of trying to perceive an underlying pattern in Nature. Species were reduced to individual type specimens, which could then be arranged according to resemblances observed in their external properties. The habitat or lifestyle of the species was only occasionally mentioned as an aid to recognition. Yet everyone knew that the natural world was something more than an abstract collection of physical forms. Living things interacted with their environment and with one another. The eighteenth-century naturalists were well aware of this and sought to recognize another kind of order that would take these practical factors into account. The Creator must have foreseen the ways in which the species He designed would interact with one another in the real world. The origins of what we might today call an ecological viewpoint can be discerned in these early efforts to understand what was often called the 'economy of Nature'.

There were differences of emphasis within this concern for the economy of Nature. The very term 'economy' carries the implication that the system has been rationally ordered for someone's material benefit, and it was easy to create the impression that the human race was intended to exploit the system wherever possible. God had created the world's economy to sustain the human race, and was happy to see us modify the system if we thought fit. But other naturalists adopted a less exploitative view of the world. For them, the complexity of the system was a source of wonder that compelled respect for every species, however trivial. Humankind was a part of the system, and the adaptations we make through agriculture enhance rather than mar the beauty of the whole. As real-life agriculture became ever more mechanized, this sense of the respect for the traditional form of the landscape was extended to a nostalgic longing for a more traditional way of life.

Rural Harmony

A classic expression of this desire to understand the ways in which animals and plants interact with their environment is *The Natural History of Selborne* by Gilbert White (1720–93). Published in 1789, this was a collection of letters originally written to fellow naturalists. White's observations of the wildlife to be found in his parish in the south of England have established his reputation as founder of our modern respect for undisturbed Nature. Apart from his descriptions of the elusive inhabitants of the woods and fields, White also celebrated the important role played by humble creatures such as the earthworm. Here was an idyllic image of the countryside in which humankind seemed part of a stable natural order maintained by the ceaseless activity of other species. Yet White's England was already an artificial economy, drastically modified by humankind over a thousand years or more, and it was now facing an increasing assault from the more efficient farming techniques being developed to feed the population of the world's first industrial nation. White's Arcadian image of country people in harmony with their environment was not typical of his own time – indeed, his *Natural History of Selborne* did not begin to attain its status as a classic until the early decades of the following century, when growing nostalgia created a market for literature encapsulating the virtues of a pre-industrial age.

Even White was aware of the practical advantages that might be gained from better knowledge of plants[5]:

> Instead of examining the minute distinctions of every various species of each obscure genus, the botanist should endeavour to make himself acquainted with those that are useful ... The botanist that could improve the swerd of the district where he lived would be an useful member of society: to raise a thick turf on a naken soil would be worth volumes of systematic knowledge, and he would be the best commonwealth's

man that could occasion the growth of '*two blades of grass* where *one* alone was seen before'.

Horticulturalists had already begun to apply an increased awareness of the relations between species to practical ends. In the early eighteenth century Richard Bradley (1688–1732) had noted how each species of insect tended to feed upon a particular plant. He warned farmers not to kill the birds in their turnip fields, because they fed upon the insects that were damaging the crop. Like many of his contemporaries, Bradley realized that all the species of animals and plants are dependent upon one another in a complex web of what we should now call ecological relationships. Like White, though, he was convinced that humankind had the right to interfere with the network in order to destroy the pests that plagued the crops. The fact that the pests themselves may have been encouraged to breed by the farmers' tendency to concentrate on single crops did not occur to anyone at the time.

This image of Nature as a network of relationships that was complex, yet designed to withstand human interference, emerged within the world view of natural theology. John Ray and others had developed the claim that a benevolent Creator had endowed each species with instincts that gave it a lifestyle compatible with its physical structure. Ray was certainly aware of the fact that some species were designed to get their food by preying upon others, and were thus dependent upon the prey species maintaining its numbers. His follower, the Boyle lecturer William Derham, developed the theme of interdependence at length in his *Physico-Theology* of 1713. Each species had its natural habitat, and maintained its numbers because its reproductive capacities were exactly in balance with its lifespan and the threats from predators and other natural hazards[6]:

The Balance of the Animal World is, throughout all Ages, kept even, and by a curious Harmony and just Proportion between the increase of all Animals, and

the length of their Lives, the World is, through all Ages
well, but not over-stored.

If Nature was indeed this stable, people need not worry about
the prospect of damaging it with their interference.

Maintaining the Balance

It was Linnaeus who developed this theme of the divinely
preordained balance of Nature to its highest pitch. Whatever
his fame as a taxonomist, Linnaeus was deeply interested in
the way in which species subsist within their natural environments, and his vision of a divine created order required
him to believe that ecological relationships are essentially
stable. In a thesis entitled 'The Oeconomy of Nature' –
attributed to a student, Isaac Bilberg, but probably written by
Linnaeus himself – he provided a comprehensive account of
how the Creator ensured that the balance was sustained.

The world of Linnaeus and the natural theologians was
a mechanical system, but one that had been designed
by an all-wise engineer who had taken into account all
possible relationships between the component parts. 'The
Oeconomy of Nature' begins with the hydrological cycle, in
which evaporation from the oceans provides the supply of
water that will fall as rain to nourish the earth. The image
of a perfect balance between rainfall and drainage provides
the model that Linnaeus then extends to the world of living
things. Each species feeds upon its own prey, and is in turn
fed upon by others. Here we have the concept of a food chain,
but it is linked to the assumption that divine providence will
ensure the continuation of every link in the chain[7]:

> Thus the *tree-louse* lives upon the plants. The fly called
> *musca aphidovora* lives upon the *tree-louse*. The *hornet*
> and *wasp fly* upon the *musca aphidovora*. The *dragon
> fly* upon the *hornet* and *wasp fly*. The *spider* upon the
> *dragon fly*. The *small birds* on the *spider*. And lastly the
> *hawk* kind on the *small birds*.

Predators are always less numerous than prey, so that the supply of prey does not become exhausted, the balance being maintained by the natural reproductive capacities of the species. Only a divine Creator could have thus ensured that every species would be provided for in perpetuity.

In conclusion, Linnaeus' tract emphasized that humankind had a duty to do more than merely observe the functioning of this exquisitely contrived machine. The fact that so many species were of use to us suggested that the Creator intended us to take charge of the machine and manipulate it for our benefit. The balance of Nature was stable by God's decree – yet it was flexible enough to allow us to multiply the numbers of those species we value and to destroy those which are pests. Here we see a residue of the belief that God has created Nature to serve humankind still influencing those naturalists who were most aware of the bewildering complexity of the natural world. The possibility that human interference might upset the balance of Nature was still unthinkable, although the increasing pace of industrialization in the late eighteenth century would soon begin to bring this assumption into question.

The naturalists of the Enlightenment were, however, aware of those plagues of rats, mice and insects which continued to threaten humankind's food supply. At first these were dismissed as divine punishments, but as the century progressed it became less fashionable to think in such terms. The balance of Nature was not absolutely uniform, and one species might occasionally increase its numbers beyond what would be needed simply to maintain the balance. Many species had been given great reproductive powers to allow them to sustain their numbers against enormous predation – but that fecundity meant that the balance could be upset from time to time by accidental factors. It was assumed, however, that natural compensating mechanisms would soon come into play to restore harmony. In his discussion of the hare in volume 6 of his *Histoire naturelle*

(1756), Buffon noted that the fecundity of Nature allowed such explosions to occur, but argued that the increase of predators rapidly cut the numbers back down to size. He also suggested that a species that multiplies excessively experiences a bout of temporary sterility, which helps to reduce its numbers back to their normal level. Although frequently critical of natural theology, Buffon shared the view that species have been endowed with characteristics that help to preserve the natural balance.

No one was as yet prepared to admit that some plagues might be an unexpected, but natural and inevitable by-product of human interference with the balance of Nature. And yet the sheer fecundity of some species could not be overlooked, and a growing awareness of this factor may have played a role in breaking down the overconfident image of stability projected by the natural theologians. Living things were not machines – they were systems with reproductive capacities that were far easier to grasp if one switched to an image in which life was an active power, constantly striving to overcome the natural limitations placed upon it by the environment. 'Vitalism' (the belief that life is a distinct force that cannot be explained in purely mechanistic terms) did, in fact, begin to flourish within late-eighteenth-century physiology. The purely mechanical world view was at last beginning to break down. For the thinkers of the Darwinian era, this point would be brought home most clearly by a book published at the end of the century, Thomas Malthus' *Essay on the Principle of Population* of 1797. In an effort to check the optimistic predictions of radicals calling for social reform, Malthus proclaimed that the human race's capacity to breed ensured that the population would constantly tend to outstrip the food supply. When applied to the animal kingdom, this insight would usher in a debate that would threaten the very foundations of natural theology.

The static character of Linnaeus' world view is illustrated by his concept of the 'station' or habitat characteristic of each species. The economy of Nature was obviously broken up into local systems, each consisting of species adapted to that particular environment. The very close adaptation of structure to function in every species ensured that most would be, of necessity, confined to an area to which their mode of life was suited. Yet it was obvious that not every species was confined to a particular location. Gilbert White was well aware of the fact that many birds migrate from Britain to warmer areas during the winter, and suggested that a route via the Straits of Dover and Gibraltar would allow birds to reach North Africa without crossing extended areas of ocean.[8] Of greater significance still was the growing recognition that species could be transported to new areas and still survive. Botanical gardens were now full of exotic species, although these were maintained under artificial conditions. But it was well known that the rabbit, for instance, had been spread across Europe by the Romans and had established itself very well in many new areas. The common frog was supposed to have been introduced into Ireland by the fellows of Trinity College, Dublin, in the late seventeenth century. Everyone was aware that the potato and other New World plants flourished in Europe.

The Geography of Life

It was thus becoming obvious that the environment to which a species was adapted did not coincide with the actual area it inhabited. In many cases there were other areas potentially available, if a few members of the species could be transported to establish a breeding colony. In some areas the actual environment could be seen to change through time, with major consequences for the inhabitants. Linnaeus knew that, as bogland gradually dried out, new species established themselves in the area in a definite sequence that could be observed whenever the same change of conditions took place. Since he was a supporter of the retreating-ocean theory, he must have realized that over a long timescale the earth must have witnessed many such changes in the inhabitants of particular areas. The growing awareness that the earth itself was subject to change created a problem for those naturalists who wanted to visualize the living world as a static system. If the conditions changed, then at the very least species must have been forced to move around the world in order to maintain themselves in a suitable habitat.

The question of permanent migrations had been tackled by seventeenth-century thinkers concerned about the biblical story of Noah's ark. If all modern species were descended from parents who survived the great flood aboard the ark, then there must have been an extensive migration across the globe after the flood waters had receded. The discovery of America raised additional problems for the ark story, partly because it vastly increased the number of species that Noah would have had to take on board, and partly because it was difficult to see how the American species could have migrated there from Mount Ararat in the Middle East (the ark's traditional resting place). Either a land-bridge between Eurasia and America must have existed in the past, or the whole concept of a universal deluge became unacceptable. It was also pointed out that, with such small numbers of each species on the ark, the carnivores would soon have eaten the only surviving members of the prey species. Even in the seventeenth century, some bold thinkers had concluded that the inhabitants of America – including the native Indians – were separately created and had not been affected by the flood. Linnaeus realized that a strict interpretation of the retreating-ocean theory made the concept of a relatively recent deluge unacceptable. Yet his strong belief in a divine Creator led him to modify the biblical story of the garden of Eden in a way that gave rise to similar problems in the area of biogeography.

In his dissertation 'On the Increase of the Habitable Earth' Linnaeus argued that God might originally have created all the species in a single area, from which they would have spread out around the globe as the lowering of the sea level opened up more land area. The original site of creation must have contained habitats suitable for all species. It must, therefore, have been located in the tropics, but must have contained a high mountain whose elevated slopes would have provided the cold conditions suited to those species which would eventually migrate to the polar regions. This was an ingenious way of reconciling the traditional notion of a single divine creation with current ideas about adaptation and the changing state of the earth's surface. But Linnaeus paid little attention to the problems that would have followed from any attempt to explain how and why the species had migrated to their present locations. The arguments against the belief that all species had spread out from Noah's ark were equally valid when applied to this updated version of the garden of Eden. If anything, they were even more damaging, since now both plants and animals would have had to be transported over vast distances.

The difficulties attendant upon any theory postulating a single source for all animals and plants were magnified as Linnaeus' students and disciples went out to collect and classify the inhabitants of all the regions of the globe. Daniel Solander's trip to the South Seas with Joseph Banks and Captain Cook was but a single example of this process – Linnaeus' students also collected in regions as diverse as the Americas, Siberia and Japan. The specimens they sent back ensured that Linnaeus' collection at Uppsala became a much sought-after prize after his death. But it became increasingly obvious that the animals and plants found in the different regions formed recognizably distinct groupings. The fact that similar physical conditions were present in different areas did not mean that the animals and plants were the same. There were superficial similarities because of the need for species to be adapted to the environment, but the plants and animals of America, for instance, were quite distinct from the equivalents to be found in the Old World.

As Buffon noted, the puma and jaguar are not misplaced lions and tigers – they are distinctly American species of big cats. It gradually became fashionable for naturalists to write books devoted to the flora or fauna of a distinct region. The actual term 'flora' began to take on a distinctly regional connotation, denoting a biological unit composed of all the plants typical of the region.

The notion of distinct biological regions or provinces thus began to emerge in the late eighteenth century, and some naturalists began to think about the processes that determined the distribution of the species. Johann Reinhold Forster (1729–98) was a German naturalist who settled in England and accompanied Cook on his second voyage around the globe. Despite falling out with both Cook and the Admiralty, he eventually published his *Observations made during a Voyage round the World* in 1778. He noted that the plants and animals of a region form a unit defined by the environment, particularly by the prevailing temperature. The units succeeded one another in a regular sequence as one passed from the tropics to the frigid zones, with the tropics always having the most numerous and the most spectacular species. In addition, though, there were biological provinces defined by longitude rather than latitude. The Pacific islands, for instance, saw an intermingling of the typical Asian and the American species. Islands in general had species similar (but often not identical) to those of the nearest continental land mass. Forster extended these generalizations to include the human inhabitants of the biological provinces.

The notion of distinct regions was refined by the German botanist Karl Willdenow (1765–1812) in his *Principles of Botany* (1792, translated 1805). By eliminating the kind of anthropological speculations favoured by Forster, Willdenow helped to establish botanical geography as a distinct scientific specialization. He recognized the interaction of geographical, geological and biological factors in determining the character of a region's inhabitants, but also noted the existence of distinct provinces that could not be explained purely in terms of adaptation. As a supporter of the retreating-ocean theory, he modified Linnaeus' suggestion to postulate several widely scattered mountains upon which the Creator had formed distinct sets of species. As the land surface expanded, the species had spread out to colonize the surrounding territory and to interact at the margins. To account for anomalies where a species characteristic of one province was found in another part of the world, he argued that catastrophic geological events must have transported some species over considerable distances.

THE POSSIBILITIES OF CHANGE

Natural theology encouraged the belief that the system of Nature was perfectly stable. The various species could be assembled into a rationally ordered pattern, while at the mechanical level the whole system worked like clockwork. If a species became extinct or changed its structure, there would be a gap both in the pattern of creation and in the complex web of predator–prey relationships that upheld the balance of Nature. Any change would represent a departure from the Creator's intentions.

Yet by the mid eighteenth century geology was already providing evidence that the earth itself had undergone massive changes in the course of time. Linnaeus' appeal to the retreating-ocean theory confirms that he was aware of this trend, but he had merely adapted the idea to his own belief in divine creation. The notion of botanical or zoological regions was also an attempt to modernize the idea of creation by postulating several centres of divine activity. But the implications of the new geology could not be evaded so easily. The rival theories of the earth each postulated its own pattern of radical change in the environment to which the planet's inhabitants would have had to adapt. The fossils suggested that some species had either become extinct or changed into something new. Some eighteenth-century naturalists and philosophers began to suspect that life itself had a history of change and development. As radical social thinkers challenged the conservative ideology that supported natural theology, they became willing to cast doubts upon the static, hierarchical model of Nature.

The later eighteenth century saw the emergence of new attitudes, which seemed to challenge the dominance of natural theology. Even conservatives such as Linnaeus made limited concessions to the evidence for change, while radicals rejoiced in the construction of a world view in which nothing was stable because all material structures were constantly being destroyed or modified by natural forces. The claim that there were 'precursors' or 'forerunners' of Darwin in the eighteenth century is based on the assumption that some naturalists, at least, must have been sufficiently aware of the evidence for change to begin constructing an evolutionary world view. In the writings of the materialists one can find passages that seem to anticipate various aspects of the modern theory of evolution. Yet we must be very careful not to take such passages out of context. It is all too easy for the inexperienced historian to be misled into thinking that a certain eighteenth-century figure must have *almost* discovered the theory of evolution, because all the necessary components can be found scattered somewhere in his writings. The modern historian of science seeks to understand the context within which the eighteenth-century naturalists developed their ideas, and in most cases the context turns out to be incompatible with our own. The various anticipations of Darwinism are artefacts of the historian's imagination, not genuine insights into the discoveries of the radical Enlightenment.

Filling in the Gaps

Concepts of change are certainly more prevalent in the later eighteenth century, but it is possible to conceive of change taking place in many different ways. Often, a limited appreciation of natural change was admitted into a system whose overall framework was still designed to maintain an image of the universe as an essentially static system. Charles Bonnet eventually conceded that the higher levels of the chain of being must have become manifested in physical reality only in the later epochs of the earth's history. But as the divine pattern of creation, the chain itself was unchanging: if not a model of what *is*, it was a pattern laid down at the beginning to predetermine what *could be*. If some boxes in the pigeon-hole system were empty to begin with, they were nevertheless there as concepts or possibilities from the moment the divine plan was conceived. Such a concept of change is quite different from the open-ended, opportunistic model of evolution projected by modern Darwinism. Even a materialist such as Buffon managed to create a view of the earth's history that preserved the concept of species as eternally fixed types – structural possibilities simply awaiting the conditions that would allow them to become manifest in the material world.

Only at the very end of the eighteenth century do we find the first serious suggestions of mechanisms that could change a species into something entirely different. But these were not always produced by thinkers in touch with the latest evidence of change from the fossil record. Hypothetical mechanisms of transformation could be constructed from new ideas that were emerging in physiology, ideas about the inherent activity and creativity of organic matter that seemed to imply an ability of living things to adjust themselves to changes in their environment. Concepts of historical development and mechanisms of day-to-day change were not as yet being integrated in the way we take for granted in the modern theory of evolution. The final breakdown of the static image of Nature would take place only when these two levels of investigation were integrated in the nineteenth century.

In his early writings Linnaeus had declared that no new

species had appeared since the creation. Later on, however, he began to suspect that the Creator had allowed some possibility of change through time. It had always been agreed that species may contain well marked 'varieties' – populations adapted to different local conditions encountered across the species' range. But such varieties were still members of the same species because they could interbreed when brought together. Linnaeus now began to suspect that, in at least a few cases, the process of adapting to the local conditions had gone so far that the new form could not interbreed with the old. A new species had thus been formed, closely related to, but distinct from, its parent.

On a very small scale, this idea anticipated the modern concept of evolution, but it was not Linnaeus' main way of acknowledging the possibility of change. He had become aware of new species appearing within his botanical garden, apparently as the result of hybridization or cross-breeding between parents of two distinct species. Building upon this discovery, he eventually postulated that in the beginning God had created only a single species in each genus – all the others we now observe had been created in the course of time by hybridization between the original forms. Here was a theory of the multiplication of species without evolution, and without any real threat to the basic idea of a stable divine plan. New species are never genuinely *new* – they merely represent new combinations of characters that were already present in the original species created by God.

Buffon

The majority of the naturalists who used Linnaeus' system of classification paid little attention to these later speculations. But for those influenced by the new radical philosophy, any compromise with the idea of divine creation was unacceptable. Of all the radicals, the most active opponent of Linnaeus and the most popular exponent of the new philosophy of Nature was the superintendent of the Jardin du Roi in Paris, Georges Leclerc, comte de Buffon. Buffon's immense *Histoire naturelle* was intended to provide a complete description of the animal kingdom. It was written in a style that would appeal to the sophisticated intellectuals of the Parisian *salons* who were the arbiters of eighteenth-century thought and taste. Yet Buffon had experienced naturalists as his collaborators, and offered radical comments on the implications of the new information flowing in from around the world.

In the early volumes of his *Histoire Naturelle* (1749) Buffon proposed that the earth had begun as a globe of molten matter struck off from the sun by a colliding comet (chapter 4). But he was also determined to extend his radical programme to the history of life on earth. He attacked Linnaeus and the other taxonomists for subordinating science to their belief in divine creation. By treating species as abstract units in a divine plan, they ignored the natural processes that sustain the structures we observe in the material universe. Even if such a plan actually existed, Nature is so complex that naturalists could never be sure that they had correctly interpreted the relationships they observed – the system might be a figment of their own imagination. For Buffon, a species was not an element in an abstract pattern of creation, it was a group of animals or plants in which reproduction sustained the same basic structure through time.

In the fourth volume of the *Histoire Naturelle* of 1753 Buffon described the ass and addressed the question of what it meant to say that the ass is 'related' to the horse. In Linnaeus' system, 'relationship' implied only a similarity in the divine plan of creation, but for Buffon this was mere verbiage. If the term had any meaning in the real world, it must imply a physical connection, i.e. that the ass species was composed of horses that had degenerated through time. In effect, the notion of relationship must carry the same implication as it does in human families: if two people are related, it is because they share a common ancestor at some point in the past. Some historians have argued that Buffon wished to demonstrate his own support for what we would now call the theory of evolution. But in fact he took great pains to show that the ass is *not* related to the horse; they are two distinct species and the relationship between them exists only in the mind of the classifier who has chosen to concentrate on certain superficial similarities. All such alleged similarities are purely arbitrary, products of the human mind, not of Nature itself. Buffon demonstrated his contempt for the whole procedure by returning to the old technique of starting his description of Nature with those species which are useful to humankind. If all 'relationships' are equally imaginary, this was as good a system to use as any other.

Reproduction was crucial because it preserved the form of the species from one generation to the next. But how was this achieved – how did Nature guarantee that the offspring had the same basic structure as their parents? Buffon was a leading opponent of the preformation theory, in which God was supposed to have created tiny miniatures of every organism to serve as the basis for reproduction. To him, such an idea did not explain the generation of new organisms, it banished true generation from Nature by claiming that all structures come directly from the Creator. Buffon proposed that the embryo is formed from 'organic particles' collected by both parents in their reproductive systems. When combined in the mother's womb, the particles somehow arrange themselves to form the body of the new organism. Explaining how material particles driven only by natural forces could arrange themselves into the complex structure of a living body was one of the great problems facing the mechanical philosophy. Explaining how the new structure always copied the basic form of the parents' species raised even more difficulties, which Buffon simply evaded by claiming that each species was based on an 'internal mould' that somehow constrained the organic particles to form themselves into the same basic structure. Nature contains only a limited number of these moulds, corresponding to the distinct species we observe.

As Buffon steadily expanded his coverage of the animal kingdom in successive volumes of the *Histoire naturelle*, he gradually changed his mind about the number – but not about the status – of the distinct species. In an article on the 'Degeneration of Animals' in volume 14 (1766) he accepted that closely related forms such as the horse and the ass, or the lion and the tiger, had descended from a common ancestor. Each Linnaean genus was, in effect, a single species that had developed a number of well marked varieties that naturalists have mistakenly called true species. Buffon was encouraged in this view by reports – presumably mistaken – of fertile hybrids being produced between species of the same genus. He had originally ignored the mule because it is sterile, but if fertile mules were produced, then the horse and the ass could not be distinct species; they were merely varieties of a single form. This considerably widened the scope for natural change within each species, but it did not alter the basic fact that the species themselves were distinct. The horse species and the big cat species were totally separate, each with its own internal mould, and there could never have been any connection between them.

Buffon's move towards 'evolutionism' was thus very limited, since it did not involve rejection of the belief that Nature contains a number of absolutely distinct types of organization. What is perhaps more significant about his later opinions is his effort to understand the multiplication of 'species' within each type as the result of geographical dispersal. He argued that if a population of animals was forced to live in a new environment, changes would appear in individual animals and would eventually become hereditary. The physical manifestation of the 'internal mould' would change because the supply of organic particles was different under the new conditions. This was not an adaptation to the new conditions, but a direct, and often degenerative, effect of the unfamiliar environment upon the organism. Such changes would take place when species were able to migrate into new areas as a result of geological alterations of the earth's surface. Buffon accepted that there were a number of distinct species found only in the New World. At first he speculated that they might be the degenerate remnants of Old World types that had crossed to America on a now-sunken land-bridge across the Atlantic. The claim that the New World offered only an inferior environment leading to inevitable degeneration was resisted by American naturalists and scholars anxious to demonstrate the superiority of their own territory.

In his supplementary volume entitled *Les Epoques de la nature* of 1778, Buffon linked his concept of modification by migration to his theory that the earth was gradually cooling down. Elephants such as the mammoth and mastodon had once lived in the northern regions at a time when the climate was much warmer than it is today. As the earth cooled, the animals were forced to move south in an effort to maintain themselves in the conditions they favoured, although even so the modern tropical forms had degenerated when compared to their giant northern ancestors. Since there had once been a connection between Asia and North America, the ancestral forms had roamed freely between the two continents. But when they were forced southwards, the populations of the New and the Old Worlds had become separated, and each type had become modified to give the apparently distinct 'species' we recognize today. Some types had become extinct on one continent, while surviving on the other. Here was a bold effort to provide a natural explanation for the existence of zoological provinces in the modern world.

Materialism and the Origin of Life

How had the ancestral forms of the north come into existence, if it was not by divine creation? Buffon was determined to eliminate this last vestige of the biblical story, and to do this he invoked the ancient concept of 'spontaneous generation' – the belief that living things can be produced directly from non-living matter under certain conditions. Early in his career he had joined with an English emigré, John Turberville Needham (1713–81), to perform experiments that seemed to show the production of micro-organisms in sterilized jars of meat gravy. To Buffon, this was clear evidence that organic particles could spontaneously organize themselves into living structures even outside the womb of a parent organism.

Needham's experiments were violently criticized by the supporters of the preformation theory, who held that matter had no such creative power. Eventually Lazarro Spallanzani (1729–99) would demonstrate that Needham had not sterilized his material well enough. But in the meantime a number of materialists, including Buffon, took the experiments as a justification for arguing that Nature could produce living structures without divine interference. They speculated that under certain conditions (such as the warmer environment of the primitive earth) much larger creatures might have appeared directly from unorganized matter. Here, then, was Buffon's theory to explain the origin of life in the northern continents. He also speculated about an earlier act of spontaneous generation that had produced forms adapted to the intensely hot conditions preceding our own epoch. Such forms would have become extinct as the planet cooled, leaving their fossilized remains to puzzle us in the rocks.

Spontaneous generation, not evolution from a common ancestor, was thus Buffon's explanation of how the ancestral members of each basic organic type had been formed. He remained convinced that each type was eternally distinct from the others – there was no possibility, for instance, of a common ancestor for all the mammals. Although each species was malleable up to a point, its basic structure was determined by the laws of Nature themselves, which somehow guaranteed the eternal stability of the internal moulds. Buffon argued that, as each planet of the solar system cooled to the appropriate point, species identical to those now on the earth would be produced by spontaneous generation. The form of each species is thus built into the very fabric of Nature, representing a potentially stable organization of material particles that will achieve physical manifestation whenever the physical conditions are suitable. Nothing could more clearly illustrate how difficult it was for an eighteenth-century naturalist to throw off the assumption that Nature is constructed around an eternally stable pattern. For all Buffon's materialism, his contempt for the biblical creation story and his recognition of the earth's changing environment, he could not shake off the conviction that he was describing species that are permanent features of the natural order.

It was the more radical materialists of the Enlightenment who came closest to abandoning the idea of an underlying natural order. Writers such as Denis Diderot (1713–84) and the baron d'Holbach (1723–89) were active in criticizing the existing social order and saw the image of an ever-changing Nature as a powerful weapon in their fight against orthodoxy. In a number of works issued clandestinely from the 1750s onwards, they argued that Nature exhibits no stable, hierarchical pattern; instead, the ceaseless activity of natural forces is forever destroying those structures that already exist and creating new ones in their place. Diderot was fascinated by the appearance of monstrosities, seeing such disturbances of the normal reproductive process as evidence that Nature contained no power capable of guaranteeing the stability of species. Was it not possible that an occasional monstrosity might be capable of surviving and reproducing, thus establishing a new species? Invoking Empedocles' vision of the origin of natural forms, he postulated a trial-and-error process of spontaneous generation in which only those combinations lucky enough to be viable would survive to become the parents of the species we know. D'Holbach's *Système de la nature*, published under an assumed name in 1770, became known as the 'Bible of atheism' for its portrayal of a world in which there was no stable order, only the ceaseless activity of material Nature. Here materialism revived the pantheist tradition that had been driven underground in the previous century, proclaiming Nature itself as the source of all creative power.

This radical philosophy threatened the very foundations of natural theology and the belief in divine creation. It seems to contain the seeds of an evolutionary viewpoint in which natural developments are open-ended and unpredictable. Yet philosophers such as Diderot and d'Holbach proposed no detailed theory of organic change. Although passionately interested in the achievements of the new science, they had other fish to fry. The transmutation of species was, in any case, only a minor part of their system. Since they believed that Nature could spontaneously generate even complex living structures, there was no need for them to postulate an extended process of evolution to explain the diversity of modern species. A theory of continuous natural development would only become plausible once the idea of spontaneous generation was abandoned, or at least restricted to a role in which it explained only the origins of the simplest forms of life. By the last decades of the century, moves were already being made in this direction, paving the way for the controversies that would rack the natural sciences of the nineteenth century.

Natural Progress

One of the earliest of these theories came from the pen of Erasmus Darwin (1731–1802), grandfather of the man whose name would later come to symbolize the evolutionary movement, Charles Darwin. Erasmus Darwin was an eminent doctor, a representative of the rising middle class that was bidding for influence within the newly industrialized world of late-eighteenth-century Britain. He wrote a number of poems expounding an image of Nature as an ever-developing system in which life was constantly struggling against the limitations imposed upon it. His theory of transmutation was contained within a chapter of his *Zoonomia* (1794–6), a comprehensive account of laws governing the world of living things. Although often hailed as an anticipation of his grandson's theory, Erasmus Darwin's view of natural development had very different foundations. He saw life originating in its simplest form and then advancing through the efforts of individual organisms struggling to cope with an ever-changing environment. Each generation built upon the efforts of its predecessors and added its own increment of self-development to the evolutionary process.

In his own time Erasmus Darwin was best known for his poetry, in which his vision of an endlessly active Nature was expressed in typical eighteenth-century rhyming couplets. Part II of his *The Botanic Garden* (1789–91) was devoted to 'The Loves of the Plants', describing the Linnaean system of classification by the sexual organs in anthropomorphic and indeed downright suggestive terms. *The Temple of Nature* of 1803 brought in the theory of evolution[9]:

> Organic life beneath the shoreless waves
> Was born and nurs'd in Ocean's pearly caves;
> First forms minute, unseen by spheric glass,
> Move on the mud, or pierce the watery mass;
> These, as successive generations bloom,
> New powers acquire, and larger limbs assume;
> Whence countless groups of vegetation spring,
> And breathing realms of fin, and feet, and wing.

Darwin also described the struggle for existence resulting from Nature's superfecundity, although there is no evidence that he saw it as a creative evolutionary force. His poems provide a classic illustration of the intersection between science and the arts, although his classical style would soon be swept away by the Romantics.

Erasmus Darwin's radical philosophy of natural development certainly attracted the criticism of conservative thinkers, but its roots lay in the medical and psychological theories of the time and it had little direct influence within natural history. But similar ideas were developed only a few years later by a French naturalist whose work seems to represent a concluding synthesis of all the trends affecting the eighteenth century's view of Nature. J. B. Lamarck (1744–1829) began his scientific career as a botanist, but was appointed to the Muséum d'Histoire Naturelle in Paris (reorganized from Buffon's old Jardin du Roi after the French Revolution) to describe the invertebrate animals. In fact, Lamarck produced major advances in the classification of the invertebrates, thus guaranteeing himself a place in the history of biology. But he became best known for the comprehensive theory of transmutation that he advanced only in the later part of his career, especially in his *Philosophie zoologique* of 1809.

Lamarck's mechanism to explain how species adapt to their environment attracted the attention of later naturalists seeking an alternative to Charles Darwin's theory of natural selection. But the 'Lamarckism' of the late nineteenth century (see chapter 8) exploited only a single component of his overall theory. In fact, Lamarck combined a number of ideas that had been circulating in the eighteenth century to create an evolutionary world view that was in many ways quite different to that which would become popular in the following century.

Lamarck's starting point was the concept of spontaneous generation. He assumed that electricity had the power to vivify globules of matter to give the most primitive forms of life. More complex living structures were produced by progressive evolution over vast numbers of generations. In principle, progressive evolution would generate a single continuous pattern of organic forms linking the most complex (humankind) to the most simple. Here, perhaps, we see the influence of the 'chain of being', although Lamarck insisted that there must be two parallel hierarchies for the plant and animal kingdoms. But as an experienced naturalist he knew that the species we observe do not fit into a linear pattern, and it was to explain this 'deviation' that he invoked the necessity for living things to adapt to an ever-changing environment. The actual structure of a given species is determined both by its position in the hierarchy of complexity and by the adaptations it has undergone.

The mechanism that Lamarck proposed to explain how species adapt is the one part of his theory remembered by later naturalists under the name 'Lamarckism'. The essence of the mechanism is a process known as the 'inheritance of acquired characteristics' because it supposed that characters acquired during the organism's own lifetime can be transmitted to its offspring. If this were true, a man who developed strong muscles through weightlifting might hope to see his children born with stronger muscles than they would have had if he had not engaged in this activity. Lamarck supposed that, when animals are exposed to a new environment, they must change their habits to adapt, and must thus exercise some parts of their bodies more than they did before. Those parts will grow in size and the next generation will inherit some of the increase, and will build upon it by continuing to exercise the same parts. The classic illustration of this is the giraffe, which Lamarck supposed to have developed its long neck through generations of its ancestors reaching up to eat the leaves of trees. In this theory there is no selection – all members of the population acquire the new habit (e.g. the new feeding pattern of the giraffe) and all participate in the process of self-adaptation, which they transmit to their offspring.

Modern geneticists insist that acquired characters cannot be inherited because there is no mechanism by which they can be imprinted on the genes that transmit characters from one generation to the next. But most biologists in Lamarck's time – and for a century afterwards – believed that the process could work. Ordinary people have always tended to believe that important effects in their own life can be transmitted to their children, and it took biologists a long time to realize that much of the evidence for this assumption is faulty. Lamarck's great innovation was to realize that, if the process could be continued indefinitely, it would provide an alternative to the traditional concept of design by a wise and benevolent Creator. Species were adapted because there was a natural process that enabled living things to change their behaviour and, as a consequence, to change their bodily structure.

At one time historians supposed that Lamarck was ignored or ridiculed by his contemporaries, but we now know that there was a current of radical thought about Nature that remained active throughout the early nineteenth century, and which drew upon the idea of evolution as a means of challenging the traditional image of a static world designed by God (chapter 7). Lamarck was thus one of the vehicles by which the materialism of the Enlightenment was transmitted into the following century.

Yet, for all its power to influence later thinkers, Lamarck's theory failed to address some of the key issues that would become crucial in early-nineteenth-century science. The evidence he presented was unsatisfactory because it never occurred to him that another process (natural selection) might produce similar results. More important, he did not link his concept of progress to the evidence that was now beginning to emerge from the fossil record. Lamarck was convinced that Nature would allow no species to become extinct, and interpreted the supposedly extinct fossils as evidence that old species have changed into something quite different in the modern world. There was also no attempt to explore the geographical dimension of evolution. Lamarck was a museum naturalist, not a traveller, and his concept of adaptation was devised from theoretical principles, not from a close observation of how animals and plants actually survive in the world. If some of the ideas he proposed had the capacity to stimulate later thought, they did so within a conceptual environment that would be drastically changed by the effort to incorporate new lines of evidence emerging from geology and biogeography.

The Heroic Age

The nineteenth century has been called the 'heroic age' of geology, because it witnessed the creation of a complete outline of the earth's history. By the middle of the century, the sequence of geological periods familiar to any modern scientist had already been established. But this was not only the heroic age of geology – it was an age that saw the whole surface of the earth exposed to scientific scrutiny. The discoveries of the earlier explorers were consolidated and extended so that by the end of the century the interior of every continent – even the 'dark continent' of Africa – had been investigated by western geographers. The depths of the oceans were beginning to yield up their secrets, and the pattern of the weather itself was being studied by scientists organized on an international scale. This was an age in which the outline of our modern understanding of the environment and its changes began to take shape. It also saw the emergence of the modern form of the scientific community, with its societies, journals and competition for government and private funding.

The outlines of what happened are clear enough, but there is plenty of room for disagreement among historians over details. As science began to take over an increasingly large area of conceptual territory, it was caught up in the social upheavals of the time. The potential for geology to come into conflict with traditional religious views on the earth's creation had already become apparent in earlier centuries. Now that conflict was to be resolved as science established itself as the sole source of authority on matters of physical fact. Yet the image of a 'war' between science and religion has been criticized by many modern historians as misleading. The emergence of the theory of evolution was certainly one of the great revolutions in the history of western thought, and geology played a role in establishing the foundations upon which the new world view would be established. But it has proved all too easy for us to concentrate on the areas of obvious controversy, oversimplifying the issues to fit in with the metaphor of conflict, and ignoring a vast range of developments that took place independently of the great public debates.

National preconceptions have determined which areas of science have attracted historians' attention. English-speaking historians have tended to concentrate on those areas where the debate with religion was most active, since here the intellectual issues seem most exciting. Such an approach may puzzle historians from continental Europe, since in their own countries religion played a much less decisive role in the nineteenth-century debates. Areas of science where there was little room for conflict with traditional values have been ignored, except by specialists whose work deals with complex technicalities of little interest to the outsider. We know far more about the history of geology than we do about the development of oceanography and meteorology – even though the same people and the same organizations were often involved in all three enterprises. Although science was becoming increasingly specialized, the interactions between the conceptual issues raised by the various areas ensured that links would be retained. It is the modern historian who chooses to concentrate on one area at the expense of another, thereby giving a distorted and compartmentalized image of the past. Our growing awareness of environmental issues may now focus more attention onto the less fashionable areas of science's attempts to understand how the terrestrial system functions.

Even within individual sciences, distortions have arisen and are only slowly being corrected by modern historical scholarship. The search for 'heroes and villains' has led to a skewed interpretation of nineteenth- as well as eighteenth-century geology. Anyone who opposed the claim that the earth has been subject only to gradual changes has been branded a 'catastrophist' and accused of trying to retain a role for the biblical story of Noah's flood. Modern historians have been able to correct the bias introduced by hindsight and have shown that the once-despised catastrophists played a major role in constructing the outline of earth history we still accept today. Equally significant has been the recognition that the much-publicized debate over the rate of geological change was not necessarily the most fundamental issue at stake. The techniques that established the sequence of geological periods, and the attempts made to understand the formation of mountain ranges, represent areas of geology that produced major results without necessarily raising the hackles of religious thinkers. Here the history of science is at last confronting the technical issues that mattered to the people who established our modern view of the natural world.

Historians have also been active in studying the social development of nineteenth-century science. Although a rudimentary scientific community had come into existence in the previous century, it was only now that the professional 'scientist' in the modern sense of the word began to appear on the scene (the word itself was coined by William Whewell in 1840). Scientists began to organize themselves both into specialized societies for the promotion of individual disciplines and into general organizations designed to promote the role of science in society. Governments were forced, some of them reluctantly, to support scientific research and education. The scientists argued that, in an increasingly industrialized world, they alone held the key to progress and hence to national development. In order to make this case, however, they had to stress the practical value of scientific knowledge rather than its theoretical content, often concealing their own real interests from their paymasters.

Science promoted an essentially materialistic view of Nature, which was very much in keeping with the demands of the new industrial society. It was favoured by those who wished to see the search for knowledge as a process of gathering facts about an objective world existing 'out there' independently of the human mind. The simple philosophy of materialism was modified by 'positivists' such as Auguste Comte, who saw science as the attempt to describe natural processes without reference to human speculations about ultimate causation. Such ideas fitted in well with the ambitions of those who wished to control Nature.

Yet, in an age increasingly dominated by the expansion of industry, there were many intellectuals who rejected the values of commercialism and materialism. The mechanistic view of the world was opposed by rival schemes designed to emphasize the spiritual dimension of Nature. In contrast, the Romantic movement in the arts stressed the importance of the human spirit and its desire to transcend the petty world of materialism. Various forms of philosophical 'idealism' tried to show that our image of Nature is a product not just of sensations received from 'out there' but also of the mind's efforts to interpret those sensations. Some philosophers went on to suggest that Nature itself was a manifestation of spirit, not of matter – a direct expression of the divine will. Far from being antagonistic to science, such ideas could be incorporated into it, and often served as the basis for important new initiatives. Scientists actively participated in the great debates of the time, and it would be a mistake to assume that they invariably joined in on the side of materialism.

To some extent the debates between science and religion can be seen as a consequence of the emergence of a new professional and entrepreneurial class demanding a say in how society should be governed. The challenge to aristocratic authority was inevitably associated with a challenge to traditional moral values. Professional scientists were caught up in this social upheaval, mostly on the side of reform, although we should never forget that traditional sources of authority also sought to 'modernize' their value systems and could thus play a role in the development of scientific ideas. The most exciting of the recent contributions to the history of nineteenth-century science have come from those scholars who have tried to explore the interaction between technical debates, professional interests and political loyalties. They have shown that in many cases the social background played a major role in shaping the theories advanced to explain the ever-widening array of factual information.

THE ORGANIZATION OF SCIENCE

As Europeans took stock of their ever-increasing industrial power and consolidated their grip on the world through colonization, science came to symbolize the power of the human mind to dominate the material world. The scientists themselves realized that the hope of material benefits would encourage governments and private industries to provide funding for research. The early nineteenth century saw the beginnings of the modern system of state-funded scientific research organizations and state-supported scientific education. Continental Europe led the way, with Britain lagging behind. Scientists became aware of their potential influence and began clubbing together to form the first professional societies. By the later decades of the century, the organization of science had taken on a recognizably modern form, with America rapidly catching up as its industrial power expanded.

Europe Surveys the Earth

By the end of the eighteenth century a number of European countries had begun to encourage scientific research and education. The highly centralized government of France had played a leading role, but many German states also had mining colleges in which geology was taught. In the 1790s the revolutionary government in France replaced the old Jardin du Roi (of which Buffon had been the superintendent) with the Muséum d'Histoire Naturelle. This was to be a teaching institution as well as a museum, and both Lamarck and his great rival Georges Cuvier (1769–1832) were based there (see chapter 7). The École des Mines offered education in geology, and in the 1820s the government sponsored a plan to prepare a geological map of the country. Although now lacking an overseas empire, the French still sponsored exploration and oceanography, the voyage of Dumont d'Urville from 1826 to 1829 being particularly important. Paris had the first Geographical Society in Europe (founded in 1821), which played a leading role in later decades encouraging the creation of a North African empire.

In the early decades of the new century German universities created the modern system of scientific education, in which students are eventually directed to a research project that they complete in order to obtain a doctorate. German scientists were also becoming more aware of themselves as a professional group seeking to play an active role in society. A meeting of the German Association of Naturalists and Physicians in 1828 attracted scientists from all over the world to Berlin and encouraged the setting up of similar national organizations in Britain and elsewhere. The geographer Alexander von Humboldt (1769–1859), who had helped to organize the 1828 meeting, promoted international co-operation in the gathering of information about the earth's magnetic field.

Britain lagged behind the continent in many respects, partly because the free-enterprise system of early industrial capitalism did not encourage the state to become involved. If science was of value to industry, then industry should pay for it. The mathematician Charles Babbage returned from the Berlin meeting of 1828 and wrote a book on

The Decline of Science in England. Partly in response to Babbage's call, the British Association for the Advancement of Science was founded in 1831. The Association served as a means of communication between scientists and the general public, and also lobbied government for more support. These efforts bore fruit in the establishment of an empire-wide chain of magnetic observatories. Edward Sabine (1788–1883) responded to Humboldt's call by encouraging the Association to press for government participation in the global study of the earth's magnetic field. The observatories were established under the control of the Army (whose Ordnance Department was responsible for mapping), and observatories were set up in Montreal, Tasmania, the Cape of Good Hope and Bombay. Apart from encouraging international scientific co-operation, the observatories played an important role in promoting science in newly emerging nations.

The involvement of the military in Sabine's project illustrates the close links between science and imperialism. The Royal Navy had supported exploration and continued to do so in the new century. When Charles Darwin travelled around the world aboard HMS *Beagle* (1831–6), he was accompanying an expedition whose main purpose was to chart the waters around South America. If most of the world's land masses had now been discovered, it was still necessary to map their coasts if the trade routes that benefited British industry were to be operated safely. Naval officers such as Robert Fitzroy (1805–65), captain of the *Beagle* during Darwin's voyage, also played an active role in meteorology. Several expeditions were sent in search of the North West Passage through the Arctic from the Atlantic to the Pacific, sometimes with disastrous results. Sir John Franklin's expedition of 1845 was wiped out completely, largely because the Royal Navy refused to copy the native Eskimo techniques for survival in the harsh environment. Franklin became a hero back home, and the eleven-year search for the remains of his expedition led to the exploration of much of the Canadian Arctic.

At home, the government was drawn into providing support for science through the creation of the Geological Survey of Great Britain. This was founded by a rather haphazard process during the 1830s, when the geologist Henry De la Beche (1796–1855) obtained government finance to help him finish a geological map of southwest England. By exploiting his friendship with the Prime Minister, Sir Robert Peel, De la Beche was eventually able to establish himself as director of a permanent Geological Survey charged with the mapping of the whole country. Here the appeal to the practical applications of science was direct: a geological map would encourage the location of useful minerals and thus benefit the country's industry. From the 1850s the Royal School of Mines offered a technical education including geology and paleontology. Soon there were geological surveys operating in most parts of the empire, including Canada, Australia and India.

The work of any geological survey was dependent upon the existence of accurate maps of the territory whose underlying structure was to be determined. De la Beche's efforts were facilitated by the existence of the maps produced by the Ordnance Survey, which had originally been established to map the country for military purposes after the crushing of Bonnie Prince Charlie's rebellion in 1745. In remoter parts of the empire, mapping was of crucial importance to Europeans seeking to symbolize their control of disputed areas. The Trigonometrical Survey of India was established in the early nineteenth century under Sir George Everest, and in 1856 it identified the highest mountain in the world in the Himalayas, to which Everest's name was given. It would be many decades before Europeans penetrated anywhere near the mountain itself, although mapping of this remote area was carried on by secret agents sent into unconquered territory by the British government.

Specialized scientific societies were now established. As

president of the Royal Society, Sir Joseph Banks resisted the foundation of more specialized groups, but he had not been able to prevent the formation of the Geological Society of London in 1807. Over the following decades the Society became one of the most active centres of scientific debate in the country. Its membership was an elite group of specialists who set themselves up to define the professional standards of their discipline. Soon there would be a whole collection of societies specializing in the various areas of science and natural history.

The Royal Geographical Society, founded in 1830, encouraged the exploration of many parts of the world. Under the leadership of the geologist Sir Roderick Murchison (1792–1871), it patronized expeditions to Australia, Africa (including the journeys of David Livingstone) and the Arctic. Murchison saw exploration as the means by which Britain's overseas empire would be extended and called for the annexation of unclaimed territory even when the government was unwilling to commit itself. The involvement of a leading scientist in this activity once again suggests that the expansion of geographical and geological knowledge was being linked to the imperial ambitions of the western powers. Some of the early explorers were strong characters seeking to establish a reputation, but as the century progressed the demand that geography should become a science led to the sending out of increasingly well equipped expeditions.

America Catches Up

In the United States, too, there was a steadily increasing involvement in science by both state and federal governments. In 1803 Thomas Jefferson commissioned Meriwether Lewis and William Clark to set out on an expedition to explore along the Missouri river and into the western half of the continent. As well as indicating US interest in the territory concerned, the expedition brought back important scientific information of the region and its inhabitants. Significantly, the expedition had been financed from the military budget, and the US Army continued to be involved in the exploration of the west in later decades. In 1807 the Coast Survey was established to study the waters around the American coastline. After being revitalized by its second superintendent, Alexander Dallas Bache (1806–67), in 1844, the Coast Survey became one of the country's most effective scientific institutions. It was beset by inter-departmental rivalries, however, and had to fight off a constant threat of being incorporated into the Navy's Hydrographic Office. All too often the involvement of governments brought problems as well as benefits; in the search for research funds, scientists ran the constant risk of their projects being taken over by those who had purely utilitarian goals.

The same difficulties arose in the case of geological surveying. A number of states established geological surveys in the early decades of the new century, New York's being particularly effective. But state legislatures demanded immediate results in the shape of newly discovered mineral deposits. It was difficult to convince the politicians that a more abstract approach was needed if state-funded science was to yield long-term benefits. A good geological map would be useful to generations of prospectors, but the geologist could not afford to get bogged down in the search for individual mineral deposits. In some cases, however, the geologists wished to engage in purely theoretical work and were forced to conceal the true nature of their activities. James Hall of the New York Survey published extensively in paleontology even though his work had little practical value.

The federal government was particularly active in encouraging exploration of the western part of the continent. There were a number of different surveys, which gradually established their independence from military control. From 1869 the US Geological and Geographical Survey of the Territories sent out expeditions under Ferdinand V. Hayden

(1829–87). Hayden was an influential figure with many important contacts both in government and in international scientific circles. His survey generated valuable information, and yet held mining and other industrial interests at arm's length. It was Hayden who reported the scenic wonders of Yellowstone and campaigned for the establishment of a National Park there. Even hard-headed scientists could see the importance of preserving such areas for posterity, and the National Park was created by Act of Congress in 1872. To the Romantic philosophers of the early nineteenth century, mountains had been symbols of the sublime. By the end of the century, improved transportation facilities were allowing ordinary people to seek refreshment in the wilderness. Scientists would increasingly be drawn into the controversies that arose as the desire for conservation came into conflict with the desire to exploit (see chapter 8).

In America these conflicts were compounded by the existence of a number of rival scientific institutions. The US Geological Survey was founded in 1879 under Clarence King, who was soon succeeded by John Wesley Powell (1834–1902). Powell had made his name by exploring the Grand Canyon of the Colorado river in 1869. His survey functioned independently of Hayden's group and actively collected both geological and anthropological information in the west. In 1886 the survey was investigated by the Allison Commission, which had been set up by Congress to evaluate government support for science. Powell defended the survey by insisting that government science was needed to investigate issues that were of long-range value but were too broad for any private enterprise to tackle. Powell warned that it would be impossible to irrigate large areas in the arid lands of the west and protested against the destruction of forests. He thus came into conflict with the developers and had to face a restriction of the Survey's funding in the 1890s. Here again, the transition to 'big science' created problems by encouraging scientific and political rivalries centred on matters of increasing public concern.

There was a close link between the growing power of the industrialized western nations and the expansion of nineteenth-century science. Yet we should beware of oversimplifications in this area. Some scientists enthused over industrialization and imperial expansion, others were indifferent or even hostile. People with a variety of different interests could now become scientists, and we should not generalize too rigidly about their motivations. The Romantic movement encouraged the exploration of mountains because they seemed to express a wildness akin to the creative spirit of humankind, and there can be little doubt that many geologists were inspired by such sentiments. In Germany, nature-philosophers such as F. W. von Schelling opposed the materialist philosophy and urged that Nature be seen as a system based upon spiritual foundations. Poets such as William Wordsworth and Samuel Taylor Coleridge articulated the image of Nature as a harmonious system animated by spiritual forces. In the modern world we find it difficult to believe that art and science interact, but in the early nineteenth century the Romantic movement inspired many scientists with the urge to look beyond the details of observation and create wide-ranging theories that would allow us to see the universe as a coherent whole. Some of the greatest achievements of the age arose from the interaction of the practical demand for knowledge and this Romantic search for a global synthesis.

A NEW GEOGRAPHY

Opposition to a purely mechanistic philosophy of Nature is evident in the expansion of geographical theory in the early nineteenth century. In Germany, Karl Ritter produced a synthesis of geographical knowledge based on the assumption that the earth was a divinely created system with a coherent plan that humankind could understand. This approach was later imported into America by Arnold Henri Guyot's *The Earth and Man* of 1849. For some scientists, natural theology encouraged the hope of tracing out the pattern of natural interactions, but this explicitly religious factor is less obvious in the writings of the geographer who had most influence on early-nineteenth-century science, Alexander von Humboldt.

Humboldt's impact was so great that some historians have coined the term 'Humboldtian science' to denote the study of the environment in the early nineteenth century[1]. Humboldt allied himself with those philosophers who challenged the materialistic view of Nature, but he was far from being a pure 'Romantic'. He wanted to collect factual information and to process it with a view to understanding the complex relationships binding all aspects of the natural world together. Humboldt gave a great impetus to the systematic study of the earth's physical and organic environments, but he hoped to found a new science that would emphasize the interactions between geological agents, living things and human activity. His vision of a unified science of the environment could not be realized in an age of increasing specialization, but he pioneered techniques that would be used by geographers, meteorologists and oceanographers, and he encouraged his followers to look for connections between these areas wherever possible.

Humboldt and the Cosmos

Humboldt came from an aristocratic Prussian family, although his own views were decidedly liberal and he frequently campaigned for the rights of oppressed peoples. He studied geology under Werner and worked for a time as an inspector of mines for the Prussian government. He was influenced by biogeographers such as Karl Willdenow and George Forster, who were attempting to explain the differences between the plants and animals of the different regions (chapter 5). This would become one of Humboldt's chief goals, but to tackle the problem he was inspired to undertake careful measurements of all aspects of the physical environment. On his travels he took a vast array of instruments for measuring temperature, humidity, air pressure and other variables.

Humboldt's attitude towards the facts he collected was shaped by the intellectual revolution that took place in late-eighteenth- and early-nineteenth-century Germany. Immanuel Kant had stressed that the human mind did not receive information passively from the senses – in order for the world to be intelligible, we must organize the information and impose a framework of understanding upon it. The Romantic movement also sought to unite the human mind with the spiritual dimension of Nature. Humboldt met the writer and philosopher J. W. von Goethe, and was clearly influenced by the image of wild Nature as a symbol of the sublime. He saw the artist's effort to present an overall impression of a landscape as a guide to the scientist seeking to understand the natural relationships involved. He was also convinced that the landscape and the vegetation exerted a formative influence upon the human inhabitants of any area. The new, anti-materialistic philosophy gave Humboldt his passion to search for the relationships between natural phenomena, but unlike the Romantic artists he did not repudiate measurement and experiment. His was to be a new kind of empiricism in which measurement would be incorporated into a science that sought to uncover the unity of Nature.

Humboldt spent the years 1799–1804 exploring in South and Central America with the botanist Aimé Bonpland (1773–1858). Much of the territory was as yet unknown to science, so there was almost unlimited scope for collecting animals, plants and information on the geology and climate of the area. Humboldt confirmed that there was a link between the Amazon and Orinoco river systems, throwing doubt on Bruache's concept of river basins as isolated geographical units. He climbed several volcanoes in the Andes, including Chimborazo, and temporarily held the record for the highest altitude reached by any European explorer. Mountain-climbing allowed him to see a cross-section through the vegetation zones ranging from the tropical forest to the snowline. Humboldt also studied the economic life of the region, commenting angrily on the injustice of slavery. On his return to Europe, he settled in Paris until returning to Berlin in 1827. It was in Paris that he prepared the results of his journey for publication – a task that would ultimately use up the whole of his inherited fortune.

Thirty volumes were published covering all aspects of South American geography, natural history and economics. Humboldt made important contributions, especially in the study of plant geography. He also wrote a *Personal Narrative* of the voyage, which stimulated many later naturalists (including Darwin) to visit the tropics. After his return to Berlin, he conceived the plan of writing a universal survey of the earth and its inhabitants that would reveal the connections between the physical, organic and human factors controlling the environment. The first volume of his *Kosmos* appeared in 1845; three more volumes appeared during the remainder of Humboldt's life and a fifth was completed from notes left behind after his death in 1859. From the beginning, Humboldt made his intention clear[2]:

> In considering the study of physical phenomena, not merely in its bearings on the material wants of life, but in its general influence on the intellectual advancement of mankind, we find its noblest and most important result to be a knowledge of the chain of connection, by which all natural forces are linked together, and made mutually dependent upon one another; and it is the perception of these relations that exalts our views and ennobles our enjoyment.

Humboldt made no effort to explain the harmony of Nature as an indication of design by a supernatural Creator, an omission for which he was sometimes criticized. For purely philosophical reasons he wanted to use empirical science in the service of a world view that would stress the interaction between all natural phenomena.

Humboldt was the last scientist who could seriously think of providing an overview of the whole world of Nature. In an age of increasing specialization, his ambition was sometimes criticized as unrealistic, a throwback to the old days of amateur enthusiasm. He founded no coherent school of thought, and no one took up the task of completing the universal survey. But Humboldt's influence was pervasive throughout the rest of the century. He provided many of the conceptual tools that would be used by others seeking to understand the global pattern of the environment. He inspired many younger scientists with the enthusiasm to search for those global patterns, and helped to create the framework of international co-operation that made this kind of study possible. Many of the best thinkers of the time – Darwin is a prime example – developed their insights precisely because they were willing to look for the kind of connections that Humboldt had stressed.

Humboldtian Science

Humboldt introduced techniques for representing geographically diverse information in a comprehensible form. He drew vertical sections through the Andes to illustrate how vegetations changed with altitude. He also pioneered the use of lines on a map to link all points with the same value for a variable such as temperature. An 'isothermal' is a kind of contour line joining all places with the same average temperature. A global map marked with isothermals allows one to see at a glance how geographical factors influence the prevailing temperature, thus facilitating the detailed study of global weather patterns. By plotting isothermals on a global scale, Humboldt was able to realize that the distribution of continents and oceans had a significant effect upon the climate. The fact that the major land masses lie more in the northern than the southern hemisphere means that the zones of latitude do not have the same climate north and south of the equator. Continental interiors experience greater extremes of heat and cold than oceanic regions.

One product of Humboldt's efforts was the creation of an international network to study variations in the earth's magnetic field. Eventually this study would reveal that 'magnetic storms' (rapid changes in the intensity of the field) are linked to the appearance of sunspots. This was just the kind of correlation between apparently distinct phenomena that Humboldt anticipated. Similar developments took place in meteorology. The introduction of the telegraph allowed the collection of simultaneous information on the weather from a wide area, making it possible to show how storms and other weather systems move across the globe. After his return from the *Beagle* voyage with Darwin, Robert Fitzroy became a leading figure in this area. He was put in charge of the Meteorologic Office when it was founded in 1855 and began to publish daily weather forecasts. The cup anemometer for measuring wind speed was invented in the 1850s by Thomas Romney Robinson, director of the Armagh Observatory in Ireland. Like many astronomical observatories, Armagh also kept track of the weather, although it was sometimes impossible to process the masses of data thus generated. By the 1870s an international scientific framework for studying the weather had already begun to emerge, with the first International Meteorological Congress meeting in Vienna in 1873.

Oceanography also benefited from the new desire to study phenomena on a global scale. Many scientists collected information on the chemical composition, temperature and pressure of the ocean at various depths and in different regions. The difficulty of gathering information about the ocean depths was immense. At first it was believed that the temperature in the depths never fell below 4°C, until it was shown that the figures were distorted by the effect of pressure on the thermometers. There was intensive study of tides and ocean currents, and a number of physicists discussed the forces responsible for the movements of the water. James Rennell (1742–1830) provided the first accurate map of the currents in the Atlantic Ocean, and the US Coast Survey made extensive studies of the Gulf Stream. The zoologist Edward Forbes argued that no life existed below a depth of 300 fathoms (about 600 metres), a view widely accepted until disproved by the voyage of the British research vessel HMS *Challenger*, in 1872–6. The *Challenger* expedition provided valuable information about the sea bed, including the discovery of manganese nodules that are now being seen as a potentially valuable source of minerals.

The first detailed map of the deep sea bed was provided for the Atlantic by the American geographer Matthew F. Maury (1806–73), director of the US Naval Observatory and Hydrographic Office. Maury's pioneering *Physical Geography of the Sea* of 1855 was based on the assumption that the earth had been designed for the benefit of humankind by a wise and benevolent God. He devised new techniques for sounding the depths and his work proved of great value in the laying of the first transatlantic telegraph cables. He also studied global wind patterns and was able to provide sailors with guides that significantly reduced the time taken on many routes. Some oceanographers believed that the winds were responsible for producing ocean currents such as the Gulf Stream, but Maury disagreed. He argued that currents were produced by changes in the density of sea water due to temperature, which set up systems of circulation between warm and cool regions of the world. Maury believed that the upwelling of a warm current would produce open sea around the north pole, a claim not disproved until Fridtjof Nansen allowed his vessel the *Fram* to be carried in the ice to within a few degrees of the pole in the years 1893-6.

Many areas of Humboldtian science remain unstudied as yet by historians, but this is not true of geology. As a student of Werner, Humboldt himself realized that the earth's present environment is the product of a long sequence of historical changes. He wrote on the geological succession of rocks and identified a new formation, the Jurassic, named after the Jura mountains in Switzerland. One of the great intellectual triumphs of the early nineteenth century was the completion of what we now call the 'stratigraphical column', the sequence of rock formations arranged according to their order of formation in time. By the 1840s the modern sequence of geological periods had been established, and every rock exposed on the earth's surface could be assigned to its position in the sequence. Historians are drawn to this topic because it serves as a prelude to the 'Darwinian revolution', and their interpretations have often been distorted by a desire to emphasize those issues which most obviously seem to anticipate the radical components of Darwin's vision of the history of life on earth.

As with Werner and Hutton (chapter 4), the tendency to identify heroes and villains has produced an oversimplified view of the issues. Charles Lyell revived Hutton's gradualism and has been hailed as one of the founders of modern geology. His opponents, the 'catastrophists', have been vilified as bad scientists who tried to defend the biblical story of Noah's flood. Recent scholarship has exposed the hollowness of these interpretations. Like Werner, the catastrophists exerted a major influence on the geological thinking of the time and on the development of modern stratigraphy. There were apparently sound empirical and theoretical reasons for postulating earth movements on a catastrophic scale, and the link to Noah's flood (which had been stressed only in England) was soon abandoned.

Many areas of the science were extended almost without reference to the debate on the rate of geological change. This is most apparent in continental Europe, where the traditions established by Werner survived the demise of the retreating-ocean theory. As Humboldt and others began to recognize the significance of earth movement in shaping the present surface, they nevertheless saw themselves as maintaining the Wernerian tradition of trying to establish the sequence in which the formations had been deposited. What is misleadingly called 'catastrophism' represents an extension of the Wernerian programme because the underlying trend of a decline in the sea level was replaced by the cooling-earth theory. There was still a *direction* in the earth's history, a point of origin followed by a cumulative change leading towards the present. It was only to be expected that geological agents would have been more active in the past, when the earth's internal temperature was much higher than today. Like Hutton, Lyell tried to replace 'directionalism' with a cyclic or steady-state model of change. He forced the catastrophists to scale down their image of past violence, but he was unable to shake the basic assumption that the earth has changed consistently since its original formation.

Historians are now more aware of the social framework within which the geological debates took place. In early-nineteenth-century Britain, a group of 'gentlemanly specialists' used the Geological Society of London to define themselves as an elite group entitled to speak with authority on disputed questions[3]. Amateurs now became mere fact-gatherers, foot soldiers to be directed by the generals who determined strategy at the theoretical level. Some of the specialists had appointments at the very conservative universities of Oxford and Cambridge and were thus forced to project the new science as an initiative that was compatible with religion. Others had no such encumbrance, however, although to begin with there was little effort to forge a link between the new geology and industry or government. This was in contrast with the situation on the continent, where state involvement in science was already well established. In America and eventually in Britain, the age of the gentlemanly specialist began to give way to the age of the professional scientist earning a living from education or from applied research.

Fossils and Stratigraphy

The aforementioned triumph of early-nineteenth-century geology was the establishment of the stratigraphical column, allowing every stratum of rock to be assigned to a unique period of deposition in the earth's history. Werner had pioneered the recognition of distinct formations, each identified with a particular episode of deposition. But the Wernerians' ability to date the rock formations was limited by their assumption that each period of deposition had its own characteristic mineral types. By 1800 it was becoming obvious that mineralogy by itself was not enough, because the same kind of rock had been laid down at different periods in the earth's history. It was now realized that the fossil content of the rock stratum, not its mineral composition, best identified its period of deposition. In effect, this implied that the population of animals and plants had changed systematically through time. Certain fossils were characteristic of particular epochs and could be found in all rocks formed at that time, whatever their mineral composition. The fossils allowed rocks of equivalent period of origin to be recognized from all over the world, and by correlating information from different areas a complete sequence of rock formations could be established. This corresponded to a complete sequence of the geological periods in the earth's history.

Geology thus gave a new sense of the time dimension in the study of Nature. This in itself modified the underlying assumptions of the mechanical philosophy: it was obvious

Fig: 6.1 *Building the stratigraphical column*

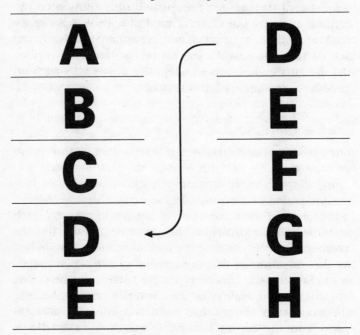

Although very extensive sequences of strata are sometimes visible, as in the Grand Canyon of the Colorado river, no part of the earth's surface has had sedimentary rocks deposited upon it throughout the planet's history. The stratigraphical column is an idealized representation of what a continuous sequence of deposition would look like, but it must be composed of information drawn from a vast number of different locations.

The correlation of this geographically diverse information is made possible by the method of dating rocks by fossils. If a continuous sequence A B C D E is found at one point, and another sequence D E F G H somewhere else, the identification of D and E as being equivalent strata (on the basis of their fossil contents) allows the two sequences to be correlated to establish the whole pattern from A to H. By repeating this process to link information from many locations, the complete sequence can be established. The geological systems listed in the accompanying diagram each consist of a whole series of strata whose fossils are fairly similar to one another, but are significantly different from the fossils of adjoining systems.

that, however constant the laws of Nature, the earth itself had changed through time and could only be studied by methods analogous to those of the historian who seeks to reconstruct the human past from fragmentary records. The Victorian era saw a new awareness of the historical dimension in human affairs, associated with a growing interest in the idea of progress. Archaeologists and geologists extended this dimension into the ever-deeper past. Geography was also important, since geological sections (corresponding to historical sequences of rocks) were linked to maps showing how the different formations outcropped onto the surface.

English-speaking historians have tended to associate the new methodology with the work of William Smith (1769–1839), dubbed the 'father of English geology' by later generations of stratigraphers. But recent research has cast doubts on the relative significance of Smith's role, suggesting that his elevation to the status of a 'hero of science' was an artefact of later British geologists' desire to evade the debt they owed to continental pioneers. Smith was an engineer and canal-builder who recognized the importance of fossils for identifying rocks in the course of explorations undertaken for purely practical reasons. His geological map of England (1815) was based on several decades of work, and was certainly an impressive achievement. But Smith had no access to the work of the Wernerian school, and his humble origins made it unlikely that he would be taken seriously by the gentlemanly specialists. However much they may have wanted to deny it later on, the British stratigraphers were most impressed by the new continental methods. Smith's map was the culmination of a long tradition of practical, local geological work, but the new generation of specialists were actively trying to separate themselves from this tradition. There was almost certainly some leakage across the social boundary, but it was unfashionable to admit a link with 'trade' until nationalist sentiments at last made it necessary to recognize the local British tradition that had paralleled the continental developments.

The most active pioneers of fossil-based stratigraphy were Georges Cuvier and Alexandre Brongniart (1770–1847), who worked on the rocks of the Paris basin. Cuvier achieved fame for his reconstructions of fossil vertebrates (chapter 7) and had become convinced that there were a number of distinct populations that had succeeded one another in the course of geological time. In collaboration with Brongniart, he realized that the successive layers of rock deposited in the Paris area could be identified by the fossil invertebrates they contained. Wherever the same stratum appeared on the surface, it could be identified by its fossil contents. A surface map could thus be correlated with a vertical cross-section illustrating the distribution of the underlying strata. An idealized section of such a cross-section would then represent the stratigraphical column for that area, with every stratum assigned to a unique position in the sequence. Cuvier and Brongniart's *Descriptions géologiques des environs de Paris* of 1811 sparked off an explosion of activity as other geologists rushed to extend their method to the rocks of Europe and, ultimately, of the world.

Fossils – hitherto dismissed as incidental features of the rock – were now the key to its precise identification. But this precision was only plausible if one assumed that the earth's population changed systematically through time, so that the rocks of any one period displayed an absolutely unique collection of fossils. The technique also seemed to depend on the fact that the fossil contents changed abruptly from one stratum to the next. If the population changed gradually, as Lamarck's theory of evolution implied (chapter 5), then there would be no clear-cut breaks in the record and the precise identification of a rock's age would be impossible. Cuvier rejected Lamarck's theory, justifying his claim that the fossil populations changed abruptly by arguing for wholesale extinctions caused by geological catastrophes. New animals and plants then appeared either by migration from unaffected areas or by creation. Cuvier's *Discours sur les révolutions de la surface du globe* of 1812 became a pioneer source for catastrophist opinions, the term 'revolution' taking on the meaning of abrupt change that it had now acquired in the political history of France.

The Ancient Rocks

Cuvier and Brongniart worked with the geologically recent strata of the Tertiary era, exploiting an area where the strata were superimposed quite regularly with only minimal later distortion. But elsewhere in the world the strata had often been folded and split by extensive earth movements occurring after the layers of rock had been deposited. The older the rocks, the more likely they were to have been affected in this way. In some areas, massive quantities of older rock had been eroded away before new deposits were laid down, giving an apparent jump from an old to a new formation. The older fossils were often bizarre creatures that took all the skills of the naturalist to identify. Some rocks contain no fossils at all, and must be dated by fossil-bearing deposits above and below. A series of heroic efforts covering the period of the 1820s and 1830s succeeded in establishing the outlines of the geological sequence we accept today, but only after intense activity and debate.

Some of the most important contributions were made by British geologists. They had access to vast areas of what Werner had called the Secondary and the Transition rocks, and were in an ideal position to establish the earlier phases of the sequence. The Carboniferous rocks were particularly important because these were known to be the principal source of coal deposits. But beneath (i.e. older than) the Carboniferous was a vast sequence of strata that had never been adequately surveyed by the Wernerians. Some of the geologists who unravelled these rocks had come into the science by very roundabout routes. Adam Sedgwick (1785–1873) was elected professor of geology at Cambridge despite knowing nothing at all about the subject (a nice comment on the state of scientific education in Britain at the time). Roderick Murchison was an ex-soldier whose wife persuaded him to take up geology as an outdoor substitute for fox-hunting. Between them they worked out the sequence of Cambrian and Silurian rocks in Wales, and played an important role in establishing the Devonian system. The latter was a particularly tricky piece of work, since it involved the use of fossils to identify the rocks of Devon with the mineralogically quite different Old Red Sandstone of Scotland.

Because of the complex state of the older strata, the major divisions of the column such as the Devonian, Silurian and Cambrian systems were not natural entities waiting to be discovered. In principle, a new system (equivalent to a new geological period) could only be established if a group of strata could be identified by a sequence of related fossils that were quite clearly distinct from the types found elsewhere. Apart from the problem of misidentifying fossils, the boundaries between the systems were not as clearly demarcated as catastrophist theory would imply. Recent study of the process by which the divisions were established has shown that they were the result of protracted – and often acrimonious – debate between the specialists involved. Murchison later tried to incorporate Sedgwick's Cambrian into the lower regions of his own Silurian system. The Devonian was created as a means of resolving an argument between Murchison and De la Beche, the founder of the Geological Survey. These debates also reflected the changing character of geology; De la Beche was anxious to protect his own status in the Geological Survey, while Murchison's position as De la Beche's successor in the Survey gave him a temporary advantage in his campaign against Sedgwick. The geological periods we take for granted today were the product of social interaction between the specialists who were trying to make sense of a bewilderingly complex array of rocks.

TABLE 6.1 *The sequence of geological formations*

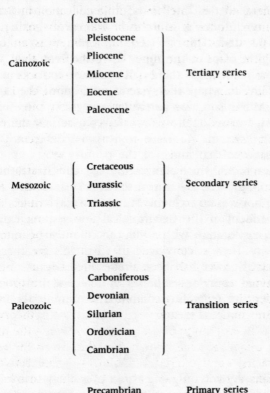

Cainozoic	Recent	
	Pleistocene	
	Pliocene	
	Miocene	**Tertiary series**
	Eocene	
	Paleocene	
Mesozoic	Cretaceous	
	Jurassic	**Secondary series**
	Triassic	
Paleozoic	Permian	
	Carboniferous	
	Devonian	
	Silurian	**Transition series**
	Ordovician	
	Cambrian	
	Precambrian	**Primary series**

By the late nineteenth century the basic subdivision of the stratigraphical column into formations (equivalent to periods of deposition in the earth's history) had taken on its modern form. The major divisions of the Wernerian theory (Tertiary, Secondary, Transition and Primary) had been divided into distinct formations with the names shown above. (The very extensive and much older rocks of the Precambrian were to remain a problem well into the twentieth century.) The three great eras of the history of life were given the names Cainozoic, Mesozoic and Paleozoic (new life, middle life and old life) by William Smith's nephew, John Phillips, in 1841. Although identified by Phillips on the basis of invertebrate fossils, these three eras correspond roughly to the ages dominated by mammals, reptiles and fish in the history of vertebrate life.

Murchison's attempt to take over the Cambrian was part of a campaign of intellectual imperialism – he wanted as much territory as possible to be represented as belonging to 'his' system. This reflected the important role played by geological maps at the time. Once the geological systems were agreed upon, the earth's surface could be mapped according to where the systems outcropped onto the surface. Murchison was particularly aggressive in trying to get large areas of the world depicted as belonging to his Silurian system. He went to Russia in the early 1840s and showed that much of the country was composed of Silurian rocks, which allowed him to claim that the area had been 'conquered' by British science. This use of a military metaphor was no accident. Murchison played a leading role as president of the Geographical Society in encouraging world exploration with a view to colonizing 'unoccupied' territory. He was convinced that Britain's greatness as an industrial power had been predestined because her rocks contained abundant supplies of coal and iron ore. Here geology and geography combined to promote the image of Britain's imperial status.

CLIMATE AND TIME

Stratigraphy was only one aspect of geology, and left many questions unanswered. What had the conditions been like when the various formations of sedimentary rocks were laid down? What forces had altered the rocks since their deposition, elevating them to form mountains and eroding them to form valleys? These questions were related because it came to be widely assumed that the earth was cooling down from an initially molten state. It was thus expected that the earlier geological periods must have experienced both more violent earth movements and much warmer conditions than those of today. But there was another set of phenomena that appeared to have quite different implications for the same questions. During the middle decades of the century it was recognized that many northern areas had been affected by extensive glaciation in the recent geological past – hardly what one would expect if there had been a steady decline in the overall temperature. To explain the ice ages, geologists switched their attention from internal to external sources of heat. Whatever the earth's original state, its climate in more recent times was governed by changes in the amount of heat received from the sun.

The Cooling Earth

Buffon supported his cooling-earth theory by appealing to the remains of 'elephants' found in Siberia and North America (chapter 4). If tropical creatures had once lived in these northerly latitudes, the climate of the whole earth must have been warmer. This particular line of evidence did not last very long. Cuvier showed that the mammoth and the mastodon were not identical to the living elephants, and it was evident that the mammoth might have lived in a temperate or even a cold climate. But the general principle that one might use the extinct creatures of the past as evidence for changes in the climate remained plausible. Naturalists could usually tell if a particular species was adapted to a warm or a cold climate, and if most of the fossils from a particular period were 'warm' species, then it seemed reasonable to assume that the climate had indeed been tropical.

There was some evidence that the climate of the earlier Tertiary had been warmer than the present, but the clearest evidence for extensive tropical conditions in the past came from the fossil plants of the coal beds. The founder of paleobotany (the study of extinct plants) was Adolphe Brongniart (1801–76), son of Cuvier's co-worker on the Paris basin. In 1828, Brongniart summarized the history of plant life on the earth and argued that the Carboniferous period had witnessed tropical conditions even in northern Europe[4]. He deduced from this that there had been a general decline in the earth's surface temperature, a claim that seemed to fit in very well with the popular assumption that the earth had originated as a globe of molten rock and had gradually cooled down.

The physicist Joseph Fourier (1768–1830) argued that the earth's central heat (revealed by temperature readings in deep mines) was best explained by the assumption that the whole planet had once been much hotter. Fourier's arguments were widely used to support the cooling-earth theory, and many geologists assumed that the heat emanating from the interior still had an appreciable effect on the climate. If this were so, the earlier geological periods would have been warmer, as Brongniart's Carboniferous fossils implied. Brongniart also argued that there had been an extensive change in the quality of the atmosphere. The air had originally contained a high proportion of carbon dioxide, but it had gradually been purified as the carbon was absorbed into plants and then embedded in the crust as coal. The higher animals appeared only when the condition of the atmosphere had improved towards its modern state.

The assumption that the generally warmer climate of the Carboniferous must have been a consequence of the earth's higher internal temperature was challenged by Charles Lyell (1797–1875) in his assault on the catastrophist position. Lyell's *Principles of Geology* (1830–3) revived Hutton's vision of a steady-state earth in which there had been no cumulative change in the overall conditions. Lyell was anxious to attack the cooling-earth theory because it seemed to uphold his opponents' claim that geological activity had been on a more violent scale than anything witnessed today. He accepted Brongniart's evidence for extensive tropical vegetation in the Carboniferous, but argued that the generally higher temperature might have been the consequence of other factors that had nothing to do with a cooling of the whole planet. Humboldt had suggested that the distribution of the land masses had a major effect on the climate. Lyell believed that earth movements were constantly elevating or depressing the surface, thus gradually creating and destroying continents. If the cycle of elevation happened to create a concentration of land masses around the equator, the earth as a whole would warm up because land absorbs the sun's heat more effectively than ocean. Conversely, a concentration of land masses in polar regions would produce an era of universal cold. By applying this theory to the Carboniferous, Lyell could argue that the higher temperature was merely an extreme point in a never-ending cycle of climatic fluctuation.

Lyell's cyclic theory of earth history won few adherents, but his arguments did have some effect. Geologists realized that they could not simply assume that a higher surface temperature in the past was the result of heat conducted up from the earth's interior. Changes in the earth's geography might be equally effective in modifying the climate. There was, in any case, a growing doubt that the surface temperature might depend on internal heat. Even if the centre of the earth was very hot, physicists now calculated that the amount of heat conducted up to the surface was trivial compared to that received from the sun.

The leading advocate of the cooling-earth theory in the later part of the century was the eminent physicist William Thomson (1824–1907), better known under his later title of Lord Kelvin. From the 1860s onwards, Kelvin attacked Lyell's steady-state theory on the grounds that a hot earth must be cooling down. But even Kelvin acknowledged that the cooling could have had little effect on the climate after the first, very rapid, lowering of the surface temperature. An equilibrium would soon be reached in which the amount of heat conducted to the surface was so small in comparison with the sun's heat that the surface would remain at the same temperature even as the interior cooled down[5]. The period in which the fossil-bearing rocks were laid down represented only the later phase of the earth's history, and any changes in the climate during this phase must thus be explained by some other factor. Kelvin tried to show that the sun's heat must be diminishing since there was no known process that could maintain its temperature.

Ice Ages

This point was driven home by the emergence of the ice age theory. If the recent past had witnessed conditions of extreme cold, then the gradual cooling of the earth did not determine the surface temperature. Modern geologists accept that the northern regions of both Europe and North America have had their superficial features shaped by extensive glaciation. There are great beds of gravel and massive 'erratic boulders', which are found at a considerable distance from the nearest outcrop of similar rock. But the significance of the evidence was missed at first because geologists preferred to see these phenomena as the result of massive tidal waves caused by catastrophic earth movements. The gravel beds and erratics were 'diluvium' produced by a great flood that had swept across Europe in the recent geological past. They certainly could not be explained by any cause that could be observed in operation today.

Only the people who lived high in the Alps were in a position to observe that there *was* a modern process that could move the rocks and gravel, because they were familiar with the activity of glaciers. There was a widespread belief among local people that the glaciers had once extended far beyond their modern limits. Apart from the gravel and erratics, there were rocks with parallel scratches or striations that must have been caused by moving ice. Yet few of the geologists who were now exploring the mountains would take these claims seriously. Only in the mid-1830s did Jean de Charpentier (1786–1855) and Louis Agassiz (1807–73) begin to accept the possibility that much of the 'diluvium' had in fact been moved by extensive ice sheets that had covered Europe during an 'ice age' in the recent geological past.

Agassiz lectured on the ice age theory in 1837 and published his *Studies on Glaciers* in 1840, narrowly ahead of a rival book by Charpentier. Thus Agassiz gained much of the publicity, although his own version of the theory was far less realistic than Charpentier's. The latter postulated only a substantial ice sheet that had extended from the mountains after a long period of poor weather. But Agassiz postulated a catastrophic lowering of the temperature, which had covered much of Europe with a thick layer of ice and had wiped out all life on earth (he was, of course, a creationist). The movement of the ice that had produced the debris scattered over a wide area was a result of the subsequent uplift of the Alps. By modern standards much of Agassiz's interpretation was in error, but in the eyes of his opponents the theory also flew in the face of all the evidence for a steady cooling of the earth through geological time. Agassiz himself accepted the cooling-earth theory, but argued that the process was not continuous. Like a good catastrophist he postulated occasional very sharp drops in the overall temperature interspersed between long periods of stable conditions. Nevertheless, the claim that the last drop in temperature had created conditions that were *colder* than those of today seemed to violate the basic logic of the cooling-earth theory.

Humboldt and many continental geologists argued strongly against the ice age theory. In 1840 Agassiz travelled to Britain and found evidence of glaciation in Scotland. He gained few converts, though, only William Buckland (1784–1856) converting from diluvialism. Lyell toyed briefly with the idea of a non-catastrophic ice age, since this fitted in with his own theory of climatic fluctuations. But he soon returned to his old position in which the diluvium was explained as debris dropped by melting icebergs during a time when Europe was covered by a shallow ocean. One partial convert was Charles Darwin (then better known as a geologist than as a biologist), who records how the ice age theory opened his eyes to the significance of phenomena he had hitherto ignored in the valleys of North Wales[6]:

> Yesterday (and the previous days) I had some most interesting work in examining the marks left by extinct glaciers – I assure you no extinct volcano could hardly leave more evident traces of its activity and vast powers . . . The valley about here, & the Inn, at which I am now writing, must once have been covered by at least 800 or 1000 ft in thickness of solid Ice! – Eleven years ago I spent a whole day in the valley, where yesterday everything but the Ice of the Glacier was palpably clear to me, and I then saw nothing but plain water and bare Rock.

The ice age concept was, however, to upset one of Darwin's own pet theories. He explained the 'parallel roads' of Glen Roy in Scotland as shorelines inscribed onto the mountainside when the area was submerged beneath the sea. It was Agassiz who noted that a glacier blocking the mouth of the glen would produce a lake having the same effect – an explanation that Darwin himself eventually accepted.

Most geologists rejected Agassiz's views, Murchison proving an especially vigilant champion of the diluvialist alternative. Only in the 1850s and 1860s did the majority of geologists began to take the ice age theory seriously, and then only after it had been separated from Agassiz's catastrophist interpretation. Studies of glaciers by many scientists, including the physicist John Tyndall (1820–93), revealed their power as agents of geological change. A. C. Ramsay (1814–79) of the British Geological Survey began to argue that many features seen in the mountains of Scotland suggested that the valleys had been eroded by ice rather than running water. If Hutton and Lyell's theory of *gradual* erosion was to be accepted, both ice and water would have to be invoked to explain the surface structure of northern Europe. The classic distinction between the V-shaped valley formed by water and the U shape formed by a glacier was recognized. The signs of extensive glaciation were also studied in North America. Ramsay extended the ice age theory even further by pointing to evidence suggesting that there may have been episodes of glaciation much earlier in the earth's history, especially in the Permian era.

A number of effects were now seen to be associated with the ice age. The sea level would have been lower because of the amount of water transferred to the continents as ice. The land itself may have sunk under the weight of the ice sheets – and would only gradually rise to its original level after the ice had melted. This would produce a temporary rise in the apparent sea level immediately after the ice age. The gradual lowering of the Baltic Sea used by Linnaeus and others to support the retreating-ocean theory was a manifestation of this effect. The modified climate of the ice age had produced effects even outside the area of actual glaciation. American geologists showed that the Great Salt Lake of Utah had been much bigger, owing to the generally wetter climate. Wind-borne earth or 'loess' had covered other parts of the world during the same period.

What could have caused such an extensive – yet temporary – lowering of the temperature? One suggestion made by the American geologist James Dwight Dana was that the elevation of mountain ranges might affect continental climates by deflecting the prevailing winds. But this was difficult to reconcile with the comparative rapidity with which the last ice age had ended. Indeed, no geological process based on non-catastrophic effects could account for the dramatic warming of the earth in the post-glacial epoch.

Attention now switched to external factors as a means of explaining the climatic fluctuations. One hypothesis came from James Croll (1821–90), who began as an amateur but achieved fame for an 1864 paper outlining a theory of astronomical causation. Over the next decade Croll refined his theory, culminating with the publication of his *Climate and Time* of 1875. The theory represents a fascinating exercise in

Humboldtian science, since it links a variation in the sun's heat to the formation of ice sheets via a complex climatic mechanism involving both meteorology and oceanography. On astronomical grounds, Croll suggested that changes in the earth's orbit would produce fluctuations in the amount of heat reaching the surface. Although there would be no variation in the *total* amount of heat received, the seasonal fluctuation between summer and winter would vary so as to give periodic extremes. During a period of extreme fluctuation there would be a series of cold winters, which would alternate between the northern and southern hemispheres on an 11 000 year cycle, the last such episode occurring between 250 000 and 80 000 years ago.

Croll argued that a series of cold winters allowed a steady build-up of ice on the surface. Once the ice began to accumulate, various effects would begin to reinforce one another. The ice attracted fog and cloud, which would in turn reflect more of the sun's heat away. Eventually the circulation of the winds would be disturbed, in turn upsetting the circulation of the ocean currents. Croll attacked Maury's theory that currents were due to variations in the density of the sea water. He argued that the Gulf Stream was produced by the winds blowing across the Atlantic. At a critical point in the sequence of cold winters, the winds responsible for the Gulf Stream would be deflected and the current would cease to flow, breaking down the circulation of water in the ocean. With no warm water being brought up into northern latitudes, the effects of the cold winters would be doubly reinforced and an ice age would result. Only the return of warmer winters would break the cycle and restore the pre-glacial conditions to the north. The earth was susceptible to minor changes in the heat received from the sun because the climate could be disrupted by a complex interaction of factors. In such circumstances the earth did not behave as a self-regulating system; the climate was subject to major fluctuations because variations reinforced one another rather than cancelling each other

TABLE 6.2 *Sequence of Pleistocene glaciations as accepted by geologists in the early twentieth century (most recent at top)*

Europe	North America
Würm	Wisonsin
Tidd	Illinoian
Mindel	Kansan
Günz	Nebraskan

The assumption that there were four glaciations affecting both continents has been undermined by later evidence suggesting that there were many more glacial epochs in this, the most recent period of climatic fluctuation.

out. On the other hand, Croll pointed out that, without the ocean currents to distribute the sun's heat, the earth would be permanently uninhabitable.

If Croll was right, the ice age should not have been a single episode, but a series of glaciations alternating with warmer conditions. Archibald Geikie (1835–1924) had already noticed evidence of plants in the middle of glacial debris, and, in response to Croll's prediction, his brother James Geikie (1839–1915) began a systematic search for evidence of warmer 'inter-glacial' periods. In his *Great Ice Age* of 1874 Geikie postulated four major glaciations followed by a less severe cold spell. In the early twentieth century Albrecht Penck and Eduard Bruckner gave the four main glaciations names derived from characteristic Swiss deposits, while American geologists introduced names based on the states where the local evidence was most obvious. It was assumed that the four periods coincided, although modern work has established a larger number of glaciations.

There were problems, however, with Croll's dating of the ice ages. Most geologists, especially the Americans, became increasingly suspicious of the claim that the last ice age should have ended as long ago as 80 000 years. The field evidence suggested a much more recent pattern of glaciations ending at the most 10 000 or 15 000 years ago. If this dating was valid, Croll's astronomical–climatic theory must be wrong – despite its success at predicting inter-glacials. European geologists retained the hope that something might be salvaged from the astronomical theory, but the century ended on a note of confusion. The more complex ice age theory was triumphantly established, but no one could be sure what had caused such tremendous fluctuations in the climate. It would be left for twentieth-century geophysicists to launch a renewed assault on the problem of causation (chapter 9).

MOUNTAINS AND CONTINENTS

Ice and water have shaped the present surface of the earth, but they have done so by excavating material from areas elevated by some other means. The retreating-ocean theory was abandoned in the early decades of the new century in the face of growing evidence that the crust of the earth is unstable. Mountains and continents are created not by the disappearance of the ocean, but by the elevation and depression of the surface itself. The oceans constitute a fixed volume of water whose location at any one time is determined by where the surface is most depressed. In the traditional interpretation of the history of geology, acceptance of this point was followed immediately by a controversy over the rate of elevation and depression. The 'catastrophists' argued that the earth movements were sudden and violent. Erosion, too, was often catastrophic, caused by the tidal waves generated as a result of sudden uplift. Against this theory Lyell revived Hutton's gradualism under the name 'uniformitarianism', arguing that both uplift and erosion are slow processes. Historians used to assume that Lyell was largely successful in eliminating catastrophism, thus paving the way for Darwin's theory of gradual evolution in the organic world.

The situation was actually far more complex. Although overt appeals to catastrophic uplift declined, most geologists continued to believe that past levels of activity were higher than those of the present, as predicted by the theory of the cooling earth. Concentration on the uniformitarian–catastrophist debate in Britain has led historians to ignore major traditions of geological theorizing in continental Europe and in America. For the Europeans, the Alps provided a crucial problem, focusing their attention on the existence of mountain ranges. Americans were similarly concerned with the complex structure of the Appalachians – which was very different to that of the Alps. Geological theorizing took on a distinctly nationalistic flavour, as the scientists of each country sought to explain the phenomena that were most obvious in their local territory.

The Shrinking Earth

The retreating-ocean theory collapsed as Werner's students travelled to parts of the world where the evidence for uplift was more apparent. Humboldt's experiences in the Andes convinced him that volcanic activity was immensely important in creating mountains. Leopold von Buch (1774–1853) published studies of the Alps between 1802 and 1809 that were dedicated to Werner – yet they confirmed that the mountains had been elevated by massive earth movements. Von Buch also studied the extinct volcanoes of central France and accepted that there had been extensive volcanic activity throughout the earth's history. Humboldt and von Buch saw themselves not as opponents of Wernerianism, but as supporters who were ridding the programme of the unnecessary burden of the retreating-ocean theory so that the work of establishing the sequence of development could go forward. Geologists now had two jobs to do instead of one: they had to establish the sequence in which the formations of sedimentary rock were laid down and also work out the sequence of earth movements that had elevated the various chains of mountains.

Werner's followers at first assumed that chemical reactions could provide the energy needed for melting and elevating rocks. The chemist Humphry Davy (1778–1829) suggested that extensive local heating could be caused where highly reactive metals such as the newly discovered sodium combined with water or oxygen. If the earth had originally contained large underground deposits of such metals, episodes of violent activity would occur from time to time whenever air or water penetrated into the depths to begin the reaction. Humboldt noted that volcanic activity was often accompanied by an uplift of the surrounding countryside, suggesting immense pressures deep in the earth. Von Buch drew upon this idea to develop his 'craters of elevation' theory, proposed after a visit to the Canary Islands in 1815. He supposed that extensive – but localized – underground heating would either eject molten rock onto the surface, or build up areas of pressure underground that would elevate all of the overlying rock to form a mountain range. The subsequent collapse of the middle section of the elevation would often give a crater-like depression in the centre of the range. Von Buch was certainly a catastrophist: he thought that erratic boulders might be thrown considerable distances through the air by the sheer force of the uplift.

The chemical theory was gradually abandoned in favour of the idea that the whole centre of the earth was intensely hot, presumably as a result of the planet having been formed as a molten mass. The mid nineteenth century saw an outburst of theorizing designed to link the cooling-earth concept to the formation of mountains and continents. The pioneer in this effort was the French geologist Léonce Elie de Beaumont (1798–1874), who outlined his views in an important paper of 1829. Here the theory of a gradually cooling earth was linked to the idea of sudden uplift via the assumption that each major system of mountains represented a single episode in the systematic crumpling of the crust. As the earth cooled, its volume decreased while the surface area of the crust remained constant. The resulting stress could only be relieved by folding, which took place in sudden episodes when the resistance of the solid crust gave way.

One of Elie de Beaumont's followers, Constant Prévost (1787–1856), proposed an analogy to explain the process of mountain-building: the earth could be compared to an apple whose skin wrinkled as the interior shrank through loss of moisture. On this theory there was no overall uplift as in von Buch's craters of elevation – the force applied was horizontal rather than vertical. The wrinkling was as much down as up, and since the overall volume of the earth was decreasing it could actually be said that the mountains were left standing as the regions around them were depressed. During the 1850s Elie de Beaumont achieved considerable influence within French geology and his theory was accepted as dogma. Elie de Beaumont himself went on to refine a suggestion made originally by Humboldt that mountain ranges of the same age might have the same geographical orientation. He now tried to depict the earth as being divided into geometrically regular pentagons by the various mountain ranges.

In America the idea of lateral folding was used to explain the complex structure of the Appalachians by James Dwight Dana (1818–95). As an exponent of the cooling-earth theory, Dana argued that an area of sea bed such as the Atlantic Ocean formed a gigantic depression or 'geosyncline' created by shrinkage pressure during the earth's early history. The oceans covered those areas of the surface which had cooled fastest and had hence contracted more. This meant that the continents and oceans were permanent features of the earth's geography, preserved since the earliest phases of cooling. The margins of the ocean were constantly filled with sediment derived from the eroding continent; this hardened into rock and was then periodically folded to create a range of mountains along the edge of the continent. The Appalachians represented, in effect, the eroded remnants of a series of extensions added onto the east coast of North America. Since the later episodes of folding were also attributed to lateral pressure caused by later cooling, Dana's theory was seen as a continuation of Elie de Beaumont's approach.

The theory of permanent continents was rejected by one of the most influential European geologists of the time, Eduard Suess (1831–1914). The structure of the Alps was quite different to that of the Appalachians, and seemed to require a different style of explanation. In his books *The Origin of the Alps* of 1875 and *The Face of the Earth* of 1883–1904, Suess developed a global synthesis based on the concept of a gradually cooling earth subject to a steady diminution in the rate of geological change. Although he agreed that the horizontal pressures created by cooling were crucial to mountain-building, Suess used evidence from the Alps to argue that the pressure could be relieved not only by folding, but also by an 'overthrust' where the crust splits and one side slides horizontally over the other. The Alps as a whole were the result of a massive northwards thrust in the comparatively recent geological past. Such events were fairly abrupt – they were not catastrophes in the old sense of the word, but they did represent distinct episodes of mountain-building interrupting the normally stable conditions of the earth's surface. Suess left no room for elevation by internal pressures and argued strongly against the view that the Andes, for instance, had been elevated by plutonic forces.

Suess did not believe that the division of the earth's surface into continents and oceans was permanent. He invoked major episodes of crustal collapse to explain the production of ocean basins, but argued that the basins would gradually fill with sediment, thus raising the sea level so that it could invade the continents. Eventually another episode of collapse would drain the water away into a new basin. Suess believed that a massive ancient continent that he called 'Gondwanaland' had been broken up by the collapse of its interior to give the separate continents of today. The possibility of land-bridges between the continents in the distant past was widely used to explain similarities between the fossil populations of the continents up to a certain point in the geological record.

Floating Continents

The supporters of Dana's 'permanent continents' theory explained the presence of marine sediments in continental interiors by supposing that the land might occasionally sink enough to allow invasion by a shallow sea. Deep-sea sediments were never found on dry land. But there were two possible explanations of the permanent continents. Perhaps they were relics of the irregular shrinkage of the earth, locked in position since ancient times. But an alternative possibility emerged when it was realized that continental rocks tended to be less dense than those of the ocean floor. The continents might be blocks of lighter rock that remained permanently elevated because they rested on a denser substratum girdling the whole earth.

The rival theories pioneered by Dana and Suess both came under pressure towards the end of the century. Apart from the growing evidence that the continents were indeed composed of lighter material than the deep ocean bed, physicists now suspected that the assumed cooling of the earth could not possibly have generated the level of contraction demanded by the geologists. This point was stressed by the British geophysicist Osmond Fisher in his book *The Physics of the Earth's Crust* in 1881. The permanence of the continents was soon widely accepted (although it was still felt that ancient areas of land could have been sunk to break the original connections between the continents).

The possibility that continents might be visualized as rafts of lighter material floating on denser rocks beneath was popularized by the American geologist Clarence Dutton (1841–1912). In 1889 Dutton coined the term 'isostasy' to denote the process by which continents find their own level due to gravity. On this model, if erosion stripped material from a continent, the 'raft' became lighter and would rise upwards to compensate. Conversely, an area where sediment was being deposited would sink as the weight accumulated. The concept of isostatic equilibrium seemed to fit in with a growing body of evidence suggesting that the continents were indeed composed of a less dense material than the ocean beds. The term 'sial' (silicates of aluminium) was coined for the lighter material, 'sima' (silicates of magnesium) for the denser.

In the early years of the twentieth century, the American geologist and cosmologist Thomas C. Chamberlin (1843–1928) proposed a new synthesis in which cycles of change took place on an earth that had permanent continents and did not contract. He assumed that continents were gradually eroded until they were almost flat, the resulting sediment filling the oceans and allowing the sea to invade the land. Under these conditions the earth as a whole enjoyed a warm, humid climate. Isostatic forces would then raise the continents, creating mountain ranges that broke up the weather patterns to give sharp distinctions between dry and humid zones. The cycle of erosion and sedimentation would then begin again.

Chamberlin's theory was the product of a time when certain key assumptions of nineteenth-century geology had been abandoned. The contracting-earth model had enjoyed considerable popularity but was no longer tenable. The continents were now assumed to be permanent, but as yet there was no conceivable reason to suppose that they might move as units across the face of the earth. Chamberlin's theory was widely discussed, but it did not achieve the kind of dominance once enjoyed by the cooling-earth model. There was no overall geological consensus – but not everyone thought that such a consensus was desirable. In many areas it was still possible to do perfectly adequate detailed work without worrying about the lack of a global picture. The geology of the early twentieth century continued on an erratic course for some time before a new synthesis emerged (chapter 9).

THE RATE OF CHANGE

The uniformitarian-catastrophist debate used to form the centrepiece of any history of nineteenth-century geology. Catastrophism was depicted as a theory whose main purpose was to defend the reality of Noah's flood. This backward-looking approach was challenged and soon defeated by Charles Lyell's uniformitarianism, which laid the foundations for modern, scientific geology. Geology emancipated itself at last from religious preconceptions, the debate between 'Genesis and geology' representing another victorious campaign in the warfare between science and theology. We have already seen enough to persuade us that this was a very short-sighted interpretation. Outside Britain there was no attempt to link catastrophism with the biblical story, and even the British catastrophists were careful not to allow their science to be straitjacketed by their religion. Nor was Lyell's campaign for the uniformitarian alternative a complete success. Although there was less emphasis on the sheer violence of past geological activity, hardly anyone accepted the cyclic or steady-state alternative that Lyell was offering.

Catastrophist Geology

Cuvier, von Buch and others promoted the view that the 'revolutions' affecting the earth's surface were sudden and violent. But this was no simple-minded return to biblical geology. Before the emergence of the ice age theory, many of the superficial phenomena seemed inexplicable except in terms of violent events such as a tidal wave or an enormous earthquake. How else could massive erratic boulders have been transported across considerable distances? Cuvier also noted the apparently sudden transitions from one fossil population to another in successive strata. Did this not indicate sudden extinctions accompanied by equally sudden changes in the environment? Whatever its broader implications, the theory of sudden revolutions followed by periods of stability seemed to represent the most obvious interpretation of the facts.

Catastrophism certainly had had ideological overtones –it was often supported by conservatives trying to resist calls for the gradual reform of society. By arguing for rare upheavals punctuated by periods of stability, they could undermine the model of gradual change that was preferred by the reformist advocates of social evolution. But it would be ludicrous to suppose that conservatives were obsessed by the need to link catastrophes to the literal truth of the Bible. Far from being an exponent of scriptural geology, Cuvier actively denied that the last catastrophe was a universal deluge and set up paleontology as a new science of history that would replace traditional stories about the distant past.

It was in England that the link between catastrophism and religion was forged. Its leading architect was William Buckland, who was appointed reader in geology at Oxford in 1819. Oxford was the centre of Anglican conservatism, and to make his science acceptable Buckland was forced to look for ways of minimizing its apparent challenge to the Bible. Cuvier's catastrophism offered the perfect opportunity. If the last catastrophe could be presented as a universal deluge, then the story of Noah's flood was vindicated. Buckland found his best evidence in a cave at Kirkdale in Yorkshire. The cave was filled with mud, in which were embedded the bones of animals no longer found in England. Applying Cuvier's paleontological techniques,

Buckland was able to show that the cave had been a hyenas' den – the teeth-marks on the bones exactly matched those from the hyenas' cage at the London Zoo. He assumed that the epoch in which the hyenas flourished had been ended by a violent flood that had filled the cave with mud. Since similar caves were known elsewhere, Buckland assumed that the flood was universal. His *Reliquiae Diluvianae* of 1823 announced that geology had confirmed the reality of the deluge.

Even Buckland did not subordinate his science completely to his religious views. His work on the bones was a model application of Cuvier's techniques, and a flood seemed the only explanation of how the cave had been filled with mud. Buckland's error was to assume that all the relevant phenomena had been formed by a single, violent event. Yet his deluge was only the last in a vast series of catastrophes that must have affected the earth throughout its long history. Buckland's theory intersected with the biblical story at only one point – beyond that the whole history of creation had been modified, much to the disgust of the true theological conservatives. Diluvialism was only a small part of catastrophism, soon eclipsed by the growing popularity of Elie de Beaumont's theory of violent contractions in the crust of a gradually cooling earth. The popular myth that the catastrophists appealed to supernatural causation has no basis in fact.

The Uniformity of Nature

Charles Lyell began his career as a catastrophist in the 1820s, and only began his attack on Buckland's position in the last years of that decade. He had no position in the Church and opposed the social conservatism that was linked to Buckland's theory. He became disturbed by Buckland's efforts to link geology to the Bible and looked for evidence that would discredit the universal deluge postulated in the *Reliquiae Diluvianae*. He was alerted to the possibility of a uniformitarian alternative by George Poulett Scrope's study of the extinct volcanoes of central France. Scrope (1797–1876) showed that these volcanoes had been built not by single eruptions, but by a series of lava flows interspersed with long periods of erosion. Only by assuming vast amounts of time could the geologist account for such phenomena.

Lyell himself travelled to Sicily to study the largest volcano in Europe, Mount Etna. Here he was able to show that the massive bulk of the mountain had been built by the slow accumulation of lava flows, only the last few of which had occurred during recorded history. Looking down at the slopes of the volcano covered with subsidiary cones now overgrown with vegetation, he wrote[7]:

> This volcano is placed as if to give just & grand conceptions of time to all in Europe . . . Nothing can be more beautiful than the view from many parts of Etna down into these wooded volcanoes covered with oak & pines & with their craters variously shaped . . . Their number is a clear indication of time.
>
> The number of years then which would roll away before all the great mountains in the woody region disappeared would be very great. The times of History would but little diminish their number & several repetitions therefore of that duration of time must be calculated back ere we can strip the flank of Etna of these numerous craters. But in doing so we should but slightly lessen the total bulk of this great mass.

The volcano was immensely old by human standards – yet it rested on geologically very young rock, showing that its age was trivial compared to that of the earth itself. Lyell also studied earthquakes in order to prove that they were often accompanied by an elevation or depression of the land surface. There was evidence that the Mediterranean area had been depressed significantly since Roman times and then re-elevated. If such 'normal' processes could be visualized acting over a timespan of millions of years, they could account for the elevation of mountain ranges and of the continents themselves.

Lyell thus decided that the best way of attacking diluvialism was to undermine catastrophism itself by reviving Hutton's vision of an earth that had never been subject to any other changes than those we still witness today. Mountains were not built by violent upheavals, but by ordinary earthquakes whose effects accumulated over vast periods of time. Valleys were excavated not by tidal waves but by the slow destructive action of the weather and flowing water, again acting over long periods. Time replaced violence as a means of explaining how the surface of the earth was brought into its present state. This point was made explicitly in the first volume of Lyell's *Principles of Geology* of 1830. Drawing on the reports of Humboldt and others he wrote[8]:

> We know that one earthquake may raise the coast of Chile for a hundred miles to the average height of about five feet. A repetition of two thousand shocks of equal violence might produce a mountain chain one hundred miles long and ten thousand feet high. Now, should one only of these convulsions happen in a century, it would be consistent with the order of events experienced by the Chilians from the earliest times; but if the whole of them were to occur in the next hundred years, the entire district must be depopulated, scarcely any animals and plants could survive, and the surface would be one confused heap of ruin and desolation.

Lyell explained the diluvium of northern Europe by arguing that a relatively minor depression of the surface in the recent geological past would have allowed the sea to invade the land. Melting icebergs would have deposited the erratics and other debris (he was only briefly tempted by the glacial alternative). The apparently sudden changes in the fossil populations of the sedimentary strata were explained in terms of long periods of time elapsing between successive depositions in any one area. Just because one layer of rock is superimposed directly on another, we should not assume that the deposition was uninterrupted except by a catastrophe. Conditions might change to prevent the accumulation of sediment for a considerable time, and when deposition began again the population would have changed in response to any gradual modifications of the environment. Lyell showed that there was often a certain degree of continuity between the successive popula- tions – some species survived the 'catastrophe' while others were replaced, suggesting that there had in fact been no violent extinction.

Limited catastrophes were, of course, possible as the result of unusual combinations of circumstances. Lyell visited North America and observed that the current position of Niagara Falls is determined by the point that the river has reached in its erosion of a gorge in the escarpment separating Lake Erie from Lake Ontario. The falls are gradually extending the gorge, and when they eventually reach Lake Erie the water will drain away quite rapidly, exposing hundreds of square miles of dry land. This might well count as a 'catastrophe', but its effect would be both predictable and strictly localized.

Lyell justified his approach by arguing that geology would only become a true science if it relied solely on observable causes in its efforts to reconstruct the past. Any other policy would open up the floodgates of pure speculation and might even tempt some to think in terms of supernatural causes. The *Principles of Geology* began with a historical introduction that has been the source of much later misrepresentation. Lyell identified the catastrophists' tendency to assume that past causes were different from modern ones with their inability to throw off non-scientific influences such as biblical literalism. Only by adhering to the rule that the past should be explained in terms of known causes operating at known intensities could geologists be certain that they were avoiding speculation.

The problem was that this rule forced Lyell – as it had forced Hutton – to adopt a steady-state theory of history in which there can be no development through time. Even the remotest periods must have been subject to conditions (earthquakes, erosive forces, etc.) exactly the same as those of today. The uniform activity of natural processes gradually alters the superficial features of the earth, but the overall picture remains always the same. If there was a 'beginning' with conditions different to those we now observe, it must be lost in the mists of time before the rocks composing the modern crust were formed.

Lyell's rigid application of the uniformitarian method thus backed him into a corner where he was forced to deny any cumulative development through time, at least in the period of the earth's history accessible to scientific study. To undermine the claim that the warm conditions of the early geological periods were a result of the earth's higher internal temperature, Lyell developed his theory of climatic modification through the rise and fall of continents. There could be fluctuations in the conditions, but no steady trend. Lyell even insisted that the fossil record provided no clear evidence of an overall progress of living things from simple to more complex levels of organization. Reptiles might once again rule the earth if there was a return to the physical conditions that had prevailed in the Mesozoic. Like Hutton, Lyell had become fascinated by the idea of a world designed by its Creator to maintain itself as a habitat for living things through an indefinitely long period of time. Far from rejecting the link between science and religion, he tried to preserve it in his own peculiar way.

Lyell's arguments for gradualism were certainly influential. One of his most important converts was Charles Darwin, who studied the elevation of the Andes first-hand while on the voyage of the *Beagle* and found clear evidence that the mountains had been raised by a long sequence of earth movements. The *Beagle* visited Concepcion in Chile in 1835, shortly after the town had been devastated by an earthquake. Darwin noted that the land had been raised slightly above the sea, leaving banks of shellfish stranded above the new high-tide mark. He also found all the material signs of a seashore – pebbles, shells, etc. – at various levels on the sides of the local mountains, showing that the elevation had proceeded in many small stages. Darwin made use of the idea of gradual subsidence to explain the coral reefs of the Pacific islands. Since the coral organisms can only live in shallow water, a catastrophic lowering of the surface would wipe them out. Only a gradual depression of the land would give the corals the opportunity to keep building up to the surface, forming an ever-widening lagoon around the slowly sinking island.

The catastrophists themselves accepted that the agents of geological change were of the same kind as those now observed, and in the face of Lyell's arguments they gradually scaled down the level of violence attributed to earlier earth movements. What they could not accept was Lyell's complete uniformitarian system in which only known causes *acting at modern intensities* should be invoked, even for the distant past. This would require the complete rejection of developmental theories in favour of a steady-state model for the earth's history. The majority of geologists argued, quite reasonably, that if there were good reasons for suspecting that geological processes had acted more vigorously in the past, then the scientist should be free to investigate this possibility. The cooling-earth theory provided a framework of plausibility for a developmental approach in which the level of activity declined along with the earth's internal temperature. Late-nineteenth-century geology was Lyellian only in certain respects. A measure of gradualism was accepted, but the majority of theories continued to be based on the assumption that the past was different to some extent from the present, if only in the overall level of geological activity.

The Age of the Earth

Perhaps the clearest indication of the weakness inherent in Lyell's position is the attack launched by the physicists in the later decades of the century. In the 1860s the cooling-earth theory was strongly defended by William Thomson, Lord Kelvin. Kelvin made the obvious point that, if the interior of the earth is hot, then it must be gradually cooling down. He assumed that there is no source of energy that can compensate for the heat radiated out into space. By making various assumptions about the earth's internal structure and temperature, he calculated that only one hundred million years had passed since the planet was a ball of molten rock. Lyell had made no estimates of the total amount of time needed to build up the present state of the crust by uniformly acting processes, but he certainly required far more time than Kelvin was prepared to allow. Various efforts were made to throw doubt onto the details of Kelvin's calculations, but other scientists independently came up with similar estimates based on different physical processes. In any case, the basic logic of the argument was incontrovertible: physics shows that all hot bodies tend to cool down, so whatever the actual timescale there could be no doubt that the earth's internal supply of energy could not be maintained indefinitely, as Lyell's theory required.

The majority of geologists accepted the timescale provided by physics. Indeed, there were geological estimates based on the rate at which sediment accumulates that seemed to support Kelvin's figure. The general assumption was that, as one penetrated further into the past, one entered periods in which the level of geological activity had been much higher than it is today, although perhaps not actually catastrophic. Cooling was routinely invoked to explain the collapse of ocean basins and the crumpling of the crust.

There were also signs of massive upwellings of igneous rock at certain points in the past. Archibald Geikie studied the evidence for ancient vulcanism in the British Isles and showed that the massive sheets of igneous rock found in Scotland and Ireland had been emitted not from volcanoes but from fissures out of which the molten rock had spread horizontally. The same effect had been noted in North America. Geikie thus accepted Kelvin's views and argued for an increased level of plutonic activity in the past.

Only in the later decades of the century did geology begin to fight back. Kelvin eventually reduced his estimate to only twenty-four million years, at which point even Geikie was forced to declare that his science could not tamely accept the physicist's demands. Geology had its own lines of evidence that were independent of physical calculations. The extent of sedimentation and mountain-building revealed by the earth's crust was too great for it to be compressed into so short a time. At the same time Thomas Chamberlin developed an alternative to the nebular hypothesis, which implied that the earth had never been molten – the planet had been formed as a 'cold' aggregate of smaller bodies. Whatever the source of the internal heat, it was not the residue of an originally molten state. By the end of the century, the foundations of classical physics were, in any case, already coming under fire. The discovery of radioactivity introduced a totally new factor into the energy equation. The early 1900s saw the logic of Kelvin's argument crumbling before the demonstration that the decay of radioactive elements in the earth's core could provide energy for billions of years of geological activity (chapter 9).

Nineteenth-century geology had lasting achievements to its credit. The establishment of the stratigraphical column and its correlation with the major episodes of mountain-building had created a framework for the earth's history that the modern era would take over almost intact. It was now clear to everyone that the present structure of the earth, with all its scenic wonders and mineral riches, had been formed by a long process of natural change. The traditional assumption that this had all come about as the result of divine forethought had quietly slipped into the background. But the actual causes of all the activity taking place within the earth's crust were still unknown. The theories that had held sway throughout the century were crumbling, and as yet there was no sign of a new synthesis in the offing. Newly discovered effects such as radioactive heating would have to be incorporated into the picture before the state of the earth's interior could be comprehended. Lyell's emphasis on gradual change taking place over millions of years would receive a new lease of life owing to the revolution in physics, but the creation of a global mechanism of geological change would require the postulation of processes that neither Lyell nor his opponents could have anticipated.

The Philosophical Naturalists

Like geology, botany and zoology expanded their territory throughout the nineteenth century. Knowledge of distant lands was consolidated and an ever-widening stream of newly discovered species was brought to the notice of European scientists. The exploration of the fossil record also forced naturalists to become more aware of the time dimension. However diverse the animals and plants of the modern world, they were but the last in a long sequence of populations that have replaced one another throughout the earth's history. Historians have treated the emergence of the theory of evolution as the culmination of scientists' efforts to explain this diversity in space and time. The 'Darwinian revolution' is seen as a watershed dividing the era of description and classification from the modern desire to explain everything as the product of natural processes. This scientific revolution provoked a dramatic shift in the values of western civilization, as belief in the existence of a benevolent Creator was replaced by a more ruthless attitude reflected by the use of phrases such as 'the survival of the fittest'.

Major changes in the way we conceptualize Nature certainly took place at this time, but historians now suspect that the emergence of Darwin's theory should not be treated as the only watershed dividing the old way of thinking from the new. The emergence of the theory of evolution has turned out to be a complex process. Earlier ideas such as those of Lamarck formed a bone of contention throughout the pre-Darwinian decades, and the aspects of Darwin's thinking that are seen as most innovative by modern biologists were not necessarily those which interested his contemporaries. Concentration on evolutionism has led historians to ignore a wide range of parallel developments affecting the way western culture visualized the natural world. The desire to study and control Nature was an integral part of the ideology of the newly industrialized nations of Europe and the United States. In a host of ways not directly connected with Darwin's theory, naturalists sought to reduce the world to a comprehensible order.

Darwin's theory linked living things to their physical environment in a new way. Adaptation was conceived as a process, not a fixed state designed by God. But some nineteenth-century naturalists tried to minimize the significance of adaptation. Their search for the order of Nature still reflected a 'modernized' version of the traditional belief that Nature expresses harmonious relationships transcending the everyday problem of coping with the environment. The idealist view that Nature is the expression of a *rational* creative force upheld many pre-Darwinian efforts to trace the relationships between living things, and the patterns of distribution in space and time.

These 'philosophical naturalists' searched for the order of Nature not in natural laws of the kind familiar to the physicist, but in universal patterns linking the diverse forms of living structure, patterns that were seen as a direct product of the Creator's rational thought. Darwin challenged this approach with his revival of the materialistic alternative. Natural selection adapts populations to their environment through a combination of observable processes that shows no sign of divine forethought. But despite Darwin's efforts, evolutionism was at first absorbed into the search for a universal order. Only in the later decades of the century was there a growing realization that to understand life one would have to take account of the complex interactions between all the living things occupying a particular territory.

There were still many different approaches to the study

of the natural world. Field naturalists collected in the local countryside or explored distant parts of the world and sent specimens back home. Some devoted themselves solely to classification, although others became interested in geographical problems and in the relationship between species and their environments. But many new species, especially the higher animals, had to be described by comparative anatomists who worked in museums and dissecting rooms and who never ventured into the field. At this level, there was a strong link between natural history and medical anatomy – and some of the most active pre-Darwinian debates on evolutionism took place in the medical schools. To the anatomists, the search for order involved the detection of underlying relationships between the structures of different species. Their approach offered little chance of understanding complex ecological relationships. The reconstruction of extinct species from fossil bones was often undertaken by museum workers, who were thus disposed to see development through time as the unfolding of purely formal relationships between successive species.

The term 'biology' did not come into use until the end of the century, and even then it did not have its modern meaning[1]. There was an increased level of specialization in the life sciences linked – as in geology – with the emergence of a body of professional scientists. The nineteenth century has been seen as a period of transition from a largely amateur 'natural history' to a professionalized science of 'biology'. This neat division does not, however, take into account a host of associated factors. Natural history certainly had a strong amateur component, but this continued throughout the century. There were a significant number of specialists even in the early decades of the century, including both the exponents of the Linnaean tradition of classification and the philosophical naturalists, who hoped to achieve something more than mere cataloguing. As the number of these specialists increased, they tried to use the idea of evolution as a means of giving their disciplines a vague unity, thus creating the first science of 'biology'. Their efforts were largely in vain, however, since the separate disciplines tended to continue on their way unaffected by the rhetoric of unity.

Professional interests often shaped scientists' attitudes towards new theories, but political rivalries were also reflected in different images of natural order. The well known debates over the religious implications of evolutionism were symptomatic of the social tensions that built up within an increasingly industrialized society. As middle- and even working-class interests challenged the established social order, science was often employed as a symbolic arena in which to fight ideological battles. Darwinian evolutionism is an example of a theory that was adopted by middle-class thinkers because it portrayed Nature in terms that paralleled their preferred framework of society. Conservative interests sought their own way of picturing the order of Nature, preferring images that admitted change only within the framework of a divinely preordained plan of creation.

KNOWLEDGE AND POWER

The Romantic enthusiasm for the wilderness co-existed with a more practical attitude in which Nature was seen as something to be explored and catalogued. The nineteenth-century approach to Nature study was influenced by the middle classes' growing commitment to hard work as the key to achievement. Naturalists took on enormous work-loads, driving themselves to the limit both physically and mentally. The urge to number and name things was an integral part of the Victorians' desire to dominate Nature. They wanted to use Latin names for species as a symbol of science's ability to impose order on Nature's apparent variety. They also liked cabinets of specimens, neatly labelled and arranged. On a larger scale, museums and exhibitions became popular as public manifestations both of world-wide power and of the passion for order. The search for knowledge was also of practical benefit to humankind. Just as geology helped the mining industry, so the study of animals and plants would help farmers, fishermen and hunters.

Professionals and Amateurs

The Muséum d'Histoire Naturelle in Paris provided a model for teaching and research that was the envy of the civilized world. Here we see the origins of the professional body of zoologists and botanists that would eventually flourish throughout Europe, America and the colonies. But at the level of field studies there was a much less rigid distinction between professional and amateur. Outsiders could not hope to participate in the learned debates on comparative anatomy at the great museums, but field naturalists relied on local amateurs for specimens and surveys, thus ensuring a much greater degree of interaction. There was a passion for information among the general public, creating a vast demand for books and a framework of local organization in the form of natural-history societies and national bodies devoted to particular areas such as ornithology. The introduction of the steam press after 1810 reduced the price of books, allowing experts to earn a living by writing for the ever-expanding market. Cheaper ways of printing illustrations were also developed, although colour plates remained the preserve of the wealthy. Animals and plants were increasingly represented in realistic poses, and works such as J. J. Audubon's *Birds of America* (1827–38) remain classics even today.

Other technical developments helped to change the pattern of interest in Nature. Entomology became more fashionable once better killing bottles provided the squeamish with a less offensive method of collecting specimens. The railways allowed ordinary people to visit the seaside and the country, so that natural-history excursions became social events. Access to the seashore created a passion for collecting seaweeds and shells. Better dredging techniques allowed even amateurs to collect bottom-dwelling species. For botanists, Nathaniel Ward invented the 'Wardian case' in which plants could grow sealed under glass. When introduced in the 1840s this created new possibilities for shipping exotic species around the world, protected from changes in the environment. Since ferns grew best in the Wardian case, there was a fern craze in which every middle-class house displayed a collection of greenery. The vivarium and aquarium soon followed, with parallel changes in fashion.

Changing fashions in popular natural history did not affect those who were doing research in botany and zoology directly, except through the demand for books. But indirectly the growing public awareness of science and its potential shaped the professionals' world by exposing it to the rough and tumble of political life. 'Big science' on the model provided by Paris required funds that were only likely to be provided by governments, and some governments were more willing than others to get involved. Where there was reluctance – as in Britain – middle-class pressure had to be brought to bear in order to convince the politicians that something ought to be done. The societies that organized research and publication could also become arenas for political debate. The challenge to the traditional social order manifested itself in a struggle for the control of science, and historians have begun to recognize how this struggle shaped the development of scientific institutions. More importantly, we have also begun to appreciate that the ideological struggle influenced people in their choice of which theories were seen as acceptable. Theories acquired ideological labels, and social interests thus determined what a particular group would admit as a legitimate contribution to scientific knowledge.

The Paris museum (colloquially known as the Jardin des Plantes) was much more than a museum for the display of exotic specimens. It had been formed from Buffon's old Jardin du Roi (the royal zoological and botanical garden) by the revolutionary government, and had been turned into a major institution for promoting the study of all the non-physical sciences. It still incorporated both a zoological and a botanical garden, but now had an extensive research and teaching function. There were twelve professorships, each with a staff of highly trained assistants. Several of the professors were internationally known figures who lectured to large audiences and engaged in public debates over controversial issues. Lamarck had been put in charge of the invertebrates, while Geoffroy Saint Hilaire (1772–1844), another transformist, dealt with the vertebrates. Most influential of all was the professor of anatomy, Georges Cuvier, who gained considerable political power through his role in the reorganization of French scientific education. Cuvier wanted to turn natural history into a science that would rank with physics and chemistry, and to establish the empirical foundation for this new science he was determined to build a collection of specimens second to none. The expanding power of Napoleonic France gave him access to the whole range of material available in Europe.

In the middle decades of the century, it was increasingly the German-speaking biologists who led the way. Scientific research and teaching were important at many German universities, although the system was rigidly structured and gave the professor a great deal of power. When able men occupied the professorial chairs, there was scope for immense development. The Germans were particularly strong in morphology and embryology, using the increasingly more sophisticated microscopes to explore and describe the structure of a wide variety of living forms. When studying this period, English-speaking historians have concentrated on the work of Darwin and the early careers of the biologists who supported and opposed the theory of evolution. But we now realize the extent to which these scientists looked to France and later to Germany for a lead. Professional biology flourished in continental Europe at a level that it would not achieve in Britain or America until the last decades of the century. The small number of influential British biologists, men such as Richard Owen and T. H. Huxley, often modelled themselves self-consciously on the Germans. Darwin's originality lay in the fact that, as a wealthy amateur, he could study in the field as well as in the dissecting room and could thus synthesize the information from biogeography and morphology.

Science and Politics

Cuvier was influential, but he did not have everything his own way, and rival groups such as the transmutationists were able to make their case with a fluency that historians have ignored until recently. New ideas such as the transmutation of species acquired political overtones because they symbolized a challenge to the existing power structure both within science and within society at large. The Paris museum was a centre of both research and passionate debate over the principles upon which the new science should be built. It is small wonder that Paris became a Mecca upon which naturalists converged in droves once the disruptions of the Napoleonic wars had ended in 1815. Conservative and radical scientific ideas were disseminated from Paris by naturalists and anatomists returning home inspired by what they had heard. If the intellectual lead in the area of biology gradually switched to Germany in the course of the mid-century, Paris had already established the ideal framework through which to promote scientific research into animals and plants.

In Britain, the French model became a focus of political debate between conservatives who wished to keep science under the control of the aristocracy and radicals who pressed for increasing public involvement. The British Museum had natural-history collections, but these were poorly housed and were considered less important than the museum's artistic holdings. The governors saw the museum as a storehouse for the nation's treasures, not as a research institution, and even public access to the museum was restricted. The space problem was relieved to some extent when the museum moved to its modern site in the late 1820s, but natural history still rubbed shoulders with art and archaeology. A public enquiry in 1835 aired the radicals' demands for a Paris-style research museum, but traditional forces ensured that the scientists were kept subordinate to public figures who were more concerned with the arts. The research function did increase, with the anatomist Richard Owen (1804–92) being appointed superintendent of natural history in 1856. Owen had an international reputation and close connections with political figures, but even he was subordinate to the principal librarian. Owen campaigned long and hard for a new building with more space and supervised the move to the present Natural History Museum in the 1880s (chapter 8).

A Zoological Society was founded in London in 1826 to act as a showcase for Britain's colonial possessions. It appealed directly to aristocratic sporting interests, and the Zoological Gardens in Regent's Park were at first restricted to members and their guests. Hunting was an aristocratic pursuit at home, and the hunting of big game symbolized the white man's superior status in colonized territories. The display of exotic animals in zoos was a public manifestation of the industrialized nations' ability to dominate the world. But working naturalists were interested in the whole range of exotic species, not just big game, while the radicals pushed for more research and public access to the zoo. The supply of exotic animals for dissection was of great importance to anatomists hoping to make their name through exciting discoveries or looking for ammunition to use in theoretical debates. Richard Owen gained part of his reputation through the description of new species shipped home from Britain's colony in Australia. He was able to checkmate the French evolutionists' efforts to use the duck-billed platypus as a link between reptiles and mammals because he alone had access to a good supply of specimens.

The debates between radicals and conservatives were not concerned merely with the quality of science and access to the necessary resources. Opinions on a whole range of theoretical issues were polarized in ways that can be related to positions taken up in the struggle for control of the social framework of science. Middle-class reformers and those with more radical opinions used the newest ideas to discredit the 'old-fashioned' science favoured by the traditional sources of power and influence. Conservatives sought to update their theoretical perspective without giving way to ideas that would threaten the notion of a divinely structured natural hierarchy. To them, the new theories smacked of materialism, and during the early decades of the century they were quite successful in blocking the spread of radical ideas. Control of the existing social machinery counted for quite a lot in this battle; the reformers forced some concessions but failed to topple the existing hierarchy. Recent historical research has shown that the battle that would be won by the Darwinians of the 1860s and 1870s had been fought and lost by an earlier generation of reformers. Far from being a new threat, evolutionism was something that the conservative forces had been battling with for several decades.

In botany the theoretical debate was less acute, although the prospect for the practical application of scientific knowledge was greater. Sir Joseph Banks had pioneered the effort to transplant useful species from one area to another, and hoped to use his position at the royal botanical gardens at Kew to further the interests of Britain's overseas colonies. Yet after Banks' death the British government did nothing to help establish the gardens as a national centre for botanical research until a public outcry in 1839–40 led to the appointment of William Hooker (1785–1865) as the first official director of Kew. Under William and his son – and successor – Joseph Dalton Hooker (1817–1911),

Kew began to acquire facilities for studying exotic plants and soon became a centre of international importance. J. D. Hooker collected plants on expeditions to the southern polar regions and to the Himalayas, and soon plants (in Wardian cases) were flooding in from all parts of the world. A museum of economic botany was opened in 1847, and in later decades Kew played a vital role in the spread of useful plants around the British empire.

In the United States a national botanical garden was founded to exploit the seeds and plants brought back by a Pacific exploration expedition under Charles Wilkes in the years 1838–42. At the same time the US Patent Office became interested in the possibility of using tropical plants to revive the agriculture of the southern states. The Patent Office also promoted the use of chemistry in soil analysis, seeing this as a means of preventing the exhaustion of nutrients. The German chemist J. von Leibig's *Chemistry in its Applications to Agriculture and Physiology* was translated in 1841 and inspired a generation of chemists to apply their skills to agriculture. Entomology was also taking on an economic role as its application to pest control became evident. The State Agricultural Society of New York hired an entomologist, Asa Fitch, in 1854 to study the life history of insects in the hope that this would suggest new methods of control. Fitch also experimented with chemical insecticides. The foundations were thus being laid for the massive expansion of government involvement in the environmental sciences in the later decades of the century.

THE PATTERN OF NATURE

The classification of species was still of vital importance to both field and museum naturalists. Linnaeus' artificial system of working out the relationships between species was now replaced with a natural system in which as many characters as possible were taken into account. From this point onwards historians have tended to lose interest in the debates over classification, assuming that they were increasingly concerned not with the principle of how to determine relationships, but with disagreements over practical applications of the technique. In fact, though, a vast amount of ink continued to be spilled in arguments over taxonomy, and disagreements over how to classify particular groups still reflected differing philosophical positions on the significance of natural relationships.

The assumption that evolution solved the problem by showing that 'relationship' should be interpreted in a genetic sense oversimplifies the situation. Darwin argued that classification is an expression of genealogy: closely related species show a basic similarity because they share a recent common ancestor. But this still left certain questions unanswered, and many naturalists did not accept the Darwinian approach to the question. Far from welcoming the concept of the evolutionary 'tree', zoologists and botanists resisted the idea that relationships were the product of an open-ended process of divergence under adaptive pressure.

This assertion conflicts with Michel Foucault's interpretation of the history of taxonomy (chapter 5). Foucault argued that the period around 1800 saw the breakdown of the 'classical' view of natural relationships in which species were slotted into a conceptual grid whose structure was determined by the human mind. The most obvious indication of this transition was the replacement of the chain of being (the simplest possible grid in which all the pigeon-holes are arranged in linear pattern) by an open-ended branching system of relationships. Foucault associates this development with the work of Georges Cuvier, who rejected the chain of being and pioneered a tree-like model of relationships. In this sense Cuvier anticipated the modern, Darwinian view of Nature even though he rejected the idea of transmutation. Unfortunately, Foucault did not realize the extent to which nineteenth-century naturalists rejected both the Cuverian and the Darwinian perspectives.

In fact, there were numerous efforts to resist the idea that natural relationships were unpredictable. The concept of the chain of being was translated into a more flexible notion of a hierarchy of organization, while other naturalists replaced the linear chain with an equally closed system based on a circular pattern.

Even the transition from an artificial to a natural system was not completed in a uniform manner throughout the scientific community. British botanists still resisted the natural technique of de Jussieu into the early decades of the new century. But the natural system was not without its own problems. Some naturalists argued that all characters should be taken into account when determining relationships, but others claimed that some characters were more fundamental than others and should thus be given more weight.

This question came to a head in the classification of the animal kingdom, where fundamental debates took place in the early decades of the new century. Foucault is correct to identify Cuvier as a central figure in these debates, even if he exaggerates how successful Cuvier's arguments were at the time. Cuvier pioneered the role of comparative anatomy as a guide to classification, thereby boosting the role of the museum naturalist at the expense of the collector in the field. From his position at the Paris museum, he promulgated a new approach in which animals would be classified not by their external appearance but by internal resemblances that could only be revealed by dissection. The significance of internal structures had already been accepted in some cases (otherwise whales and dolphins would be classified as fish rather than as mammals), but Cuvier made this central to the classification of the whole animal kingdom. He insisted on the principle of the 'subordination of characters' – some characters were more fundamental than others and should be given pride of place in assessing degrees of relationship. In theory Cuvier saw the nervous system as the most important structure, although in the case of the vertebrates he concentrated in practice on the skeleton.

Chains, Trees and Circles

In 1812 Cuvier used this technique to break up the chain of being. The animal kingdom does not display a linear pattern of species linking the lowest forms to humankind. Instead, there are four completely unrelated 'embranchments' or basic types of organization. Each of these can be recognized by a common pattern of internal structure, all the individual species within a type being but variations on the same basic pattern. The four types were the vertebrates (animals with backbones), the molluscs (snails, octopus, etc.), the articulates (segmented animals including insects, worms and crustaceans) and the radiates (animals with a circular body plan such as sea-urchins). Each type was divided into classes, orders, families, genera and species as in the Linnaean system, the types constituting the most fundamental divisions in the taxonomic hierarchy. Cuvier insisted that the four types were simply different and equally successful body plans – they should not be ranked with the vertebrates at the top, since this would smack too much of the old chain of being. The species within each type represented different modifications of the basic pattern, each adapted to a particular way of life. Because the range of possible adaptations was unlimited, the relationships could be represented as a branching tree-like pattern, with new branches being added whenever some totally new (and unpredictable) form of animal life was discovered.

The four types are the foundations of the modern phyla, the most basic subdivisions of the animal kingdom, although the articulates and radiates have been broken up as the true range of diversity has become apparent. Cuvier thus anticipated the kind of branching relationship implicit in the Darwinian version of evolution, in which species diverge from a common ancestor under different adaptive pressures.

But although Cuvier won immense prestige for his work as a comparative anatomist, he was unable to convince many naturalists that they should give up all hope of being able to predict the overall pattern of natural relationships and hence the 'missing' species still waiting to be discovered. The concept of a single chain of being was abandoned – even Lamarck had two parallel hierarchies for the plant and animal kingdoms. But Lamarck's followers were far more numerous than historians once believed, and they continued to argue that a serial relationship was apparent in the animal kingdom, at least in the arrangement of the main classes. Even if the arrangement within each class was a branching tree, the classes themselves could be set one above the other to form a continuous sequence.

This position was argued in France by J. B. Bory de Saint-Vincent (1778–1846) and in Britain by Robert E. Grant (1793–1874). Those who defended the idea were supporters of the evolutionary hypothesis who saw the linear arrangement as the ladder by which life had ascended to the highest form, humankind. Cuvier's rigid distinction between the four types was intended to discredit this interpretation, as was his theory of catastrophic extinction (chapter 6). But the radicals kept the spirit of Lamarck's theory alive through into the 1830s by searching for ways of preserving the linear hierarchy. They looked for intermediate forms to bridge the gaps that Cuvier insisted lay between the known classes and types. The debate over the classification of the duck-billed platypus – which lays eggs but also gives milk – arose out of this issue, with the serialists trying to use it as a link between the reptiles and the mammals. Owen was a staunch supporter of Cuvier's views on discontinuity, and he successfully emphasized the milk glands at the expense of the egg-laying in order to keep the platypus firmly within the mammals.

An important addition to the concept of a linear scale was the 'law of parallelism' advocated by J. F. Meckel (1781–1833) and Etienne Serres (1786–1868). The early nineteenth century saw major developments in embryology, which challenged the mechanical concept of generation and overthrew the preformation theory. The law of parallelism was a means of linking comparative embryology into the search for a unifying pattern in the organic world. Meckel and Serres believed that the scale of organic complexity used to rank the major classes of living organisms corresponded to the pattern of individual development. The human embryo, for instance, passes through stages in which its structure approximates to that of an adult fish and then an adult reptile, before acquiring its final mammalian form. On this model of organic relationships, the lower animals are merely immature versions of humankind: they develop along the same scale but mature at an earlier point in the process.

The Lamarckians used the idea of a progressive scale to define the relationships between the main groups. Adaptation might distort this scale (as Lamarck himself supposed) but it was a secondary factor, not the sole determinant of relationships. The linear scale was not, however, the only possibility now under consideration. If the chain of being had been taken over by the materialists, those naturalists who believed that the order of Nature revealed the existence of a rational plan of creation preferred to invoke more complex patterns.

The most striking expression of this approach is the circular or quinary system proposed by the entomologist William Sharpe MacLeay (1792–1865) in 1819. MacLeay was influenced by the idealist movement now flourishing in Germany, and he was determined to show that a rationally ordered pattern underlay the apparent diversity of species. He argued that all the higher taxonomic categories can be divided into five subordinate units that fall naturally into a circular pattern. The animal kingdom was divided into five (not four) types, each consisting of five classes. Each class is divided into five orders, each order into five families, each family into five genera, and finally each genus into five species. Here was an immensely complex yet rigidly structured model for the plan of Nature, which – like the chain of being – was closed rather than open-ended. Since the relationships were all in fives, one could tell if a particular category was missing from our present knowledge and predict roughly what kind of new organism would eventually be discovered to fill the gap.

MacLeay's view of natural relationships presupposed that the living world is constructed according to a rational plan. His system achieved some degree of popularity within entomology, which was in a flourishing state at the time. It was applied to the birds by the ornithologist William Swainson (1789–1855), who wrote extensively and produced a number of well illustrated works. Although the quinary system was not widely taken up, especially outside Britain, it was not dismissed out of hand as a figment of MacLeay's and Swainson's imaginations. Most naturalists thought that there was some sort of underlying pattern in Nature, even if they found the circles a little too artificial a representation of the pattern. By the 1840s naturalists were depicting relationships in two dimensions but without the orderly pattern of circles, producing images of relationships similar to geographical maps. This technique was compatible with a more open-ended view of natural relationships, although most naturalists were as yet unwilling to concede that life could have developed solely in response to adaptive pressures.

Form and Function

Cuvier's system was open-ended because he considered each species to be a variation of the underlying type adapted to a particular way of life. The form or structure of the species was determined by its function, this in turn being determined by its environment and way of life. It was impossible to predict what new variations on the theme might be discovered, because there were endless conceivable ways of gaining a living that animal species might adopt. Any system in which adaptation becomes the sole determinant of structure will be incompatible with the idea that Nature has been created in accordance with a harmonious or regularly structured grid of relationships. Cuvier's commitment to this adaptationist or utilitarian approach was complete. Although he distinguished four basic types of organization, he rejected any suggestion that the underlying unity within each type had any mystical significance. The types existed because certain basic patterns of animal organization were more efficient than any other conceivable forms, so all species were necessarily based upon one or another of these patterns.

Although Cuvier himself made no use of the theological implications of this approach, it was taken up with enthusiasm by British exponents of the argument from design. A century earlier, John Ray had used the adaptation of form to function to proclaim the existence of a benevolent God who designed His creatures to enjoy life. The threat of the French Revolution led conservative forces to close ranks against the godless atheism of the Enlightenment materialists. Intellectual arguments to demonstrate the existence of a Creator were once again in demand among those who wished to preserve the existing social hierarchy. In this emotional environment the argument from design gained a new lease of life. Its most eloquent exponent was an Anglican clergyman, William Paley (1743–1805), whose *Natural Theology* appeared in 1802. Although there was little originality in Paley's argument and little new in his approach to natural history, his emphasis on the *usefulness* of organic structures struck a renewed chord in the minds of a social elite whose wealth was increasingly based on trade and commerce. Since many English naturalists were also clergymen, they were only too willing to describe the species of animals and plants they studied as examples of divine craftsmanship.

Like exponents of the argument from design in earlier

generations, Paley had to confront the fact that some species possessed structures that were useful for the killing of prey. He was convinced that, in a world governed by natural law, life must inevitably be accompanied by reproduction and death. In these circumstances, the predators actually reduced the suffering of their prey[2]:

> A brute, in his wild and natural state, does every thing for himself. When his strength, therefore, or his speed, or his limbs, or his senses fail him, he is delivered over, either to absolute famine, or to the protracted wretchedness of a life slowly wasted by the scarcity of food. Is it then to see the world filled with drooping, superannuated, half-starved, helpless and unhelped animals, that you would alter the present system of pursuit and prey?

In this way Paley excused the violence of Nature and linked the species together not in some abstract formal pattern but in terms of how they interacted in the food chain.

Cuvier's emphasis on adaptation added a veneer of scientific authority to Paley's utilitarian argument from design. Serious naturalists could sprinkle their descriptions of the internal organization of new species with references to the skill of the Creator who had designed such complex living structures. This approach flourished through into the 1830s, when a series of illustrations of the argument from design was commissioned in the will of the Earl of Bridgewater. The eight *Bridgewater Treatises* were each written by an expert who could speak with authority on the evidence of design to be seen in his particular area. Unfortunately, the *Treatises* tended to bore their readers through the sheer number of examples cited. Some of the authors were also aware of new approaches to the study of organic relationships that were being discussed in France and which were leading radical anatomists to challenge Cuvier's authority.

This new approach centred on the 'transcendental anatomy' of Etienne Geoffroy Saint Hilaire. Although sympathetic to Lamarck, Geoffroy attacked Cuvier from a very different position. He abandoned the concept of a linear scale in favour of a more flexible method of seeking an underlying unity that would rationalize the apparent diversity of living forms. In his *Philosophie anatomique* of 1818–22, Geoffroy argued that the basic similarity that allowed all the vertebrates to be assigned to a single type of organization reflected something more than mere engineering efficiency. The unity of type suggested that all the vertebrates were merely superficial modifications of a single ground plan – the same basic animal form has been modified in a host of superficial ways. For Geoffroy, the superficial adaptations were less significant than the underlying unity of structure – the laws governing the modification of form were more important than the function that the resulting structures would serve in allowing the animals to live within their environment. Eventually Geoffroy challenged Cuvier head-on by asserting that the four types could themselves be linked by appealing to yet more fundamental unities of form. The resulting debate over the relative priority of form and function polarized anatomists throughout Europe.

Geoffroy's challenge had a significance beyond the purely intellectual level. By the 1820s Cuvier was associated with a conservative political attitude that favoured a static image of Nature in order to uphold a static model of the social order. Lamarck's theory of the gradual transformation of species had threatened this image, much to Cuvier's annoyance, and Geoffroy's very different approach to the significance of structural resemblances reopened the same debate. By the late 1820s Geoffroy was arguing that one animal form could be transformed into another by natural processes – not by the accumulation of slight modifications, but by a sudden switching of the growth process so that the development of the individual proceeded in a new direction to mature as a new modification of the basic pattern. He suggested that extinct crocodiles found in the fossil record could have been transformed into their modern relatives when a modification of the environment triggered off the transition to a new growth pattern. He also hinted that the transformation of reptiles into birds was produced by the same mechanism. Transcendental anatomy thus revived the case for materialism by suggesting that the 'relationship' between similar species reflected a real genetic link resulting from natural processes of transformation. Geoffroy had popularized a new concept of evolution by 'saltations' or sudden transitions from one structure to another.

The debate between the supporters of Cuvier and Geoffroy raged throughout the 1830s and 1840s. Transcendental anatomy was seized upon by radicals who were determined to show that the scientific thinking of the conservatives was outdated. In Britain, Geoffroy's position was supported by Robert Grant and by another Scottish anatomist, Robert Knox (1793–1862), who was later discredited by his role in the Burke and Hare scandal, in which the corpses of murder victims were sold for dissection. The arguments over the significance of organic relationships that we normally associate with the debate over Darwin's *Origin of Species* were being fought out in the natural-history museums and anatomy schools just when Darwin himself was first beginning to develop his evolutionary theory in the late 1830s. Although the conceptual foundations of the challenge to the old order were different, transcendental anatomy raised issues that would be associated with the theory of evolution through into the era of Darwinism. The rival anatomists debated the implications of transitional forms such as the newly discovered platypus and the degree to which the structure of the human body indicated a link to the apes.

In Britain the new anatomy was seen as especially threatening because the traditional souces of power and influence had committed themselves to Paley's argument from design. This added a religious dimension to the debate, with the radicals being branded as atheists who wished to subvert the traditional foundations of morality. Yet the new approach attracted too many adherents for it to be dismissed out of hand. Conservative forces in science and medicine looked for ways of modernizing their own approach so that the intellectual foundations of their authority could be preserved. In 1836 the Royal College of Surgeons, one of the bastions of traditional power within the medical establishment, appointed a young and ambitious anatomist, Richard Owen, to a professorial chair. Owen was also put in charge of the College's museum, an important source of specimens for comparative anatomy. It was Owen who took up the challenge of modernizing the argument from design in a way that would allow it to incorporate the innovative aspects of transcendental anatomy.

Owen's task was made easier by the fact that the idealist strand in German thought promoted an alternative view of natural unity. While the Romantic artists challenged materialism at the emotional level, the exponents of *Naturphilosophie* argued that the material world was merely a projection of a deeper, spiritual reality. *Naturphilosophie* was already being applied to the unity of type in the organic world by Lorenz Oken (1779–1851) and others. Owen soon began to appreciate the potential value of this rival continental influence as a means of checkmating the materialists. To the idealist, the whole world is a manifestation of a rational pattern in the mind of its Creator, and thus it is to be expected that natural forms should reveal a harmony and an underlying unity. Where MacLeay saw this unity as a circular system of relationships, Owen began to argue that the unity of type within each group displayed the rational activity of the Creator. The relationships were not physical ones produced by transmutation, but ideal ones existing in the mind of God.

In his book *On the Archetype and Homologies of the Vertebrate Skeleton* of 1846 Owen postulated a basic vertebrate pattern or archetype that the Creator had modified in different

directions to create the various species. Paley was quite
right to insist that the adaptation of each individual species
to its environment was an indication of divine forethought,
but it was equally important to demonstrate the existence
of an underlying pattern that showed that all species could
be seen as elements within a rational plan. In another book,
On the Nature of Limbs, Owen made clear the origins of his
position in German idealism[3]:

> I have used therefore the word 'Nature' in the sense
> of the German 'Bedeutung', as signifying that essential
> character of a part which belongs to it in its relation to
> a predetermined pattern, answering to the 'idea' of the
> Archetypical World in the Platonic cosmogony, which
> archetypical or primal pattern is the basis supporting
> all the modifications of such part for specific powers
> and actions in all animals possessing it . . .

Owen introduced the term 'homology' to denote cases in
which the same structure had been adapted to a range of
different purposes. Thus the wing of a bat and the paddle
of a porpoise are homologous because they show the
same underlying pattern of bones typical of a mammalian
forelimb. Form was more important than function because
in some cases the efficiency of the adaptation was impaired
because the Creator wished to maintain the consistency of
the basic pattern.

Owen could take over the transcendentalists' search for
underlying patterns, but, since the relationships were ideal
rather than physical, he could follow Cuvier's refutation
of transmutation by emphasizing the gaps between the
different forms of vertebrate life. Humankind and the apes
shared the vertebrate form because both were created by
God, but Owen could stress the anatomical differences in
order to make it clear that the one could never have evolved
from the other. The world is unified at the ideal level,
but the physical manifestations of the vertebrate archetype
are each a distinct product of the Creator's will. As with

Cuvier's system, this approach to the unity of Nature is open-ended – there is always room for another newly discovered variant of the basic pattern to be fitted in. The best model for organic relationships was a branching tree, not a linear scale.

In the 1850s Owen drew upon the challenge to the law of parallelism developed by the German embryologist K. E. von Baer (1792–1876) to illustrate this point. In 1828 von Baer showed that the human embryo never passes through a stage equivalent to an adult fish or reptile. The development of every organism starts from a very generalized structure, with the more specialized features that distinguish the particular species being added on in the course of growth. At a very early stage of growth, the embryos of a fish, a reptile and a mammal may be virtually indistinguishable, but their development proceeds in different directions and the class to which they will belong is defined as soon as they start to acquire more specialized features. The early embryo is a physical manifestation of the basic archetype, while the addition of specialized features represents a gradual adaptive modification. Since there are many possible adaptive forms, the patterns of development can be represented as a tree in which many lines branch out in different directions from the same starting point, not as a linear scale.

The model of branching relationships would eventually lead Owen towards a limited kind of evolutionism (see below), but he carefully suppressed this implication in the pre-Darwinian era. He never wavered from the view that the whole sequence of vertebrate life on earth was an expression of a divine plan, unified into a coherent whole by the rational power of the Creator's mind, but divided into distinct sections that could never be linked by natural processes. In the highly charged atmosphere of the 1840s, Owen served his conservative masters well by steadfastly blocking the radicals' attempts to postulate relationships close enough to justify the theory of transmutation.

The anatomists who described and classified species were only peripherally interested in how the organisms functioned within their environment. Work in a museum did not encourage the study of what we should now call ecological relationships. Concern for the role of the environment came from those naturalists who followed Humboldt in searching for the interactions between the physical and the organic realms. Some taxonomists wanted to go beyond mere classification and investigate the geographical distribution of the various species. Humboldt showed them how to tabulate data to reveal the effect of the environment in determining what kind of species could live in a particular area. As part of the drive to impose an appearance of order on the world, nineteenth-century science created a vast body of numerical information all carefully arranged so that geographical variations could be ascertained at a glance. Yet this approach to the question was merely descriptive. A few naturalists, inspired by developments in geology, used information on recent changes in the environment to explain a number of apparently anomalous facts revealed by the numerical data. Here at last was the starting point for a science that would attempt to explain the present state of the earth and its inhabitants in terms of natural processes acting over long periods of time.

Botanical Provinces

Although some efforts had been made to study the world-wide distribution of plants in the eighteenth century (chapter 5), the real founder of botanical geography was Alexander von Humboldt. Drawing upon the vast fund of information gathered on his trip to South America (1799–1804), Humboldt began to study how the physical environment determined the geographical distribution of plants. In his *Essai sur la géographie des plantes* of 1807, Humboldt noted that there was an equivalence between the belts of vegetation found at different heights ascending a mountain and the geographical zones that girdled the earth. The plants in high altitudes were the equivalent of an arctic flora, both adapted to cold conditions. The zones of vegetation around the earth were determined by temperature and rainfall, and within each zone the plants all possessed similar adaptations to the conditions. They were not, however, the same: different continents were inhabited by different species adapted to the same sets of conditions. The existence of distinct geographical provinces thus could not be explained in terms of climate.

The concept of geographical provinces may have gained some currency because it could be understood as a biological equivalent of the nations of humanity – and the early nineteenth century was a period of strong nationalist feelings. Humboldt himself had shown an interest in the native inhabitants of the regions he visited, and his views on how the environment affects plants and animals may have been influenced by his studies of the ways in which human societies adapted to local conditions. Social thinkers were now deeply involved with the collection of information about the human population, with censuses and various kinds of statistical studies gaining in popularity as efforts were made to reduce economic and social forces to a semblance of rational order. Humboldt introduced similar techniques to display the characteristics of the flora to be found in different regions. He used tables that allowed the reader to see at a glance what proportion of the population in any zone was made up of grasses or any other major type of plant. Such tables required the collection of masses of information on the numbers of plants to be found in any area. During the first half of the century efforts were made to extend the same technique to the various kinds of animals as well as plants, so that naturalists had access to a vast body of information about the geographical distribution of living things.

In botany this technique was extended by the Swiss naturalist Augustin de Candolle (1778–1841) and his son Alphonse (1806–93). Augustin de Candolle's *Essai élémentaire de géographie botanique* of 1820 emphasized the distinction between a species' 'station' or habitat and its 'habitation' or range. The concept of habitat focused attention on the forces confining a species to areas with a certain environment, and would become particularly important in later ecological studies. But the range of each species was also limited by factors that did not depend upon its habitat, and which were related to the existence of distinct geographical provinces. De Candolle realized that to explain the existence of these provinces would require naturalists to tackle the question of the actual origin of species, although at this point he thought the matter was beyond the scope of scientific investigation. He studied geographical variations in the number of species belonging to a genus, showing that in France, for instance, there was an average of 7.2 species per genus, while in Britain the corresponding figure was 2.3. The question of why islands were deficient in this way (but not necessarily in the number of genera themselves) would later be explained in terms of the limited opportunities for migration across ocean barriers. Alphonse de Candolle's *Géographie botanique raisonée* of 1855 provided masses of tabulated information on all aspects of the geographical distribution of plants.

In Britain, Hewett C. Watson (1804–81) worked on the variation of the flora with altitude in the Highlands of Scotland and pioneered a national census of plants, which allowed the country to be divided into a number of botanical provinces. He eventually fell out with another pioneering figure in the study of distribution, the Manx naturalist Edward Forbes (1815–54), who appropriated Watson's provinces for his own very different study. Forbes was one of the bright stars of early-nineteenth-century British biology, whose death at the age of thirty-nine was regarded as a tragedy for science. He was deeply influenced by German idealism and sought the kind of patterns in Nature that were typical of that school of thought. Following an 1841 expedition to the Aegean Sea, Forbes proposed that the fauna of the oceans could be divided into zones by depth, exactly equivalent to the zones of vegetation ascending a mountainside. Forbes proposed that no life could be found below a depth of 300 fathoms (about 600 metres), a view not conclusively disproved until the *Challenger* expedition in the 1870s. He also proposed global zones for marine life defined by ocean temperatures. In 1846 he turned his attention to the botany and zoology of the British Isles, using a modified version of Watson's geographical provinces in an effort to explain the present distribution in historical terms.

The argument between Watson and Forbes was more than a mere priority dispute – it reflected a basic disagreement over the purpose of gathering geographical information. Although Watson wanted to go beyond mere taxonomy, he adopted an essentially descriptive approach to geographical studies and saw no point in advancing speculative explanations of how things came to be in their present state until much more data were available[4]:

> The ingenious minds that take their delight in inventing causal hypotheses to account for these present seeming relations, through reference to supposed conditions of the past, will work more easily, and perhaps with more truthful results, if supplied with properly arranged data truly illustrating the present. They may then find out, instead of only feign, phyto-geographical and phyto-geological histories.

This was an implicit critique of Forbes, but Forbes saw no reason to be apologetic because he had used the concept of botanical provinces in an entirely new way. For him, the provinces made sense not as descriptive units but as products of a historical process that explained their existence in purely natural terms.

Historical Biogeography

Forbes was inspired in this direction by the work of Charles Lyell, whose uniformitarian approach to geology stressed the use of observable processes to explain the present state of the earth (chapter 6). In the second volume of his *Principles of Geology*, Lyell had addressed the question of geographical distribution as a means of challenging the catastrophists' notion of mass extinctions. He believed that the surface of the earth changed gradually over long periods of time, and saw that such changes would have an effect not only upon the physical environment to which living things were exposed, but upon the possibility of migration to new locations. If a geographical barrier such as a mountain range were gradually destroyed, there would come a point at which plant and animal species would be able to move into territory from which they had hitherto been excluded. The inhabitants of the two originally separated regions would have the chance to invade new areas – and might themselves face a challenge from immigrants on their home territory.

Lyell was able to draw upon the work of the elder de Candolle, who had recognized that the old idea of a stable 'balance of Nature' was no longer tenable. The ability of some species to undergo explosive increases in population had been largely ignored by exponents of the balance theory, but the significance of such phenomena had now been recognized owing to the writings of Thomas Malthus. In his *Essay on the Principle of Population* of 1797, Malthus had argued that the human race always had the potential to breed faster than the food supply could be increased. This claim was widely debated because Malthus used it to suggest that poverty and starvation were inevitable. For naturalists, his insight threatened the assumption that Nature was a divinely contrived harmony. De Candolle realized that all species have the potential to increase their numbers at the expense of rivals occupying the same territory. Each species' range is determined by the area within which it can outbreed potential rivals because it is better adapted to that set of conditions[5]:

> All the plants of a given country are at war with one another. The first which establish themselves by chance in a particular spot, tend, by the mere occupancy of space, to exclude other species – the greater choke the smaller, the longest livers replace those which last for a shorter period, the more prolific gradually make themselves masters of the ground, which species multiplying more slowly would otherwise fill.

Lyell saw that this more competitive model of relationships had implications for his historical model of biogeography. The breakdown of a barrier would allow species from both sides to compete with one another, and the winners would spread out to occupy any territory to which they were better adapted than the natives.

Lyell was no evolutionist, but he was convinced that each species was only created once. When the same species was found occupying two separate territories, it made more sense to work out how it could have migrated from one to the other than to assume that it had been created independently in each area. Lyell saw each biogeographical province as a 'centre of creation' where new species typical of that province were created from time to time, spreading out from the centre to occupy as much territory as they could. Major oceans would form the most long-lasting boundaries defining the provinces, but even these were not permanent, and migrations may have been possible in early geological periods when the distribution of land and sea was different.

Forbes drew upon this approach to explain the distribution of plants and animals in the British Isles. He distinguished five botanical zones, but went beyond Watson by suggesting that each had been established by a separate episode of migration from a different continental source. The migrations had been made possible by geological changes temporarily removing barriers such as the English Channel, or opening up new routes through the creation of land-bridges. He postulated that at one time there had been an area of land in what is now the Atlantic Ocean allowing plants to move from Spain to Ireland. Perhaps his most powerful hypothesis was intended to explain the fact that many British mountains had plants identical to those found in the Arctic regions of Scandinavia. Forbes assumed that much of northern Europe had been submerged by a shallow ocean during a period in which the climate was significantly colder than the present (Lyell's alternative to the ice age theory). Arctic plants would have been transported southwards on icebergs, and would have established themselves on the islands corresponding to the modern mountains. When the land was elevated, and the climate improved, the arctic plants retreated up the mountainsides in order to remain in an environment similar to that from which they had been derived.

There were many examples of species occupying territories separated by wide expanses of ocean, and in the 1850s it became fashionable to invoke prehistoric land-bridges to explain the migrations. The botanist J. D. Hooker postulated a massive extension of land in the Antarctic to account for the similarity of plants between New Zealand, South America and the islands of the remote south. Alphonse de Candolle followed Forbes in supposing that Europe had once extended into what is now the Atlantic Ocean. Others extended the land all the way across the Atlantic to explain how some species came to be resident in both North America and Europe, openly appealing to the legend of Atlantis to support this hypothesis, despite the fact that the land-bridge would have been submerged long before the human race appeared. The American botanist Asa Gray (1810–88) adopted a less extravagant view, preferring to believe that the migration had taken place across Asia. Sunken continents remained popular until better soundings of the ocean floor in the 1870s revealed the sheer extent of the elevation that would be needed to create dry land across the Atlantic and other major oceans.

The publication of Forbes' theory to explain the spread of arctic plants was a great disappointment to Lyell's better-known disciple, Charles Darwin (1809–82). Darwin shared Lyell's antipathy to the idea that the same species could appear independently in different areas, because his theory of evolution implied that the appearance of each species was a unique historical event. Independently of Forbes, he proposed an extension of the arctic flora to explain the species found on European mountains. In the geographical section of the unpublished *Essay* of 1844 outlining his theory of evolution, he assumed that, at a time of generally colder climate, the arctic plants would spread southwards, driving out the temperate species that were no longer adapted to the conditions. When the climate improved, the situation would be reversed[6]:

> What then would be the natural and almost inevitable effects of the gradual change into the present more temperate climate? The ice and snow would disappear from the mountains, and as the new plants from the more temperate regions of the south migrated northwards, replacing the arctic plants, these latter would crawl up the now uncovered mountains, and likewise be driven northward to the present arctic shores.

Darwin's theory was a better extension of Lyell's principles because it presented the species as engaged in an active struggle with an ever-changing environment. Instead of simply being rafted by icebergs, they expanded to occupy as much territory as possible, retreating only when changed conditions began to favour their rivals. Nevertheless, the appearance of Forbes' paper led Darwin to abandon any idea of a separate publication on the topic.

Although Darwin made no use of icebergs in this case, he was passionately interested in the possibility that animals and plants might be transported across oceans by various accidental means. He studied how seeds could be carried in the mud clinging to the feet of birds, and also experimented on how long seeds could remain immersed in salt water and still be fertile. He collected information on the natural rafts of timber that are sometimes swept out to sea from great rivers, often bearing plants and even animals with them. Darwin saw no need to postulate major elevations of the sea bed to explain migration. He was convinced that, in most cases, oceanic islands have been stocked by the accidental transportation of seeds and small animals from the nearest mainland. The same processes could explain how related species appeared on widely separated land masses. Through the 1840s and 1850s Darwin engaged in a debate with Hooker on this question, eventually winning Hooker around to his own way of thinking. There was, however, another crucial difference between Darwin's position and that of his contemporaries: he believed that, when members of a species were transported to a new location, they might sometimes establish themselves even if the conditions were not identical to those of the homeland, in which case the isolated population would adapt to the new environment and eventually be transformed into a distinct species.

THE HISTORY OF LIFE

Biogeography was the principal source of Darwin's theory of evolution, but to most of his contemporaries it seemed obvious that the starting point for any investigation of the history of life on earth must be the fossil record. This distinction highlights the originality of Darwin's ideas. The basic idea of evolution was not unknown because it had been actively supported by some radicals during the 1830s and 1840s. But the early debates were conducted almost entirely in terms of comparative anatomy and paleontology.

If there were general laws governing the history of life, it was assumed that they would be revealed by the fossil record. Darwin was one of a small group of naturalists who took the opposite route, following Lyell in the search for natural processes at work in the present and the recent past that would throw light on the distribution of life around the globe today. To understand the conceptual framework of his contemporaries, we must look first at the efforts being made to uncover a pattern of development among the fossils.

Although geologists made use mainly of invertebrate fossils in dating the rocks, it was the discovery of fossil bones that attracted most attention. In an age that was becoming increasingly fascinated with history, the prospect of extending the antiquity of life itself by uncovering the remains of bizarre and often gigantic animals had a romantic appeal outside the realms of science. But it was the scientist alone who could make sense of the discoveries, and late-eighteenth-century naturalists had already begun to puzzle over the bones of gigantic elephant-like creatures found both in America and in Europe. Vertebrate paleontology finally became an organized science in the early years of the nineteenth century, owing to the anatomical skills of Georges Cuvier.

As a comparative anatomist used to studying the skeletal structures of a wide range of living species, Cuvier had the experience necessary to reconstruct the often fragmentary and disjointed bones found in geological deposits. It was said that he could estimate the structure of the whole creature from just a single bone. This was something of an exaggeration, and paleontologists made serious mistakes when reconstructing the more bizarre species from the older rocks, but only the anatomist could hope to make sense out of the discoveries. Cuvier was determined to found a new science of the past by displaying the changes in the earth's inhabitants revealed by the fossil record. In the heyday of Napoleon's conquests he had access to fossils from all over Europe and was thus in a position to make the first comprehensive study. His papers on the topic, subsequently collected as the *Recherches sur les ossemens fossiles* of 1812, provided a starting point for the science of vertebrate paleontology.

The Extinction of Species

One of the most disturbing consequences of Cuvier's work was proof that some species have indeed become extinct. The possibility that some fossils corresponded to species no longer alive today had been noted even in the late seventeenth century, but had been put on one side until now. Cuvier investigated the bones of the ancient elephants and showed that they were not the same as the living African and Indian elephants. The mammoth was related to but distinct from the modern elephants, and therefore it must be an extinct species of the same genus. The mastodon was even more distinct, since its teeth were unlike those of the modern elephants. Such gigantic creatures could hardly be living in some remote spot unknown to European science, so they had either changed into something else or had become extinct. Lamarck proposed that transmutation was responsible for the changes, thus putting off the need to confront the possibility that Nature may allow some of its products to disappear. But Cuvier insisted that the transition from the ancient to the modern species was too abrupt, and that the intermediates would not have been viable in any case. Having eliminated transmutation, extinction became inescapable.

Cuvier tried to evade one disturbing implication of extinction by linking the phenomenon to his theory of catastrophic geological changes. He believed that abrupt changes on the earth's surface were responsible for killing off all the species over a wide area. The catastrophists' notion of mass extinction allowed them to believe that the disappearance of species was not a part of Nature's regular operations, but occurred only in those exceptional interludes when the earth's normal stability was interrupted by some gigantic convulsion. There were no causes operating in the world today that could have such devastating results. This situation would only begin to break down when Lyell introduced his uniformitarian geology in the 1830s. Since there were no catastrophes in Lyell's system, extinction had to become part of the normal operations of Nature. If species were indeed fixed, then their state of adaptation to the environment would gradually break down as the conditions were changed by geological forces. They might sometimes be able to migrate to new areas, but in so doing they would merely threaten the existence of the original inhabitants. In Lyell's system, extinction was taking place even now, as the population of species no longer adapted to modern conditions is gradually whittled down.

Instead of trying to explain the extinction and replacement of species in terms of observable causes, Lyell's opponents looked to the patterns revealed in the fossil record to explain the history of life on earth. Cuvier's work with Alexandre Brongniart on the stratigraphy of the Paris basin showed him that there was a whole sequence of geological deposits, each of a different age. The vertebrate fossils could thus be arranged in a historical sequence running through from the most ancient to the most recent geological periods. Evidently there had been a series of mass extinctions followed each time by the appearance of an entirely new set of species that formed the population of the next stable period. Cuvier did not adopt the most obvious non-evolutionary explanation of where the new species came from. He said nothing of miraculous creation and instead proposed that the 'new' populations simply migrated in from areas not affected by the catastrophe. This made nonsense out of the geologists' expectation of a world-wide sequence of characteristic fossils, and was never taken up. Most catastrophists preferred to believe that some form of supernatural agency was involved in the appearance of each new population.

Cuvier noticed that the most recently extinct creatures

such as the mammoth were closely related to living species. The hyena remains that William Buckland took as evidence of the deluge belonged to the same recent period, as did the giant deer sometimes called the 'Irish elk'. As Cuvier began to study the more ancient fossils, he found that they were increasingly unlike anything known in the world today. The oldest Tertiary rocks contained archaic mammals that bore no resemblance to the living families within the class. Cuvier called one of them Palaeotherium (ancient beast). In the older Secondary rocks there seemed to be no mammals at all, only bizarre reptiles. It was Cuvier's English followers who did much to open up what soon became known as the 'age of reptiles'. The aquatic plesiosaur was first described in 1821 and soon afterwards an almost complete skeleton was unearthed from the rich fossil-bearing rocks of Lyme Regis in Dorset. A local woman, Mary Anning, made a living selling fossils from these rocks to collectors, and was immortalized in the tongue-twister 'She sells seashells on the seashore'.

The fossils that have most effectively captured the public's attention as symbols of the outlandish character of prehistoric life are the dinosaurs. The first of these to be discovered was a gigantic carnivore from the Jurassic rocks of Stonesfield, Oxfordshire, which Buckland described in 1824 under the name Megalosaurus (great lizard). In the following year a Sussex doctor Gideon Mantell (1790–1852) found some large fossilized teeth in a pile of rocks intended for road-mending and noted that they resembled the teeth of the modern iguana. He named this enormous herbivorous reptile Iguanodon and went on to coin the term 'geological age of reptiles' in an article published in 1831. The general term 'Dinosauria' (fearful reptile) was introduced by Richard Owen in a survey of British fossil reptiles in 1841. Some years later Owen helped to design a collection of life-sized replicas of dinosaurs and other denizens of the age of reptiles erected at the Crystal Palace site in south London. They are still there today, providing a dramatic illustration of the Victorians' perception of the past. In fact, Owen misinterpreted the remains of Megalosaurus and Iguanodon, presenting them as gigantic lizards standing on four legs. Later discoveries would show that in fact both are bipedal, with massive hind legs and diminutive forelimbs.

Beneath the age of reptiles lay a series of formations containing the fossils only of aquatic creatures. The Old Red Sandstone of Scotland was particularly rich in the remains of strange fish, many of them covered with a kind of bony armour-plating. These were collected by a stonemason, Hugh Miller (1802–56), who became accepted as an authority on the subject and went on to become an outspoken critic of evolutionism. Louis Agassiz, then the world authority on fossil fish, studied Miller's specimens and helped to describe some of them. Beneath the Old Red Sandstone (Devonian) lay the Silurian and Cambrian systems that were being described by Murchison and Sedgwick. Here, no vertebrates at all could be found, the fossils consisting only of invertebrates such as the trilobites.

Progressive Development

By the 1840s there seemed to be clear evidence for a sequence of development in the history of life on earth. Animal life began with the invertebrates and then ascended through the sequence of vertebrate classes in the order: fish, reptiles and mammals. Only Charles Lyell resisted this interpretation, arguing that the apparent sequence might be the result of imperfect evidence. Lyell wished to deny any historical trend in the history of life since it would violate the steady-state principles of his uniformitarian system (chapter 6). A few controversial mammalian fossils were known from the age of reptiles, and Lyell appealed to these as evidence that the whole sequence was unsound. He argued that fluctuations in the earth's climate due to geological changes might alter the proportion of reptiles and mammals in the population. The so-called 'age of reptiles' was merely a period when there were more reptiles than mammals because the earth had enjoyed uniformly hot conditions. As the conditions changed to the more temperate climate we know today, so the proportion of reptiles had decreased – but there was no overall trend of the kind that would be implied if the mammals had only come into existence in the later geological periods.

Most geologists saw Lyell's steady-state alternative as completely unrealistic. They could not believe that large numbers of mammals would eventually be found in some of the earliest rocks. The basic sequence was clear enough, and to most nineteenth-century naturalists and geologists it was self-evidently a progressive development from lower or simpler forms of life to higher or more complex ones. Despite Cuvier's warnings, it was still widely assumed that the vertebrates are collectively more advanced than any invertebrate type. And within the vertebrates, the fish were the 'lowest' class and the mammals the 'highest' – since we ourselves are mammals and the Victorians were convinced that the human race was the pinnacle of creation. It was widely believed that there were no human fossils, so the recent appearance of humankind seemed to represent the last step in the ascent of life. The critical question was: Why has life undergone this progressive development in the course of time?

Lamarckians such as Robert Grant had no doubts about the significance of the fossil progression – it proved that animal life had advanced through the scale of development as the result of purely natural processes. Opponents of this interpretation had to stress the discontinuity of the development in order to undermine the plausibility of evolutionism. They wanted progress, but preferred to believe that the advances had taken place by a series of discrete steps, each too large to be compatible with any form of natural development. But if the advance was not the result of evolutionary laws, why had it taken place? Those naturalists who wanted to preserve a role for the

1 Woodcut of yarrow, foxglove, primrose etc. from Dioscorides, *De Medicinali Materia Libri Sex* (Frankfurt & Marburg, 1543), p. 338

2 The artist and engraver who prepared the plates for Leonard Fuchs' *De Historia Stirpum* (Basel, 1542)

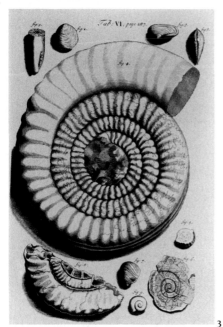

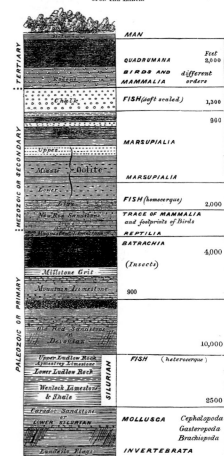

TABLE OF STRATA AND ORDER OF APPEARANCE OF ANIMAL LIFE
UPON THE EARTH.

3 Fossil sea-creatures including a large Ammonite as depicted in Robert Hooke's 'Lectures and Discourses of Earthquakes', *The Posthumous Works of Robert Hooke* (London, 1705), plate 6

4 Part of a hypothetical cross-section of the earth's crust used to depict geological structures in William Buckland's *Geology and Mineralogy Considered with Reference to Natural Theology* (London, 1837), vol. 2, plate 1

5 Stratigraphical column, from Richard Owen, *Palaeontology, or a Systematic Summary of Extinct Animals and their Geological Relations* (Edinburgh, 1860), facing p. 1

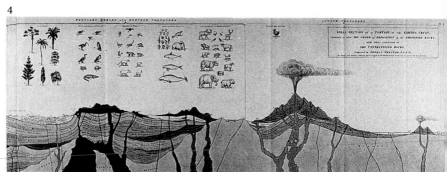

6 The Temple of Serapis at Puzzuoli, Italy, used to illustrate the nature of geological change by
Charles Lyell, *Principles of Geology* (London, 1830–3, 3 vols.), vol. 1, frontispiece. The dark bands
on the columns have been formed by the action of marine creatures, showing that since built
by the Romans the Temple has been submerged and then elevated again

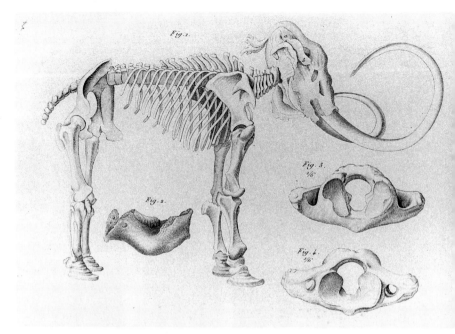

7 Skeleton of the mammoth, from Georges Cuvier, *Recherches sur les ossements fossiles* (3rd edn: Paris, 1825, 5 vols.), vol. 1, plate 11

8 Life-sized reconstruction of the carnivorous dinosaur Megalosaurus designed by Richard Owen and erected at the Crystal Palace site in South London. Later discoveries showed that Megalosaurus actually walked only on its hind legs.

9 The duck-billed platypus, the subject of much controversy when first reported to European naturalists in the early nineteenth century. This illustration first appeared in John Gould's *Mammals of Australia* (1845–63) and was widely reprinted (taken from W.H. Flower and R. Lydekker, *An Introduction to the Study of Mammals* (London, 1891), p. 121

10 Crowds at the London Zoo in the 1860s, illustrating the wide popularity of displays of exotic animals in the Victorian period; from the *Illustrated London News*, vol. 48 (1866), p. 509

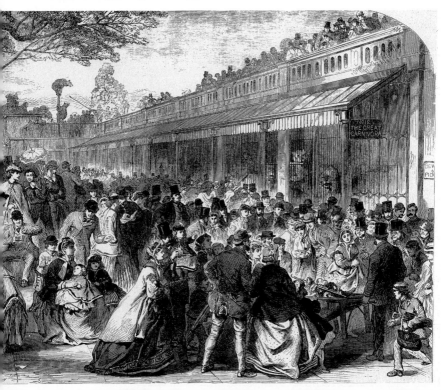

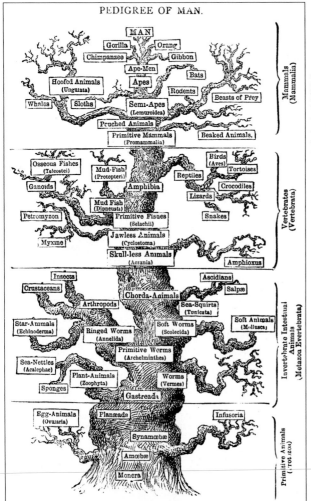

PEDIGREE OF MAN.

MAN

Gorilla — Orang
Chimpanzee — Gibbon

Ape-Men

Apes — Bats

Hoofed Animals (Ungulata) — Rodents

Whales — Sloths — Beasts of Prey

Semi-Apes (Lemuroidea)

Pouched Animals

Primitive Mammals (Promammalia) — Beaked Animals.

Mammals (Mammalia)

Birds (Aves) — Tortoises

Osseous Fishes (Teleostei) — Mud-Fish (Protopteri) — Reptiles — Crocodiles

Ganoids — Amphibia — Lizards

Petromyzon — Mud Fish (Dipneusta) — Snakes

Primitive Fishes (Selachii)

Myxine — Jawless Animals (Cyclostoma)

Skull-less Animals (Acrania) — Amphioxus

Vertebrates (Vertebrata)

Insects — Ascidians

Crustaceans — Chorda-Animals — Salpæ

Arthropods — Sea-Squirts (Tunicata)

Star-Animals (Echinoderma) — Soft Worms (Scolecida) — Soft Animals (Molluscs)

Ringed Worms (Annelida)

Sea-Nettles (Acalephae) — Primitive Worms (Archelminthes)

Plant-Animals (Zoophyta) — Worms (Vermes)

Sponges — Gastreada

Invertebrate Intestinal Animals (Metazoa Evertebrata)

Egg-Animals (Ovularia) — Planæada — Infusoria

Synamœbæ

Amœbæ

Monera

Primitive Animals (Protozoa)

11 Heads of four of the Galapagos ground finches illustrating the variation in beak-structure evolved by the different species, from Charles Darwin, *Journal of Researches into the Geology and Natural History of the Countries Visited during the Voyage of HMS* Beagle (reprinted London, 1891), chapter 17

12 Diagram to illustrate branching evolution from Charles Darwin's B Notebook, p. 36

13 The 'Pedigree of Man' from Ernst Haeckel, *The History of Creation* (New York, 1879, 2 vols.), vol. 2, facing p. 188. Note how – unlike Darwin (plate 10) – Haeckel draws a tree with a central trunk running through to mankind, all other forms being depicted as side-branches characteristic of certain stages in the main line of development

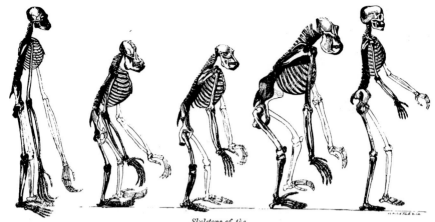

GIBBON. ORANG. *Skeletons of the* CHIMPANZEE. GORILLA. MAN.

14 Diagram to illustrate the anatomical resemblances between humans and apes (the gibbon on the left is drawn at twice the size of the other skeletons); from T.H. Huxley, *Man's Place in Nature* (London, 1863), frontispiece

DISTRIBUTION OF PRIMATES

MODERN ANTHROPOIDEA (MONKEYS, APES, BABOONS)
" LEMUROIDEA (LEMURS, LORIS, TARSIER)
E, EOCENE (AND OLIGOCENE) LEMUROIDS
O, OLIGOCENE ANTHROPOIDS
M, MIOCENE
P, PLIOCENE

	Fore-foot.	Hind-foot.	Fore-arm.	Leg.	Upper molar.	Lower molar.
RECENT. Equus.						
PLIOCENE. Pliohippus.						
Protohippus (*Hipparion*).						
MIOCENE. Miohippus (*Anchitherium*).						
Mesohippus.						
EOCENE. Orohippus.						

15 Modification of the teeth and lower parts of the limbs from the four-toed *Orohippus* of the Eocene to the modern horse. This sequence was constructed from American fossils by O.C. Marsh and was widely regarded as convincing evidence for the evolution of a modern form from a less specialized ancestor. The diagram was widely reprinted; this example is from A.R. Wallace, *Darwinism* (London, 1889), p. 388

16 Map showing the dispersal of the modern primates southwards from a Holarctic centre of origin (Eurasia and North America); from W.D. Matthew, *Climate and Evolution* (New York, 1939), p. 46

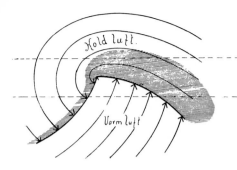

Kold luft.

Vorm luft

17 Diagram to illustrate the structure of a cyclone as warm air from the south penetrates into cold air to the north, from V. Bjerknes, 'The Structure of the Atmosphere when Rain is Falling', *Quarterly Journal of the Meteorological Society*, vol. 46 (1920), pp. 119–38, see   p. 128

18 Map to show continental drift from the Upper Carboniferous to the Lower Quaternary; the break-up of Pangaea into the modern continents is clearly visible. From Alfred Wegener, *Die Entstehung der Kontinente und Ozeane* (3rd edn; 1922), p. 4.

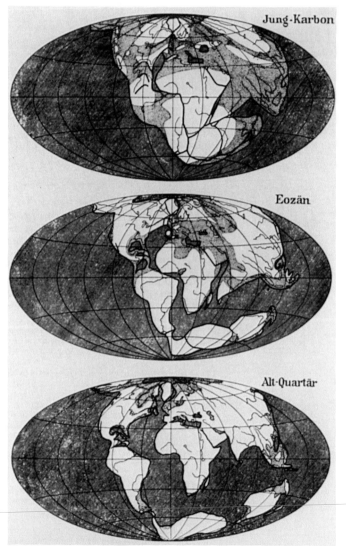

Jung-Karbon

Eozän

Alt-Quartär

supernatural in the creation of life were forced to try out a number of alternatives in an attempt to provide an explanatory framework that would serve as an alternative to evolutionism.

One possibility extended Paley's version of the argument from design to include a whole sequence of creations, each adapted to the prevailing conditions of the time. If geology suggested that the earth's climate had changed through time, it would make sense to accept that the Creator would design different species for each of the successive geological periods. Catastrophes would serve a useful function in killing off the old species just as the conditions were changing, leaving the world clear for the Creator to step in and design a whole new population adapted to the environment that would stabilize after the upheaval. The cooling-earth theory provided the perfect framework for such a model of successive adaptive stages. The supposed decline in the carbon dioxide content of the atmosphere would explain why the air-breathing animals had only been created in the later stages of the earth's history. Paleontologists such as William Buckland took up this interpretation with enthusiasm, thus combining a continued belief in the argument from design with opposition to transmutation via the concept of successive creations.

Idealist philosophers saw the human body as the most perfect expression of the vertebrate form. They visualized the human race as the goal of a developmental pattern imprinted onto history by God. Embryologists such as J. F. Meckel had tried to establish a parallel between the hierarchy of vertebrate classes and the stages of development of the human embryo. Now it seemed as if this 'law of parallelism' could be extended to the whole history of life on earth. The growth of the human embryo recapitulated the history of animal life as revealed by the fossil record. Throughout the earth's history, life had been steadily maturing, advancing through a predestined sequence of developmental stages towards the perfection of the human form. Louis Agassiz was one of the most eloquent spokesmen for this view of development as a divinely constructed pattern[7]:

> The history of the earth proclaims its Creator. It tells us that the object and term of creation is man. He is announced in nature from the first appearance of organized beings; and each important modification in the whole series of these beings is a step toward the definitive term of the development of organic life.

The apparent parallel between the development of the human individual and the history of life on earth is thus God's symbol telling us that we are the goal of His creation.

To stress the difference between his position and that of the transmutationists, Agassiz insisted that the development revealed by the fossil record was discontinuous. Whatever the overall trend, the steps from one class to the next were so abrupt that they could not be explained by gradual transmutation. God's plan unfolded by distinct stages corresponding to the sudden introduction of entirely new forms, not by a gradual ascent of the organic scale. As part of his campaign against the transmutationists, Richard Owen used his 1841 survey of British fossil reptiles to argue against the idea of continuous development. The dinosaurs were the highest form of reptile, but in Owen's time they were the earliest known members of the class – exactly the reverse of what would have been predicted by a theory of continuous development.[8]

The Scottish paleontologist Hugh Miller drew the same lesson from his study of the most ancient fossil fish. They were not the *lowest* fish, as the transmutationist predicted, but among the highest. The history of each new class could be seen as a sudden explosion of development followed by a gradual decline, just like the rise and fall of human empires[9]:

The general advance in creation has been incalculably great. The lower divisions of the vertebrata precede the higher; – the fish preceded the reptile, the reptile preceded the bird, and the bird preceded the mammiferous quadruped. And yet, is there one of these great divisions in which, in at least some prominent feature, the present, through this mysterious element of degradation, is not inferior to the past?

The cyclic model of development was intended to thwart the transmutationists, but was not a mere product of Owen's and Miller's imaginations. The fossil record seemed to indicate sudden steps rather than continuous development, although new discoveries over the next few decades would help the evolutionists' case.

The most important lesson to be learnt from the fossil discoveries of the 1840s and 1850s was that the history of life was far more complex than anyone had at first realized. As long as transmutationism was linked to the old idea of a simple ascent of a linear scale, it was doomed to failure. There was indeed an overall progression, but the ascent from fish to reptiles and then to birds and mammals was superimposed on an intricate sequence of developments that seemed to take place independently of the overall trend. There were major changes within the various invertebrate types, although these seldom caught the headlines. It did not make sense to see all the lower animals as merely immature versions of humankind, nor was it possible to assume that every extinct population had been perfectly adapted to the conditions of its own time. Any theory seeking to explain the history of life on earth would have to take into account a whole range of geographical and adaptational factors in addition to the ascent of life towards humankind.

In the early decades of the century it had been widely assumed that the distinct geographical provinces of the modern world had only come into existence quite recently in geological terms. For most of its history, the earth had enjoyed warm, uniform conditions over its whole surface. This interpretation was already being undermined in the 1830s as new evidence showed that the geographical provinces did indeed extend some way back into the past. Darwin himself contributed to this trend by bringing back fossils from the voyage of the *Beagle* showing that the past inhabitants of South America were closely related to the modern ones. The giant ground sloths and armadillos of the past were clearly South American in character, showing that this continent had been populated by unique families of animals over a vast period of time. Darwin and Owen formulated a 'law of succession of types' to indicate this geographical continuity in the fossil record. For the opponents of transmutation, this meant only that the 'centres of creation' had been active throughout the earth's history. Darwin was one of the few to speculate that the law of succession of types was best explained in terms of the constraints placed upon evolution by the major geographical barriers defining the continents.

In the 1850s the fossil record began to yield evidence for a new kind of trend within the development of each class. K. E. von Baer's attack on the linear model of embryological development inspired Owen and the physiologist W. B. Carpenter (1813–85) to look for non-linear relationships in the fossil record. They showed that each class normally appeared first in a very generalized form, from which radiated out a number of lines of development each leading towards a specialized modification of the class for a particular way of life. Branching and specialization were the key to the history of each class, not the ascent of a linear scale towards the next highest form. The German paleontologist H. G. Bronn (1800–62) also began to draw diagrams of the history of life representing the overall process as a branching tree. Owen and Bronn continued to believe, however, that the pattern of development represented the unfolding of a divine plan, the expansion of predetermined variations on a theme.

THE PROCESS OF CHANGE

With hindsight we can see how the paleontology of the 1840s and 1850s paved the way for the reception of Darwin's theory when the *Origin of Species* was published late in 1859. But to many of the naturalists actually involved, the new discoveries required only a revision of traditional ideas rather than a revolution. For all their efforts to uncover trends that would serve to unify the history of life on earth, the vast majority were unwilling to explain the trends as the result of natural processes that could be seen operating in the world today. They were still interested in patterns rather than processes, although the patterns they were now discussing were of a kind that Darwin would be able to explain.

This preference for patterns must be taken into account in any survey of early transmutationism. To the anatomists who worked in the museum or the dissecting room, processes that could only be observed in the field were of little interest. They looked for abstract relationships between the different structures, but they had little incentive to ask what kind of circumstance might lead a species to change when exposed to a new environment. The earliest accounts of transmutation fell into this trap, projecting the process as the gradual unfolding of trends that could not be explained in terms of a response to the environment. Darwin introduced a new element into the equation because he started not from the fossil record but from field studies, which forced him to consider the real-life pressures acting upon species in the course of geological time.

Transmutation before Darwin

Lamarck's theory had included the mechanism of the inheritance of acquired characteristics as a means of adapting organisms to changes in their environment. But in the *Philosophie zoologique* Lamarck had subordinated this to his belief in a progressive force that pushed living things steadily up the scale of organization. His followers tended to concentrate on the progressionist element, looking for a serial relationship between the classes of animal life and minimizing the gaps between the known forms. Little effort was made to investigate how a species might actually respond when placed in a new environment, and the Lamarckians paid little attention to biogeography. Geoffroy Saint Hilaire uncoupled transmutationism from the idea of a linear scale, but again failed to see the importance of geographical factors. For Geoffroy, a change in the environment might trigger off a new pattern of growth in the organism – but the result was determined more by the laws of growth than by the adaptive needs of the organism. To test his theory he looked at the processes responsible for producing monstrosities, because these provided the closest observable parallel to the saltative transmutations demanded by his theory.

Earlier sections of this chapter have outlined the clash between radical transmutationists such as R. E. Grant and conservatives such as Owen. The conservative forces were successful at first. By the 1840s the transmutationist challenge was ebbing, Grant was losing influence and the whole movement had been labelled as irreligious and potentially subversive. To understand the climate of opinion within which Darwin developed his own theory of transmutation during the late 1830s, we must bear in mind that Lamarckism had been branded as an atheistical theory appealing only to political revolutionaries. It is small wonder that Darwin thought twice about threatening his own privileged position by identifying himself with a new twist on this old theme.

The first attempt to break through the deadlock created by the blacklisting of Lamarckism was made not by Darwin, but by the Edinburgh publisher and amateur naturalist, Robert Chambers (1802–71). In the 1830s and 1840s Chambers issued his own periodical to reflect the interests of the middle classes. He argued that social progress towards a free-enterprise society was inevitable and saw progressive evolution as a useful foundation upon which to build a philosophy of cosmic development that would include the human race. Chambers realized that, to popularize such a philosophy in the more conservative political climate south of the border, it would be necessary to uncouple the idea of transmutation from the materialist image acquired by Lamarckism. In his book, *Vestiges of the Natural History of Creation*, published anonymously in 1844, Chambers presented transmutation as the expression of a divine plan. The progressive appearance of ever-higher forms of life was not the result of blindly operating natural laws, but the step-by-step unfolding of a preordained pattern that the Creator had built into the universe. Chambers hoped to reconcile those readers with religious qualms to the general idea of transmutation. The human race could be accepted as a product of Nature's development because that development was instituted by the Creator as an indirect means of achieving His goal.

Chambers discussed the fossil record in the hope of providing evidence for a generally progressive trend in the history of life, but his strategy depended upon distancing his theory from Lamarckism. He had no interest in the adaptation of species to their environment, nor indeed in the provision of any naturalistic mechanism of transmutation. The progress of life unfolded in the way it did because God had imposed a preordained sequence of developmental stages through which organisms must ascend towards the human form.

Chambers drew upon the law of parallelism to suggest that the actual process of transmutation from one species to the next highest in the scale took place by means of an extension to the growth of the individual. Once again the lower animals were treated as immature versions of the human race, with the growth of the human embryo recapitulating all the stages through which life had passed in order to reach the highest level of development. But now Chambers insisted that the extensions to growth took place without supernatural interference, so that one species naturally transmuted itself into a higher one at a certain point in time. The linear pattern of development was an indication that God had programmed the whole history of life to move in a purposeful direction towards higher levels of intelligence. Somewhat paradoxically, Chambers also included a chapter on MacLeay's circular or quinary system of classification, since this indicated that the whole animal kingdom was the product of a coherent, rational plan.

Vestiges was the centre of a violent controversy. Chambers had made no secret of his intention to treat the human race as the last and highest product of the ascent through the animal hierarchy. He appealed to the pseudo-science of 'phrenology', which had been built around the assumption that the shape of the brain determined a person's character. This essentially materialist approach to the mind allowed Chambers to claim that, as the brain increased in size in the course of evolution, so had intelligence expanded towards the human level. Such claims were anathema to conservative thinkers, and a host of writers led by Adam Sedgwick and Hugh Miller fulminated against *Vestiges* and its influence. It was one thing to suggest that God might have instituted a 'law of progress' to achieve His goals in Nature, quite another to suggest that human beings were merely highly developed animals.

Chambers' book nevertheless had a positive effect in encouraging people to think in terms of continuous trends in the development of life. In 1849 even Richard Owen hinted briefly that the modifications of the vertebrate archetype might have been unfolded by non-miraculous causes[10]:

> The archetypical idea was manifested in the flesh, under divers such modifications, upon this planet, long prior to the existence of those animals species that actually exemplify it. To what natural laws or secondary causes the orderly succession and progression of such organic phenomena may have been committed we are as yet ignorant. But if, without derogation of the Divine power, we may conceive the existence of such ministers, and personify them by the term 'Nature', we learn from the past history of our globe that she has advanced with slow and stately steps, guided by the archetypical light, amidst the wreck of worlds, from the first embodiment of the Vertebrate idea under its Ichthyic [fish-like] vestment, until it became arrayed in the glorious garb of the human form.

Owen never repeated this claim in the pre-Darwinian era, but liberal theologians such as the Oxford mathematician and philosopher Baden Powell began to come around to the idea of 'designed evolution' during the 1850s. A compromise was being hammered out in which the once-opposed concepts of design and transmutation would be synthesized so that the trends in the fossil record could be accepted as the gradual unfolding of a divine plan.

Chambers' book thus played an important role in creating the framework of opinion into which Darwin's theory would be received. Evolution was acceptable as long as it was presented as a purposeful process in which the human race played a central role as the goal towards which natural progress was aimed. Radicals wanted to see progress towards humankind as the central theme, but conservatives like Owen, worried about the status of the human soul, preferred more complex patterns of development that allowed for entirely new levels of organization to be introduced from time to time.

Chambers' theory offered no natural explanation of why changes took place in the way they did. The only driving force was the power of God, which somehow transcended the ordinary laws of Nature to impose a purposeful direction on the ascent of life. Far from being an anticipation of Darwin's theory, Chambers' whole approach ignored the possibility of explaining transmutation in terms of the natural laws observable in the everyday world of how organisms relate to their environment. Those who were content with this approach would never turn the basic concept of transmutation into a scientifically testable theory.

Lamarckism was the only available theory of natural evolution, and, although this was being promoted by the liberal social philosopher Herbert Spencer, biologists were too well aware of its dangerous reputation to take it seriously. A small number of naturalists, including the young Thomas Henry Huxley (1825–95), were on the lookout for a new initiative in this area. If Lamarckism was discredited, then some entirely new mechanism would have to be suggested if evolution was to be seen as a truly natural process. But Huxley condemned *Vestiges* as meaningless verbiage and was unwilling to support the basic idea of evolution unless someone could show him a natural mechanism to explain how the process could work. It was exactly this kind of initiative that Darwin would bring forward in the *Origin of Species*.

The Origins of Darwin's Theory

The circumstances leading to Darwin's discovery of the theory of natural selection form one of the most intensively studied areas of the history of science. So many books and articles have been published on the topic that we often speak of a 'Darwin industry'. Historians have been attracted by the wealth of available material: Darwin's private notebooks have survived, along with much of his correspondence, and this material is now being published so that scholars everywhere can have access to what must stand as a unique record of creative thought. But the motivation that leads so many historians to work on this material also derives from the status that the theory of natural selection has achieved within modern biology. The vast majority of working biologists now accept natural selection as the key to our understanding of how populations of animals and plants adjust to changes in their environment. Darwin is thus treated as a major figure in the history of science because he both popularized the general theory of evolution and discovered the mechanism that seems to offer the best prospect of explaining how the process works.

There is a real danger that the sheer scope of the Darwin industry will encourage us to forget that other people were thinking about organic origins at the time, but were exploring very different ways of trying to understand the development of life on earth. Darwin is important in any history of the environmental sciences because his theory focused attention onto the problem of how species become adapted to their environment. But to many of his contemporaries, adaptation was *not* the central problem, and natural selection thus failed to answer the questions that were uppermost in their minds. When we survey the reception of Darwinism (chapter 8), we shall have to beware of the temptation to assume that the success of the selection theory in the twentieth century was built upon foundations established in Darwin's own time. In fact, Darwin converted the scientific world to evolutionism despite an almost universal refusal to accept that his theory of natural selection offered an adequate explanation of how the process works. Darwin's thinking both drew upon and transcended the conventional ideas of his time.

Darwin met Robert Grant during a brief and unsuccessful spell as a medical student at Edinburgh. Although most biographers (encouraged by Darwin's own recollections) have played down the Edinburgh episode, the most recent studies have noted that he developed an interest in invertebrate zoology here that was to form a mainstay of his scientific work through into the time of the *Beagle* voyage. After abandoning the idea of taking up medicine, Darwin moved to Cambridge in order to obtain the BA degree needed if he was to become an Anglican clergyman. At Cambridge he came under the influence of both Adam Sedgwick, the professor of geology, and John Stevens Henslow, the professor of botany. Henslow was particularly important in developing the young Darwin's interest in natural history. Although Henslow's teaching was not part of the undergraduate curriculum, interested students were taken on field trips and given a good grounding in the science of the time. At Cambridge Darwin was absorbed into the Anglican way of thinking about Nature and read Paley's *Natural Theology*. At this point he accepted the argument from design, and his interest would remained focused on the question of adaptation even when his faith in divine creation began to wane.

After reading Humboldt's *Personal Narrative*, Darwin was determined to visit the tropics himself. He was also inspired by Humboldt's vision of an all-embracing theory that would explain the geographical diversity of life on earth. This resonated with his reading of Romantic poets such as Wordsworth, who also expressed a concern that science should not degenerate into a search for mere details. Although many aspects of Darwin's work reflect practical concerns, his overall vision is very much a product of the Romantic era, with its emphasis on the unity of Nature and the power of natural forces to reshape the world.

Darwin was soon offered the position of naturalist aboard HMS *Beagle*, which was being sent to chart the waters of South America under Captain Robert Fitzroy. He spent the years 1831–6 on the voyage, although much of his time was spent away from the ship exploring in the interior of the continent. While on the voyage he read Lyell's *Principles of Geology* and was soon making observations on the gradual elevation of the Andes, which converted him to uniformitarianism. Lyell's second volume raised the question of the geographical distribution of species at a time when Darwin was well placed to study this phenomenon. Edward Forbes and Darwin were inspired by Lyell to look for geological changes that would explain the present distribution of species in Europe. But Forbes was unable to shake off the belief that new species were produced by the exertion of a supernatural agent – he proposed a strange theory of 'polarity' in which the creative power was supposed to have been exerted most strongly at the beginning and at the end of the history of life on earth. Darwin's discoveries on the *Beagle* voyage had the effect of gradually breaking down his faith in the fixity of species and converting him to a dynamic view of the relationship between living things and their environment. By the time he returned home, he had already passed beyond the position of Lyell and Forbes and had begun to investigate the possibility that species might actually be modified when exposed to a new environment.

While travelling on the mainland of South America, Darwin was brought face to face with the conflict that was under way between the European settlers and the native Indians. Perhaps by analogy with this he began to see that animal species too are in a state of dynamic equilibrium with one another and with the environment, an equilibrium that can easily be disturbed by geological changes or by the immigration of new species into the area. Darwin discovered a new species of the rhea, the flightless bird of the pampas, and realized that the existence of two closely related species in the same area made nonsense out of the old idea of perfect adaptation. Even if one species was adapted to one extreme of the pampas and the other to the opposite extreme, the intermediate territory was inhabited by both and the two species must be competing to occupy as much of this territory as possible. Here de Candolle's concept of species being at war with one another, discussed at length by Lyell, must have allowed Darwin to see that the population of any species is always tending to expand into the surrounding territory, if necessary at the expense of the existing inhabitants. Given a slight change in the conditions favouring one species at the expense of its rivals, the balance of power would be shifted and the disadvantaged species might be driven to extinction.

This more dynamic model of natural relationships would become an integral part of Darwin's thinking, but it was not by itself enough to convert him to evolutionism. Here the critical developments were to take place during the *Beagle's* visit to the Galapagos islands, a small group of volcanic islands several hundred miles out into the Pacific from the South American mainland. The unique fauna of the Galapagos islands includes marine iguanas and giant tortoises, but it is the birds – especially 'Darwin's finches' – that allow the naturalist an unrivalled opportunity to study the effects of geographical isolation in the production of new species. We now know that Darwin very nearly missed this opportunity.[11] The *Beagle* was on the point of leaving when he found out that the natives could tell which island a tortoise came from merely by looking at the shape of its shell. Far from studying the distribution of the finches in the light of this information, Darwin had to sort through his collection trying to remember where each specimen had been captured. The mocking-birds were probably his best clue, since here he found that he had clear evidence showing that the populations of different islands were quite distinct.

After the *Beagle* returned to England, the ornithologist John Gould confirmed that there were several different species of finches and mocking-birds on the islands. To Darwin, this information reduced the theory of divine creation to an absurdity. He simply could not believe that the Creator had done separate miracles in order to provide each of these insignificant islands with its own species of finch or mocking-bird. It was far easier to assume that the islands had been stocked by the accidental transportation of small numbers of individuals from the mainland, after which each population had simply adapted to its new environment in a different way. The production of a group of related species on the different islands of an archipelago provided a model for divergent evolution, suggesting that, when a species was divided into separate breeding populations, each of those populations had the ability to change as it adapted to its new environment. The separate populations would first constitute mere varieties of the original species, but as the modifications accumulated they would eventually become distinct species in their own right, unable to interbreed with the parent form even if they were brought back into contact with it. Extended over vast periods of time, the same process could account for the production of all the various species of animals and plants.

Natural Selection

Darwin settled down in London for a few years and became an active member of the Geological Society. But unknown to most of his colleagues he was busily collecting information that might bear on the question of *how* a species might change when exposed to a new environment. He knew about Lamarck's theory of the inheritance of acquired characteristics, but dismissed this as a secondary process that could not account for the whole range of adaptations (he must also have been aware of the strong opposition to Lamarckism from Owen and the conservative forces in science). In search of clues he began to study the one area where species can actually be seen to change, the production of new varieties of pigeons, dogs, etc., through the activity of human breeders. Darwin's later descriptions of his theory use artificial selection as a means of introducing the reader to the possibility of a natural equivalent, and he always implied that this was how he himself discovered his theory. His notebooks tell a rather more complex story, however, and it was only after several years of work that he eventually put together the idea of natural selection.

An important clue was his reading of Malthus on population, which allowed him to see that the 'struggle for existence' caused by population pressure would act *within* the species, killing off any individuals who were not well adapted to the conditions and allowing only the 'fittest' (i.e. best-adapted) individuals to survive and breed. From the random individual variation existing within any population, the 'survival of the fittest' would pick out those with useful characters and thus intensify the adaptive feature in each generation. In the first detailed account of his theory, Darwin gave the following imaginary example to show how natural selection would work[12]:

> ... let the organization of a canine animal become slightly plastic, which animal preyed chiefly on rabbits, but sometimes on hares; let these same changes cause the number of rabbits very slowly to decrease and the number of hares to increase; the effect of this would be that the fox or dog would be driven to try to catch more hares, and his numbers would tend to decrease; his organization, however, being slightly plastic, those individuals with the lightest forms, longest limbs, and best eyesight (though perhaps with less cunning or scent) would be slightly favoured, let the difference be ever so small, and would tend to live longer and to survive during that time of the year when food was shortest; they would also rear more young, which young would tend to inherit these slight peculiarities. The less fleet ones would be rigidly destroyed. I can see no more reason to doubt but that these causes in a thousand generations would produce a marked effect, and adapt the form of the fox to catching hares instead of rabbits, than that greyhounds can be improved by selection and careful breeding.

Owing to the impact of Malthus' principle, many historians have argued that Darwin's theory of gradual transmutation through individual competition reflects the competitive ethos of Victorian capitalist society. Natural selection gives the responsibility for progress to the individuals who make up the population; in seeking to do the best for themselves, they help to guarantee the future of their species. If the population as a whole fails to adapt under pressure, the species will be wiped out by a rival.

The most original aspects of Darwin's theory were inspired by his reading of Lyell and his study of biogeography. But historians now recognize that for Darwin himself the idea of natural selection arose from a dialogue between his geographical studies and his interest in the process of reproduction by which the species is maintained.[13] Darwin's views on reproduction or 'generation' have often been ignored because they bear no resemblance to the modern theory of Mendelian genetics. It is easy to assume that there was a gap in Darwin's theory that would later be filled in by modern knowledge of heredity. But such an application of hindsight ignores the fact that Darwin's pre-genetical ideas were an integral part of the conceptual scheme within which his theory of evolution was formulated. Darwin's theory provides a classic example of how difficult it is to draw a sharp distinction between 'natural history' and 'biology': the process of evolution must of necessity mediate between the reproductive process that maintains the population and the environment to which the population must adapt.

Darwin had formulated the outline of his theory of natural selection in the late 1830s. In 1842 he wrote out a short sketch of his theory, and in 1844 a substantial essay that was intended for publication only if (as he now feared) he should die prematurely. Darwin's reluctance to publish in the 1840s can easily be understood in terms of the still active controversy over Lamarckism and Chambers' *Vestiges*. As a member of the wealthy middle class, Darwin had a social position that he might well forfeit if he became identified with the supposedly atheistical concept of transmutation.

Apprehension about the social consequences was not the only factor holding Darwin back from publication. He soon became aware that his theory was not capable of explaining some of the most exciting new developments in science. The model provided by the Galapagos islands showed how a group of physically isolated populations produced by migration might adapt to their new environments. But once adaptation to the new conditions had been achieved, Darwin assumed that evolution would come to a halt. The latest studies of the fossil record, however, suggested that many families had been subject to a constant trend towards increasing levels of specialization. Darwin saw that his theory must explain the pressure towards divergence and specialization, and it was not until the mid 1850s that he solved the problem. He now realized that specialization for a particular way of life was an advantage even in a stable environment, because it allowed a species to escape from the pressure of rivals seeking to exploit the same resource. Just as the 'division of labour' in a manufacturing process allows greater productivity, Nature can support more living organisms per square metre if they are divided into a diverse collection of specialists, each exploiting the environment in a particular way.

By the late 1850s Darwin had begun to prepare a multi-volumed account of his theory for publication. He was interrupted in this task by the appearance in 1858 of a paper outlining a similar concept of selection developed independently by Alfred Russel Wallace (1823–1913). Wallace came from a poor background, and the only way he could finance his interest in natural history was as a professional collector of rare specimens. He had worked in South America and was now exploring the Malay archipelago (modern Indonesia). He too had been influenced by Lyell, and as early as 1855 had published a paper commenting on the fact that new species always seemed to appear in the same neighbourhood as a closely related existing species. Like Darwin, Wallace had been led by a study of biogeography to see how new species might be produced when existing ones migrated into areas with different environments.

In 1858 Wallace (who had also read Malthus) hit upon the idea of natural selection and wrote up an account, which he sent to Darwin, who was widely known to be interested in the species question. There are significant differences between Darwin's and Wallace's formulations of the theory. Wallace had no interest in animal breeding and did not model his proposed mechanism on the process of artificial selection. It seems that his original conception was of natural selection operating upon varieties or subspecies, not upon individual variations within the same population. But the two ideas were similar enough for Darwin to feel very apprehensive when Wallace's paper arrived. After all, he had been working on the theory for twenty years, and did not want to see himself scooped by an outsider. On the advice of Lyell and Hooker he arranged for an account of his own theory to be published alongside Wallace's paper by the Linnean Society of London. At the same time he began writing the book-length account of his theory that was published as *On the Origin of Species* at the end of 1859. The age of evolution was about to dawn.

The Age of Evolution

The *Origin of Species* was not the first book to suggest a natural explanation of how life has developed, but its publication in 1859 precipitated the debate that converted most people to evolutionism. Darwin's theory was controversial because it challenged the conventional view that the Creator designed each species, and implied that the human race was just another animal. But the metaphor of a 'war' between Victorian science and religion has now been rejected by most historians. Evolution was accepted because it could be accommodated into a world view that implied that the development of Nature has a moral purpose. Human life was not necessarily robbed of its meaning by being incorporated into the natural system. Liberal thinkers welcomed the view that evolution came about through the accumulation of animals' efforts to conquer their environment, since this implied that the social progress resulting from commercial activity was a direct continuation of Nature's development. Even conservatives could accept evolution, provided it was seen as the unfolding of a divine plan through forces built into Nature by its Creator. Far from being perceived as the harbinger of an amoral 'social Darwinism' based on the worship of brute force, Darwin's theory became a foundation stone of the nineteenth century's faith in the inevitability of progress.

The assertion that Darwinism was a characteristic product of the time must be qualified in several important respects, however. *Evolutionism* became popular, and Darwin's name was frequently used as a symbol for the whole evolutionary movement. But nineteenth-century 'Darwinism' was not an expression of those aspects of Darwin's thought which appeal to modern scientists. In the twentieth century, natural selection has emerged as the dominant theory by which biologists explain the evolutionary mechanism. But in Darwin's own time, natural selection remained controversial and was supplemented by a range of alternative theories including Lamarck's 'inheritance of acquired characteristics' and an appeal to saltations or sudden evolutionary leaps. Even the emphasis on the interaction between a population and its environment, so typical of Darwin's approach to the question, was ignored by many late-nineteenth-century evolutionists.

Darwin stimulated the study of biogeography and other areas dealing with how organisms adapt to their environment. But these studies had to be ranged alongside a number of other disciplines that offered a very different perspective on the development of life. Paleontologists reconstructed the history of life from the fossil record, while morphologists studied the variety of animal forms in order to work out their evolutionary relationships. Neither found Darwin's detailed theory of adaptive evolution very useful. Those scientists who worked in museums and dissecting rooms could not exploit insights gained from field studies in different natural environments. The search for 'relationships' between species could be reconstituted along evolutionary lines, with the links being seen as natural genealogies rather than abstract patterns in the mind of the Creator. But those scientists who sought to reconstruct these links could do so without worrying very much about how populations of individual animals adapted to the challenges posed by an ever-changing environment.

Darwinism did not constitute a unified research programme, and by the end of the century the fragile unity imposed upon several diverse fields by the appeal to Darwin's name as a figurehead had broken down. New initiatives in the life sciences led to the establishment of disciplines such as genetics and ecology. But these new sciences were not the products of developments within either the morphological or the environmentalist study of evolution. They were reactions against what was now regarded as an old-fashioned evolutionism. A wave of enthusiasm for the experimental approach to biology meant that the field studies of the biogeographers and the descriptive genealogies of the morphologists and paleontologists were dismissed as unscientific. The experimental study of heredity led to the emergence of genetics, while ecology was presented as an extension of physiology into the study of the organism's reaction to its environment. Disciplines that we now accept as having thrown much light on the evolutionary process were thus introduced originally as alternatives to nineteenth-century 'Darwinism'.

Outside science, evolutionism replaced natural theology as the chief foundation for the attempt to understand the meaning and nature of the material universe. This was achieved by linking Darwinism to the idea of progress. Twentieth-century evolutionists such as George Gaylord Simpson have argued that Darwin's theory of adaptive evolution is incompatible with a belief in the necessity of progress, but in the late nineteenth century a philosophy of universal change could only be accepted if the changes were assumed to have a purposeful goal. The idea of progress was crucially important in an age when the west's growing technological power was allowing it to conquer and colonize most of the world. The sweeping aside of the 'primitive' inhabitants of these conquered territories could be justified by arguing that the process was necessary for progress. Darwinism was characteristic of an age of exploitation, and the idea of progress through struggle caught everyone's attention. Industrialists exploited their workers, and western nations exploited the rest of the world – but those who succeeded in the struggle were only too willing to see their success as the driving force of progress. Natural selection was not the only scientific manifestation of this spirit, but Darwinism supplied useful catch-phrases such as 'the survival of the fittest', which could be used by those seeking a scientific justification for their indifference to the losers in Nature's great race.

Darwin's theory emphasized the interaction between organisms and their environment and made it clear that species could be driven to extinction when the conditions to which they were exposed changed too rapidly for them to cope. But far from alerting people to the danger of environmental degradation, the theory was used to justify the kind of progress that increased the level of exploitation. A small number of late-nineteenth-century thinkers did begin to warn against the dangers of pollution and environmental exhaustion, but they were as yet in a minority. The majority was still not ready to abandon the belief that humankind was intended to dominate the earth. If Nature had replaced God as the power that had created the human race, the message derived from the new world view was still the same as far as most people were concerned.

Caution is also necessary when evaluating the emergence of ecology at the turn of the century. In modern times we have come to associate the very word 'ecology' with a concern for the environment – yet 'ecology' is properly the name of the science that deals with the ways in which living things interact with one another and with their environment. Like evolution theory, this science had complex relationships with the society within which it emerged, and there was no immediate link with the environmentalist movement. Some biologists saw ecology as a science that would help to support the exploitation of the environment by showing how to minimize the damage caused.

The preservation movement itself had mixed origins; the protection of endangered species in reservations or natural parks was urged by writers who wished to preserve stocks for hunting as well as by those who wished to stop the killing altogether. Like our own time, the late nineteenth century saw a constant tension between the desire to exploit and the desire to conserve, and we should beware of identifying either of these attitudes with any particular development in the scientific study of life and its environment.

Science and Empire

The idea of progressive evolution formed a natural foundation for the belief that European civilization was the high point of human achievement, destined to spread its values around the world by conversion or conquest. The last decades of the century saw the emergence of a self-conscious imperialism, in which the old system of colonial exploitation by commercial interests was replaced by an ideology of military conquest and strong control from the mother nation. The East India Company had extended British control over the Indian subcontinent in a haphazard fashion as opportunity dictated. This commercial imperialism was reflected in the liberal social evolutionism of the philosopher Herbert Spencer (1820–1903), which assumed that *laissez-faire* individualism would spread around the world by simply demonstrating its efficiency as an economic system.

The proclamation of Queen Victoria as Empress of India in 1877 heralded the emergence of the new imperialism in which European governments set out to conquer and control territory by exploiting the latest technology. The telegraph now allowed instantaneous communication with the farthest reaches of the empire, while the steamship brought everywhere in the world within a few weeks' sail of home. The conquest of tropical diseases opened up new areas to European penetration. The scramble to partition Africa into colonies – made possible once the wide availability of quinine solved the problem of fever – typified this more militaristic phase of imperialism. It also enhanced the national rivalries that would eventually touch off the First World War.

Whether the expansionist interests were commercial or military, however, science was seen as an indispensable tool for world domination. The success of the animal and plant breeders (which Darwin exploited in his analogy between artificial and natural selection) confirmed the human race's ability to manipulate organic Nature for its own ends. The transportation of useful plants from one part of the world to another had begun in the eighteenth century, and we saw in the previous chapter how Kew Gardens became the hub of the British empire's efforts to replace indigenous species with imported ones of greater commercial value. This activity continued into the later decades of the nineteenth century under successive directors: J. D. Hooker (one of Darwin's chief sources of information on the geographical distribution of plants) and his son-in-law William Thiselton-Dyer (1843–1928). Botanical stations were established throughout the empire to study native plants and identify those of potential value. Similar stations were established by other colonizing powers, and, along with colonial geological surveys, provided an important source of professional employment for the growing number of trained European scientists. The British also established an Imperial Bureau of Entomology to look into ways of dealing with insect pests in their various colonies.

The system of deliberate transplantation centred on Kew resulted in major changes in the crops grown in many countries. The availability of cheap quinine, which opened up many tropical areas to European penetration, came about when Clements Markham successfully transplanted the cinchona plant from South America to India via Kew in 1859–62. The Indian plantations that were soon established produced a vast increase in the supply of quinine, replacing the declining supply from naturally grown South American cinchona bark. Rubber trees were smuggled out of Brazil despite a government ban on export and used to create new sources of supply in Ceylon and the Malay peninsula. Tobacco, cacao plants (for chocolate) and tea were other plants that were established as cash crops in new parts of the world as a result of this programme. Botanists also influenced people's lives in other ways. The Canadian naturalist John Macoun explored the prairies in the 1850s and described them as ideal agricultural country. But Macoun was there during an unusually wet season, and when the first white settlers arrived they faced immense hardship until they developed techniques for coping with this semi-arid environment.

There was less interest in transferring exotic animal species around the world. Europeans tended to export their own familiar animals, sometimes deliberately as in the case of cattle and horses, but often accidentally and with disastrous consequences. The introduction of the rabbit into Australia offered a classic illustration of how a species could take over a new environment in which there were no natural predators. But samples of the new species discovered in colonized territories were routinely brought back to be described, classified and perhaps displayed in the great European museums and zoological gardens. These institutions played a role as imperial symbols, allowing Europeans to visualize the extent of their world-wide conquests through the display of exotic species.

Considerable excitement was generated when spectacular new species were brought home from newly explored parts of the world, as when Paul du Chaillu displayed stuffed gorillas shot during his explorations of central Africa (1855–9). The anatomist Richard Owen, now director of the Natural History Museum in London, lectured on the gorilla to large audiences. He soon came into conflict with Thomas Henry Huxley (1825–95) – known as 'Darwin's bulldog' – over the question of humankind's relationship to the great apes. Du Chaillu's discoveries did much to focus public attention on the implications of Darwin's theory, and may have helped to generate alarm over the prospect that the human race could be related to a creature whose ferocious nature had been deliberately exaggerated by its discoverer.

Museums also helped to focus attention on the time dimension in natural history by displaying spectacular collections of fossils. Once such displays were interpreted in terms of evolution, the general public could hardly fail to accept the message of the new science. The late nineteenth century saw a major expansion of paleontology as new areas were opened up for exploration. The new discoveries not only provided more detailed evidence for the development of life on earth, but also highlighted the exotic nature of the earth's earlier inhabitants. Great dinosaurs were excavated from the American west while it was still 'wild' – early fossil-hunters had to contend with hostile Indians and sometimes came to blows over possession of the richest sites. When displayed in the great cities of the east, and in the capitals of Europe, the larger dinosaurs seemed to confirm the superiority of modern industrial society by showing how the world had been conquered both in space and time.

As symbols of dominance, these displays demanded a spectacular setting, and soon every great city of the industrialized world possessed a museum whose cathedral-like setting helped to confirm the role that science had usurped as the source of moral authority in the modern world. The Natural History Museum in London moved into its present building in the 1880s, and a statue of Owen still greets visitors as they enter its main hall containing the great dinosaur skeletons. Many of the American museums were founded by capitalists seeking to proclaim both their own success and the power of industrial society. If the London museum was a product of imperial self-confidence, the American Museum of Natural History in New York celebrated the growing influence of the captains of industry in a similarly grandiose fashion.

The Professionalization of Biology

The modern scientific profession was very much a product of the late nineteenth century. Now at last governments and industries were persuaded (reluctantly, in some cases) that investment in science was a vital part of a nation's economy. France and Germany already had museums and universities with a substantial research function. Men such as Huxley and Hooker now played a significant role in the creation of the British scientific establishment by building up a system of government support for research and education. The process did not always run smoothly. As late as 1871 Hooker's position at Kew was threatened by an unsympathetic government official who wished to dismantle the garden's scientific function, and the decision was reversed only when pressure was applied by Huxley, Darwin and other eminent scientists. On the other hand, Huxley at first opposed the plans for a new Natural History Museum because of his personal hostility to Owen.

On the whole, however, this period saw a steady expansion in support for science, and the emergence of the modern system of scientific education in the English-speaking world. The concept of a unified science of 'biology' was part of Huxley's campaign to convince the government of the need for more support for modern science. Along with a number of younger protégés, Huxley established a course for high-school teachers at the Royal School of Mines (later the Normal School of Science, now incorporated into the Imperial College of Science and Technology). The older universities of Oxford and Cambridge were also persuaded to modernize their science teaching in the 1870s.

The last decades of the century also saw a vast development in America's capacity for science education. Laboratory studies in zoology had been established at the Museum of Comparative Zoology at Harvard under Louis Agassiz in the mid century. Now scientific education and research were to become a major feature of America's drive to supplant Europe as the centre of western culture. In 1876 Johns Hopkins University created a graduate school in science based on the model pioneered decades before in Germany. Other private universities including Harvard and the University of Chicago created a framework for elite scientific research freed from the immediate demands of industry. At the same time many states established land grant colleges providing instruction in the practical sciences and promoting research in areas of economic value. Agricultural institutions also began to play a major role in applying biological theory to the problems of agriculture.

But what exactly was the new science of biology that Huxley and others wanted to establish as the basis for a modern education? In order to evade the rather amateurish image associated with the kind of fieldwork done by Darwin, the new generation of scientists needed to stress the role of laboratory work and sophisticated instrumentation. Physiology was the model of an experimental science, providing a source of materialist theories coupled with the prestige of a high-technology, interventionist study of life. Evolutionism provided a parallel source of materialist concepts, but needed to be turned into a laboratory-based subject if it was to fit the new model.

Huxley's answer was to stress the role of evolutionary morphology, the detailed study of organic structures aimed at the reconstruction of the links between the major living groups. Morphology was a descriptive rather than an experimental science, but it was conducted in the laboratory rather than the field and could thus be presented as a symbol of modernization. Under its banner, biologists such as E. Ray Lankester (1847–1929) and Francis Balfour (1851–82) reformed the teaching of science at London, Oxford and Cambridge. Although only an updated version of a traditional discipline, evolutionary morphology became one of the great success stories of the late nineteenth century, providing a framework for visualizing evolution that seems quite at odds with our modern approach to Darwinism.

The problem was that evolutionary morphology ignored the central insights upon which Darwin's theory had been based. The morphologist scorned the field naturalist's study of geographical distribution, migration and adaptation – studies that were the very foundation upon which the theory of natural selection had been built. Field studies were also important for early efforts to monitor the environment, an area with which governments were increasingly concerned. There was thus a source of tension built into the very heart of the new biology, a tension that was never resolved and would ultimately divide the life sciences into a chaos of competing disciplines.

This tension can be seen in the growing interest in marine science. There were many voyages designed to survey the ocean depths in the nineteenth century, of which the most famous was that of HMS *Challenger* in 1872–6 (see chapter 6). Although the reports of this and the other expeditions provided new information on the distribution of life in the sea, much of the space was devoted to the minute description of the newly discovered species, i.e. to pure morphology. There was considerable interest in the creation of marine zoological laboratories, the first of which was established at Naples in 1872 by the German zoologist Anton Dohrn (1840–1909). E. Ray Lankester played a prominent role in the establishment of a similar laboratory at Plymouth in 1888, and in the same year the Marine Biological Laboratory was formed at Woods Hole in the United States.

Some of these marine stations were little more than dissecting rooms by the sea, but there was increased pressure from governments for scientists to study practical problems such as the effects of over-fishing. Thus while the new generation of 'modern' biologists was stressing the importance of pure morphology, governments were becoming interested in field studies because these offered the only way of monitoring the increasingly obvious changes in natural populations. The US government had established a Fish Commission under Spencer F. Baird (1823–87) as early as 1871 to enquire into the depletion of fish stocks. A decade later Baird had the sea-going vessel *Albatross* especially equipped for survey work, and he played an important role in the creation of the Woods Hole laboratory later in the decade, although this was established independently of the government.

On land, many areas of field natural history could draw upon the enthusiasm of bird-watchers and other amateurs prepared to spend significant amounts of time and money on the study of Nature. The professor of zoology at Cambridge was Alfred Newton (1829–1907), an ornithologist committed to fieldwork. Although Newton did not hinder the setting up of a programme in animal morphology under Balfour, he was suspicious of the laboratory-based discipline and played an important role in galvanizing the country's amateur bird-watchers to form a network that could provide information of real scientific value. Local observers were encouraged to report on distribution and migration patterns, thus paving the way for an interest in environmental concerns. So much information was gathered that the professionals could hardly find time to analyse it. By the early twentieth century German ornithologists took over the leading role, pioneering the use of ringing as a technique for tracing bird movements.

American ornithologists had more practical concerns that soon attracted government funding. They were worried about the spread of imported species such as the English sparrow at the expense of native birds. The American Ornithologists' Union became active in 1883, generating so much information that the government established a Division of Economic Ornithology and Mammology in 1886 under C. Hart Merriam (1855–1942). In 1896 this became the Division of the Biological Survey, which provided extensive information on the distribution of various native species of birds and animals.

Early Conservationism

The harmful effects of clearing forests had already become apparent in Europe in the early part of the century. Studies in France and other countries had revealed the extent to which forests were being cut down and showed that deforestation had a range of harmful consequences. Massive soil erosion and disruption of the natural drainage pattern destroyed the land's productivity. Drawing upon this European research and his own observations in the United States, the American diplomat George Perkins Marsh (1801–82) wrote his *Man and Nature* of 1864 to warn of the dangers. Marsh had little interest in geological changes or in evolution. He presented the earth as an essentially static, harmonious system that was being torn apart by human greed. Where an earlier generation of natural theologians had assumed that Nature would heal itself, Marsh emphasized that some acts of destruction exceeded the earth's recuperative powers[1]:

> The ravages committed by man subvert the relations and destroy the balance which nature had established between her organized and her inorganic creations; and she avenges herself upon the intruder, by letting loose upon her defaced provinces destructive energies hitherto kept in check by organic forces destined to be his best auxiliaries, but which he has unwisely dispersed and driven from the field of action. When the forest is gone, the great reservoir of moisture stored up in its vegetable mould is evaporated, and returns only in deluges of rain to wash away the parched dust into which that mould has been converted.

He continued:

> The earth is fast becoming an unfit home for its
> noblest inhabitant, and another era of equal human
> crime and human improvidence ... would reduce it
> to such a condition of impoverished productiveness, of
> shattered surface, of climatic excess, as to threaten the
> depravation, barbarism, and perhaps even extinction
> of the species.

Strong and prescient words, yet Marsh was not calling for
a halt to all human interference. He believed that sensible
policies for extracting timber would allow a balance to
be maintained, permitting humankind to harvest a permanently renewable resource. Scientific management, not
rejection of all exploitative activity, was the answer.

Marsh's book had considerable influence and played a
role in stimulating the US government to set up a Forestry Commission to manage the nation's resources. This
had little practical effect, however, until the 1890s, when
public pressure forced the government to set aside areas
of forest that could not be sold off to logging companies.
An 1896 commission by the Department of the Interior
joined the National Academy of Sciences in calling for a
doubling of these reserves, a proposal vigorously resisted
by the timber industry. During his term of office (1901–9)
President Theodore Roosevelt stressed the need for scientific
management of the nation's resources and created an active
Forest Service under Gifford Pinchot. Later presidents did
not share the same enthusiasm, however, and the slacken-
ing of government control led to increased demands from
the public that forests be protected. Roosevelt was an
outdoorsman who wanted to preserve the wilderness – but
his motivation was to maintain the stock of bear, deer and
other animals sought by hunters. Legend has it that the
first 'teddy bear' was created for him after an unsuccessful
hunting trip.

Public concern was fuelled by aesthetic rather than

economic considerations. In the 1850s Henry Thoreau expressed a new sense of delight in the experience of Nature in its original state. There was a widespread feeling that areas of outstanding natural beauty should be protected from the greed of miners and loggers and thereby preserved for the benefit of future generations. The exploration of the American west revealed many such areas, and efforts to preserve them had begun at an early stage. In 1864 the Yosemite valley was ceded to the State of California as a public park, while Yellowstone in Wyoming was established as a national park in 1872. Frederick Law Olmsted (1822–1903) played an important role in saving Yosemite, emphasizing that private selfishness would lead to destruction unless the land was controlled by the state. During the later decades of the century, scientists were active in organizations such as the Sierra Club, founded by John Muir in 1892 to help preserve Yosemite and other natural wonders in California. Yellowstone was threatened in 1883–4 by a plan to allow mining, but the park was saved after an intense debate.

Olmsted was also active in stressing the need for open spaces in the overcrowded cities, and had been instrumental in setting aside New York's Central Park in 1853. In Britain, the biologist and reformer Patrick Geddes (1854–1932) played a similar role, calling for conservation of the country-side and a better environment in the cities. From the 1860s onwards the Commons, Open Spaces and Footpaths Preservation Society campaigned to save Britain's rural heritage so that city-dwellers could be revitalized through contact with Nature (a campaign that only became realistic in the era of rapid transportation by rail).

The establishment of reserves to protect wildlife was accepted by hunters when it became clear that some species would soon be wiped out altogether. The near extinction of the North American buffalo or bison offered a chilling example of the destructive power of modern firearms. The tension between the desire to offer absolute protection, and the desire to preserve stocks for continued exploitation, was apparent in many countries throughout the late nineteenth century. America was not the only country where white men confronted a land that had originally teemed with big game. Throughout the empires established by Britain and other European countries, adventurers and colonial administrators hunted both for profit and to symbolize the power conferred by their modern technology. Famous hunters such as F. C. Selous returned home to lecture on the value of big-game hunting as a means of training the next generation of empire-builders. The novels of H. Rider Haggard offer a clear illustration of the white man's behaviour during the penetration into Africa[2]. During the middle decades of the century, vast numbers of animals were slaughtered in both India and Africa. It was often assumed that the 'natives' were unable to protect themselves from dangerous animals, thus justifying the intervention of the white man with his rifle.

By the later decades of the century it was becoming obvious that the numbers of many species were being rapidly diminished, and it was now regarded as 'unsporting' to amass the vast collections of trophies favoured by earlier hunters. Colonial administrators began to restrict the natives' access to species that could be killed for food by establishing reserves where hunting was only permitted under licence. These efforts began in India during the 1870s, while the hunting of some species was prohibited in the Cape Colony (modern South Africa) in 1886. The Society for the Preservation of the Fauna of the Empire was founded in 1903 to co-ordinate efforts to save the big-game species in British-held territory through the promotion of reserves. Here, as in America, it was the hunters themselves who realized the need to protect stocks by establishing reserves. White settlers often objected to the reserves because they were thereby denied access to farming land, and in some areas the size of the reserves was scaled down to meet their demands.

Those who opposed hunting on moral grounds were more effective in reducing the scale upon which exotic birds were slaughtered for their feathers. Evangelical opponents of cruelty to animals campaigned against the trade in feathers, along with other forms of cruelty closer to home including cock-fighting and scientific vivisection experiments. In 1885 a Plumage League was founded in Britain to protest against the trade, while the Audubon Society played a similar role in the United States. Women pledged not to use feathers for decoration, and when the Society for the Protection of Birds was first founded in Britain, men were excluded – just as women had been excluded from the ranks of many scientific societies. At home, the SPB (it became the Royal Society for the Protection of Birds only in 1904) began to set up Nature reserves during the 1880s, as the decline in some native species became apparent. Their activities thus linked up with more general efforts to protect the rural environment.

The state of the earth's physical resources was also giving cause for concern. Marsh's *Man and Nature* warned that deforestation upset the natural distribution of water and led to soil erosion. Yet opponents of the conservation movement claimed that the arid lands of the American west would attract a higher rainfall if they were ploughed and would thus be turned into valuable farmland. The geologist J. W. Powell argued against this view and sought to control all the west's water in the name of scientific management. Water must be under public control, he declared, since private exploitation would ruin these fragile lands for all time. Powell began an irrigation survey in 1888 under the auspices of the US Geological Survey and was for a while successful in convincing the government that state control was necessary. Inevitably he came into conflict with the developers, who lobbied successfully to destroy his project. Concepts of scientific management were discredited until the Dust-Bowl of the 1930s once again focused public attention onto the problem of unrestricted exploitation of land and water.

Some economists were also becoming worried about the rapid expansion in the use of the nineteenth-century's greatest non-renewable resource: coal. Geologists had been active in surveying the land to reveal new mineral deposits, but had occasionally warned that the supply was not infinite and would eventually become exhausted. In Britain, W. S. Jevons drew attention to this problem in his *The Coal Question* of 1865. Drawing upon the conclusions of the geologists, he pointed out[3] that *'there is no reasonable prospect of any relief from a future want of the main agent of industry. We must lose that which constitutes our peculiar energy.'* Jevons used exaggerated estimates of the rate at which consumption would increase in the future, and Royal Commissions in 1866 and 1901 dismissed his arguments as alarmist. The supply of coal seemed so vast that no one was willing to concede the possibility of exhaustion in the near future, and in the twentieth century oil began to offer a new source of energy that the scientists and engineers of Jevons' time had not anticipated. As yet, no one was prepared to worry about what might happen in the *distant* future. For all the warnings of the conservationists, the late nineteenth and early twentieth centuries were a period of unparalleled expansion in the consumption of resources.

THE DARWINIAN REVOLUTION

One of the main pillars of support for this optimistic progressionism was the theory of evolution. At first resisted by conservatives who feared its materialistic implications, evolutionism eventually became a central theme of late-nineteenth-century thought, the foundation for a unified philosophy of biological and social progress. The conversion of public opinion on this question was triggered off by the scientists' response to a major theoretical initiative in the field: Darwin's *Origin of Species*, published in 1859. Darwin's theory played a major catalytic role in converting the world to evolutionism, but historians have begun to realize that our conventional image of the 'Darwinian revolution' is an oversimplification inspired by modern biologists' enthusiasm for the concept of natural selection. It has become apparent that the success of evolution theory in the late nineteenth century was not based on acceptance of what we now regard as the most important components of Darwin's thinking.

Interpretations of Darwinism

Darwin's investigation of biogeography on the voyage of the *Beagle* led him to adopt an evolutionary view of how species adapt to changes in their environment (chapter 7). The *Origin of Species* was published in 1859, precipitating a debate that led most scientists to adopt an evolutionary perspective within a decade or so. The theory of natural selection conveys an image of Nature in which complex and unpredictable interactions determine the life and death of both individuals and species. Darwin stressed that pressure on one species might have unexpected consequences for others through a complex chain of interactions. From this it might have been possible to develop a philosophy in which humankind had a responsibility towards the other species with which we share the earth. But this would be a very modern reading of Darwin, a reading that extracts a message that has become popular only in the age of environmental awareness. In his own time, the ecological dimension in Darwin's thought was largely subordinated to the confident progressionism that characterized the Victorian era.

At one time it was fashionable to argue that Darwinism destroyed the traditional belief that Nature had been created by a wise and benevolent God, thereby ushering in an age of rampant materialism. The universe was reduced to a system of chaos and confusion in which all things must struggle to survive. Humankind became a product of material Nature, but had no responsibility towards the rest of the world because the process that had created humans allowed no room for compassion and encouraged everyone to seek his or her own self-interest. The age of 'social Darwinism' left no room for a sense of stewardship and encouraged no sympathy for the animal and plant species that might be threatened by humankind's dominant position.

Historians now suspect that this harsh image of Darwinism is a misinterpretation arising from the application of modern values to a past situation. Darwinism was certainly incorporated into the philosophy of progress through struggle, but the Victorians did not turn to that philosophy because they had abandoned all hope that the universe has a moral purpose. On the contrary, progress was essential to their optimistic vision of history because it seemed to guarantee that the world was moving in a purposeful direction towards a morally significant goal. The moral values of the Protestant religion that had fuelled Bacon's belief that science would enable humankind to dominate Nature were reformulated, not abandoned. It could now be argued that, throughout the history of the earth, Nature had rewarded those who were able and energetic, and punished those who could not keep up with the race towards higher things. Human progress through the conquest of the environment was merely a continuation of the evolutionary process that had encouraged the traditional virtues of industry and enterprise throughout the ascent towards humankind.

While recognizing the extent to which Darwinism was absorbed into the philosophy of progress, historians have become more aware of the variety of late-nineteenth-century reactions to evolution theory. Darwin converted the whole world to evolutionism – but this was possible only because the general idea of evolution could be exploited in so many different ways. There were many conflicting interests within nineteenth-century society, and each developed its own interpretation of evolutionism. To confuse matters for the historian, the exponents of widely differing interpretations called themselves 'Darwinians' because they acknowledged Darwin's lead even if they did not accept all the details of his theory. Less well known are the various groups of explicitly anti-Darwinian evolutionists who developed alternative theories in the later part of the century. There was no monolithic Victorian evolutionism, only a confusing mixture of different and sometimes hostile interpretations, which changed significantly as the century progressed towards its close.

The complexity of what is often misleadingly called the 'Darwinian revolution' can be judged by surveying the mixed fortunes of the theory of natural selection. Modern biologists accept that the synthesis of natural selection with Mendelian genetics has created a powerful explanation of the evolutionary process. Historians who have looked at the impact of the *Origin of Species* have thus been encouraged to assume that Darwin's introduction of the selection mechanism represented the decisive factor in the conversion of the scientific community to evolutionism. Darwin succeeded where earlier evolutionists had failed because he had the right solution to the problem of explaining how populations actually change in response to environmental pressure. The success of 'Darwinism' shows that the majority of biologists recognized this point, and the only significant area of debate remaining centred on the problem of heredity. Once Darwin's rather primitive views on this topic were replaced by Mendelian genetics in the early twentieth century, the last missing piece in the jigsaw puzzle had been inserted and the complete picture became visible to all.

Recent research has shown that Darwin converted the world to evolutionism despite the fact that his theory of natural selection was largely rejected as an adequate explanation of the evolutionary process. Even supporters such as Huxley treated it as only a secondary mechanism and looked for alternatives to supplement it. Opponents rejected natural selection outright and revived pre-Darwinian theories such as Lamarckism and saltationism as the basis for their claim that evolution could not be the totally haphazard process depicted by Darwin's theory. Many anti-Darwinians denied the significance of adaptation, preferring to believe that evolution was driven by forces arising from within the organism, independent of any changes in the environment. The late nineteenth century saw an 'eclipse of Darwinism' in which the supporters of natural selection were marginalized even within scientific biology. Recognition of these facts forces us to ask a new set of questions about the impact of Darwinism. We need to know why the *Origin of Species* had so much influence despite the relative failure of its chief explanatory tool. We also need to know why some biologists who were suspicious of natural selection called themselves 'Darwinians' – and why others were so determined to develop alternative theories.

The transition to evolutionism must be seen as a social event occurring within the scientific community. Darwin introduced new lines of evidence, but these were not enough to convince everyone of the superiority of his particular explanation, and the general idea of evolution succeeded because it could be exploited by powerful interest groups acting within science. Because of this social dimension, Darwinism was not a monolithic theory. The practitioners of different disciplines within the life sciences exploited the general idea of evolution in different ways, some of which bore little resemblance to Darwin's own approach. To the extent that the evolutionists from these different camps co-operated, they did so for social reasons rather than because there were real interactions between their areas of research.

More striking still is the variety of responses to evolutionism in different countries. In Britain, the 'Darwinians' led by Huxley succeeded in dominating the scientific community of the late 1860s and 1870s, although opposition grew in later decades. In America, an explicitly anti-Darwinian school of neo-Lamarckism emerged at a much earlier date, despite the initial success of Darwin's supporters. In Germany, Darwinism was absorbed into the developmental philosophy of Ernst Haeckel (1834–1919), who was strongly influenced by the analogy between evolution and the purposeful development of the embryo towards maturity. French biologists ignored Darwin and only turned to a Lamarkian version of evolutionism in the 1880s. These widely differing national responses suggest that we should be wary of any attempt to explain the popularity of evolutionism in terms of the self-evident superiority of Darwin's explanatory programme.

The Darwinians

The biogeographers were in the best position to appreciate the value of Darwin's new evidence for adaptive evolution, and thus to adopt a genuine 'Darwinism'. J. D. Hooker was one of the first scientists to be informed of Darwin's new theory back in the 1840s, and he now came out in public support of the *Origin of Species*. The Harvard botanist Asa Gray (1810–88), who studied the distribution of American plants, led the defence of the theory in the United States. A. R. Wallace – whose 1858 paper had prompted Darwin to publish – also made notable contributions to biogeography within an evolutionary framework. He also studied the problem of how a single original population becomes divided up into varieties (subspecies) and eventually a range of distinct 'daughter' species. By the very nature of their discipline, biogeographers could see that evolution must be an unpredictable process depending on the hazards of migration and the possibilities of dispersed populations adapting to areas with unfamiliar environments.

Other supporters had less interest in biogeography and adaptation. Thomas Henry Huxley was a young morphologist and paleontologist who was anxious to challenge the traditional role played by natural theology and establish the life sciences as independent disciplines. Huxley had been suspicious of evolutionism because he could see no plausible explanation of *how* species might change, and he welcomed Darwin's theory with open arms because it offered a new hypothesis that confirmed the scientist's right to treat the origin of species as a problem susceptible to natural explanation. Histories of the Darwinian revolution often cite his confrontation with Bishop Samuel Wilberforce at the 1860 meeting of the British Association for the Advancement of Science as a classic illustration of how evolutionary science overcame religious bigotry. Huxley's *Man's Place in Nature* of 1863 also confirmed the anatomical similarities between humans and great apes, undermining Owen's efforts to establish an unbridgeable gulf between them.

Recent studies show that Huxley's support for Darwin was by no means straightforward. Contemporary accounts do not support the evolutionists' claim that Huxley demolished Wilberforce's arguments at the British Association meeting. More important is Huxley's own attitude to the theory of natural selection. Although he saw selection as a plausible hypothesis that confirmed the scientist's right to investigate the question, he had major reservations about the mechanism and preferred to believe that evolution occurred through saltations or leaps produced by some mechanism internal to the organism. As a morphologist, Huxley was interested only in establishing the relationships between the variety of animal forms; he had major reservations about the role of adaptation in evolution. His support for Darwin was not based on his ability to take up a research programme suggested by the theory proposed in the *Origin*. He was slow to use even the basic idea of evolution in his paleontological work, and did so only in the mid 1860s after reading the work of the German evolutionist Ernst Haeckel. Huxley joined the evolutionists because he could use their theory to make a general point about the need for a naturalistic methodology in the life sciences as an essential prelude to modernization.

Like most of the early Darwinians, Huxley was also a liberal in his politics, determined to show that society could enjoy steady progress if only traditional restrictions on individual freedom were dismantled. Here he was able to make common cause with the philosopher who provided the moral underpinnings for the evolutionary movement, Herbert Spencer[4]. Spencer insisted that free enterprise was the key to social progress, and used biological evolutionism to argue that Nature itself exhibited a universal tendency towards progress. It was important for this connection that the model of biological evolution used should be one that saw change as the cumulative product of the actions of individual organisms over many generations. Progress, for Spencer, was a kind of long-range but inevitable by-product of the constant struggle of organisms – including human beings – to respond to the challenge of the environment. Since the characters that helped the individual to succeed in the battle were the traditional Protestant virtues of enterprise, initiative and thrift, Spencer could project his philosophy as a new foundation for traditional morality, despite its tendency to encourage indifference to the suffering of those who still needed to be taught a lesson by a wise but harsh Nature.

Darwin himself had stressed the long-term progressionism of natural selection, and his was certainly a mechanism that highlighted the role of the individual. Spencer coined the term 'survival of the fittest' and has been seen by some historians as the first of the 'social Darwinists'. But he saw natural selection as less important than the Lamarckian mechanism in which self-improvements resulting from the individual's response to environmental challenge are inherited and thus accumulate to change the species. The weeding out of the congenitally unfit was only a secondary process, less important than the pressure encouraging most individuals to become fitter. Spencer's emphasis on the struggle against the environment as the driving force of progress thus helped to create the popular image of 'Darwinism'. What modern biologists see as the theory's most important contributions were subordinated to a contemporary philosophy encouraging struggle and exploitation. Hooker, Huxley and the other scientists supported Spencer's evolutionary liberalism and thus united around the figurehead of Darwin despite their quite different interests at the level of scientific research.

Anti-Darwinism

The Huxley – Wilberforce debate symbolized the opposition of some conservative thinkers to the new evolutionism. But most religious thinkers accommodated themselves fairly rapidly to the basic concept of evolution. Some could even accept Spencer's philosophy of progress through struggle, because it seemed to offer an updated version of the Protestant work ethic in which thrift and industry were rewarded in this world as well as the next. In general, though, conservative thinkers chose to accept a non-Darwinian version of evolutionism. They sought alternative explanations of how the process might work, which allowed them to believe that Nature was driven by something higher than mere selfishness. Anti-Darwinian versions of evolutionism flourished in the later decades of the century because they upheld the view that natural development is a more orderly and more purposeful process than natural selection would allow.

Even the inheritance of acquired characters was taken up as the centrepiece for a self-consciously anti-Darwinian movement known as 'neo-Lamarckism'. In the 1870s some of the American disciples of the idealist zoologist Louis Agassiz accepted evolutionism, but rejected natural selection as the expression of a materialist view of life. Paleontologists such as Edward Drinker Cope (1840–97) and Alpheus Hyatt (1838–1902) argued that evolution was a more purposeful process than Darwin imagined and suggested that Lamarckism would allow living things to play an active role in determining their species' evolution. If variation were random, as Darwin supposed, then living things had no control over their own destiny – each individual's fate was sealed by the characters it inherited. But if the characters acquired as a result of purposeful changes in behaviour could be inherited, then animals could transcend their inheritance and create new avenues of evolution. When the first giraffes began to stretch their necks to reach the leaves of trees, they were choosing a new habit that would guide the future development of their species over a vast period of time.

Cope argued for a new theology in which the Creator has delegated His power to living things by giving them the ability to shape the future through their purposeful response to the environment. Lamarckism drew much of its inspiration from a moral distaste for the 'trial-and-error' process of natural selection. The opponents of Darwinism saw the implications of the selection mechanism more clearly than the 'Darwinians' who linked the theory to Spencer's progressionism. The same moral revulsion was expressed by the novelist Samuel Butler (1835–1902), who wrote a series of books and articles attacking Darwin, beginning with his *Evolution Old and New* of 1879. Of natural selection Butler later wrote[5]: 'To state this doctrine is to arouse instinctive loathing; it is my fortunate task to maintain that such a nightmare of waste and death is as baseless as it is repulsive.' Lamarckism was the only morally acceptable theory of evolution, and Darwin was all the more to blame because he had concealed the fact that better mechanisms of evolution than his own had been proposed long before the *Origin of Species* was published. When Butler first began his campaign he was ignored by the British Darwinians, but by the 1890s Lamarckism was being taken seriously even in Darwin's home country.

It has been suggested that the growth of environmentalism was stimulated by the anti-materialistic philosophy that found its expression in neo-Lamarckism. Darwinism was portrayed as a philosophy that reduced living things to automata struggling mindlessly against the physical environment. It exposed the moral dangers of the mechanistic philosophy that had begun with Bacon and had fuelled humankind's drive for dominance over Nature. Lamarckism was part of a rival world view, which stressed the ability of living things to transcend material limitations and which saw Nature as a harmonious whole rather than a scene of constant struggle. In a series of articles later collected in his book *Mutual Aid* of 1902, the Russian anarchist writer Peter Kropotkin argued that animals naturally co-operate with one another. The 'struggle for existence' was an illusion, and the driving force of evolution was Lamarckism.

The German evolutionist Ernst Haeckel developed a philosophy of 'monism' that did much to promote this view of Nature. He argued that matter and spirit were manifestations of a single underlying substance – which meant that even the most primitive form of life had some spiritual qualities. Monism thus generated a respect for life as a whole and helped to emphasize the unity of Nature. The advocates of such a holistic viewpoint would have been far more willing to accept the need for preserving the complex web of natural relationships, and would thus have gravitated towards the environmentalist movement. If this interpretation is correct, then evolutionism may have played a role in promoting a greater awareness of environmental fragility mainly through its Lamarckian rather than its Darwinian version.

Lamarckism still assumed that evolution was driven by the interaction between the organism and its environment, but some opponents of Darwinism insisted that the process must be controlled by forces arising from within the organism. These internal forces would produce new characters whatever the constraints imposed by the environment. The problem with Darwin's theory was that it reduced evolution to a chapter of accidents. Variation was entirely random, and each population evolved in response to purely local environmental pressures, which could be altered in an unpredictable way by migration or geological changes. The course of life's history was thus an open-ended, haphazard series of developments best represented by the image of an irregularly branching tree. Many naturalists were unwilling to see the living world as the product of such a haphazard process. They could only accept evolution if it were a process that gave rise to regular, predictable developments in accordance with some preordained plan that could be seen as originating in the mind of God. The solution was to argue that variation was not random, but was directed along a predetermined path by forces somehow built into the very nature of the species.

Such a view of evolution was often favoured by conservative thinkers who rejected the liberal philosophy of Spencer's cosmic evolutionism. If evolution were the unfolding of a divine plan, then there was no need to assume that human progress would come about through the summing up of individual acts of selfishness. This line of opposition to Darwin was apparent almost as soon as the *Origin of Species* was published. The astronomer Sir John Herschel (1792–1871) called natural selection 'the law of higgledy-piggledy' and in his *Physical Geography* of 1861 wrote[6]:

> An intelligence, guided by a purpose, must be continually in action to bias the direction of the steps of change – to regulate their amount – to limit their divergence – and to continue them in a definite course ... On the other hand, we do not mean to deny that such intelligence may act according to law (that is to say on a preconceived and definite plan).

In biology, Richard Owen – who had led the conservative assault on Lamarckism in the 1830s – now accepted evolution, but insisted that it worked by a preordained process of 'derivation'. In 1868 he wrote[7]:

> Derivation holds that every species changes in time, by virtue of inherent tendencies thereto. 'Natural selection' holds that no such change can take place without the influence of altered external circumstances. 'Derivation' sees among the effects of the

innate tendency to change irrespective of altered circumstances, a manifestation of creative power in the
variety and beauty of the results.

Although often dismissed as an opponent of evolutionism,
Owen can thus be seen as an opponent of *Darwinism* who
was quite happy to accept a theory of evolution based on
preordained trends.

St George Jackson Mivart's *Genesis of Species* of 1871
offered a cornucopia of anti-Darwinian arguments based
on the claim that evolution must be something more
than a haphazard process of adaptation. A preordained
trend could drive several forms in parallel towards the
same goal, thus making nonsense of Darwin's claim that
similar species must share a recent common ancestor. In
Britain the anti-Darwinians were unable to prevent the
takeover of the scientific community by Huxley and the
liberal Darwinians, but elsewhere the opposition was better
organized. Many of the American neo-Lamarckians argued
that evolution exhibited both adaptive and non-adaptive
trends, the latter being explained as the unfolding of trends
built into the very constitution of each group at its origin.
Alpheus Hyatt used an analogy with the life-cycle of the
individual organism to suggest that each group eventually
underwent degenerative evolution into a senile phase as the
prelude to extinction.

By the end of the century it was no longer fashionable
to explain such non-adaptive trends as the unfolding of
a divine plan, but the theory of 'orthogenesis' still postulated the existence of rigid trends built into the constitution of each group. These trends were supposed to be
so powerful that many groups were driven to extinction
as once-useful organs became overdeveloped to the extent
that they interfered with the animals' ability to cope with
the environment.

The popularity of theories of non-adaptive evolution
forces us to reassess the impact of Darwinism. Where

modern biologists see Darwin's theory as the expression of a new awareness of the fragility of natural relationships, his contemporaries extracted a very different message from it. Darwinism was associated with a philosophy of progress and domination over Nature, while some Lamarckians showed a greater awareness of natural interactions. But other versions of evolutionism encouraged the belief that living things are governed by forces that transcend the everyday world of the environment. Far from dying away as the century progressed, this anti-Darwinian and anti-adaptationist view of evolution became more powerful. By 1900 Darwin was acknowledged as the founder of evolutionism, but his theory of natural selection was accepted by only a minority of biologists.

THE TREE OF LIFE

Once the *Origin of Species* had converted the scientific world to a basic acceptance of evolutionism, biologists' attention was drawn to the problem of reconstructing the history of life on earth. Darwin himself had warned that this might be an impossible task, given the enormous gaps in the fossil record, which left little or no evidence for the exact course of many transitions. But many biologists took it for granted that the main purpose of evolutionism was to elucidate the precise course of life's development from its earliest origins. In the 1860s and 1870s, the evolutionists turned to morphology, the detailed study of animal structure, as a means of working out relationships that could not be traced in the fossil record. Only in the last decades of the century was morphology replaced by paleontology as the chief focus of attention in the reconstruction of what the German morphologist Ernst Haeckel called 'phylogeny' (evolutionary history). The great expansion of fossil discoveries now meant that paleontologists were increasingly confident in their ability to plot the course of each group's development. They were also able to incorporate a geographical dimension into the study of how life had developed.

Evolutionary Morphology

Darwin himself had done extensive morphological work in a detailed study of barnacles during the 1850s. He realized that the structure of each species – and especially the development of the embryo – would throw light on the relationship of one form to another. Classification had always been based on the recognition of structural resemblances, and the advent of evolutionism merely transposed this enterprise into a search for real (i.e. historical) rather than purely formal relationships. Darwin believed that very similar species, such as those grouped into a single genus, would have evolved from a fairly recent common ancestor. More distant relationships would correspond to a more distant common ancestry. Evolution was thus to be portrayed as an ever-branching tree, and a good classification system would be based on a correct identification of the crucial points at which the branchings took place. Darwin argued that the study of embryonic stages was important for working out these relationships because the embryo would be less affected than the adult form by the adaptive modifications upon which evolution was based. In some cases (as with the barnacles) forms whose adult structure seemed quite unrelated could be shown to have shared a common ancestry because their embryos were very similar at an early stage.

The evolutionary morphologists of the 1860s were determined to use comparative anatomy and embryology to work out all the crucial relationships needed to reconstruct the tree of life. Morphology was already a well established discipline, which required only a change of emphasis to become the basis for the reconstruction of phylogenetic (evolutionary) relationships. The German biologist Ernst Haeckel's *Generelle Morphologie* of 1866 pioneered the new approach. Haeckel also wrote more popular works, which were translated into English as *The History of Creation* and *The Evolution of Man*. Although counting himself an enthusiastic Darwinian, he favoured a Lamarckian view of the evolutionary mechanism and was strongly influenced by the idealist trend in pre-Darwinian biology. He saw the universe as a unified system whose development could be reconstructed by the evolutionist through the study of its modern products.

Haeckel's holistic philosophy would inspire a later generation of environmentalists, but he had been trained as a morphologist and sought other ways of displaying the unity of Nature. He believed that all living forms can be related into a single developmental sequence. The growth of the embryo towards maturity provided a model for the whole evolutionary process. Where Darwin saw resemblances between embryos as evidence of common descent, Haeckel assumed that 'lower' forms can be equated with earlier stages in the individual development of more advanced species. It was he who popularized the recapitulation theory, the claim that ontogeny (the growth of the embryo) recapitulates phylogeny (the evolutionary history of the species). Since the human race was the highest product of evolution, the development of the human embryo exhibited all of the major stages in the ascent of life. The lower animals still alive in the world today could be identified with particular stages in the overall development of life – a process that was replayed like a speeded-up movie film in the life of every human being.

This model promoted the assumption that evolution is based on a progressive trend with the human race as its goal. Darwin's 'tree' of evolutionary relationships had a radiating pattern of divergent branches so that no single branch could be said to represent the goal towards which all the others were moving. Each branch was developing its own way of coping with the environment, and progress in one direction could not be evaluated by comparison with any other. Haeckel reconstructed the tree of life to give it a main trunk or stem with the human race at the top: all developments in other directions were merely side branches of little real importance. Except for minor modifications after they branched off the main line, all living species could be treated as steps on the way towards the human form. Haeckel revived the old belief that the 'lower' species are merely immature versions of humankind. Instead of portraying evolution as an open-ended process driven by a multitude of local responses to environmental pressure, he focused on progress as the key to understanding the relationships between modern species. Evolution became a ladder rather than a tree, a linear sequence of stages through which life had advanced towards the human form.

The recapitulation theory was seductive because it seemed to offer a powerful guide to the reconstruction of those steps in the history of life hidden by gaps in the fossil record. There was a massive concentration of scientific effort in this area in the decades immediately following the general acceptance of evolutionism. Large numbers of scientists tackled some of the most basic problems, including the question of the origin of the vertebrates. It was obvious that the fish – the 'lowest' class within the vertebrate phylum – must have evolved from a pre-existing invertebrate form, yet this had occurred before fossils became plentiful in the rocks. Various hypotheses based on different kinds of anatomical evidence were offered to identify which invertebrate type had been the ancestor from which the 'main line' of vertebrate evolution had taken off.

In Britain the work of E. Ray Lankester illustrates the strength of evolutionary morphology in the 1870s and 1880s. Lankester made numerous contributions to the elucidation of evolutionary relationships and, like Haeckel, projected a model of evolution based on progress. In principle, Lankester realized that no living form can be seen as a stage in the ancestry of another. It was wrong, for instance, to treat the great apes as the ancestral form from which humankind had sprung, since the apes would have undergone changes of their own since their ancestors and ours had diverged from a single root. Yet Lankester often ignored this warning in his own work, and suggested that all forms of life can be ranked into grades defined by the point at which they branched off from the main line of progress towards humankind.

Morphology was a laboratory-based subject that did not encourage detailed study of how animals adapted to their local environment. The morphologists' interpretation of evolution invoked adaptation only in the most abstract way. Lankester suggested that the driving force of progress was the struggle of organisms to cope with environmental challenge, and warned that degeneration threatened any species that took up a less stimulating environment. This fitted in neatly with Spencer's philosophy of self-help as the driving force of evolution, and Lankester drew the obvious implication from his theory by warning that the human race might degenerate if it made life too easy for itself[8].

This approach was carried to its furthest extreme by the embryologist E.W. MacBride, who argued that all the different invertebrate types were degenerate offshoots from various points in the main line of vertebrate progress[9]: 'It is therefore broadly speaking true', he wrote, 'that the Invertebrates collectively represent those branches of the Vertebrate stock which, at various times, have deserted their high vocation and fallen into lowlier habits of life.' The challenge of the environment was the key to progress – yet neither Lankester nor MacBride was interested in the details of local adaptation. All that mattered was the retention of an active way of life that would maintain the stimulus to individual self-development.

Fossils and Ancestors

MacBride was an active Lamarckian and one of the last great exponents of the recapitulation theory. By the time he wrote the survey quoted above, evolutionary morphology was beginning to fall into the background as new sciences such as genetics took over the forefront of scientific research. But the same kind of eclipse did not affect that other great area devoted to reconstructing the history of life: paleontology. Fossils were now being collected in ever-greater numbers and from an ever-wider area of the earth's surface, and it was inevitable that paleontology would be used to throw light on the course of evolution. As its influence expanded, it began to promote a model of evolution that took far greater account of geographical factors. The fossils showed that new groups originated in particular parts of the world, and focused attention on the process by which they expand around the globe, often exterminating more primitive forms as they go. The continuous model of development was thus replaced by a more complex theory that highlighted certain crucial episodes in the history of life on earth.

Opponents of evolutionism claimed – then as now – that the fossil record revealed no sign of intermediates between the major groups of animals and plants. Fortunately for the evolutionists, there were important discoveries during the 1860s and 1870s that helped to confirm their prediction that such 'missing links' were merely hidden by the imperfection of the fossil record. Most important was the reptile–bird intermediate *Archaeopteryx* from the Jurassic rocks of Germany. Here was a form with reptilian characters and a mouth with teeth (not a beak) – yet the imprint of feathers was clearly preserved in the rocks surrounding the fossilized skeleton. Many of the other major transitions were not filled in, however, and paleontologists thus concentrated on developments within well defined groups. In America O.C. Marsh discovered a sequence of fossils illustrating the evolution of the modern horse from a more generalized four-toed ancestor. As the number of fossils increased, more and more examples of fairly continuous fossil sequences were discovered, allowing paleontologists to analyse the historical development of some groups with a fair degree of confidence.

Although supporting evolution, these discoveries did not always seem to confirm the Darwinian interpretation of how the process worked. In Britain, Owen and Mivart argued that many cases of evolution consisted of parallel lines of development within the same group, all driven in the same direction as though by an internally programmed force. Darwin's image of an irregularly branching tree thus did not fit the evidence. The American neo-Lamarckians made the same point, arguing that the fossils seemed to support a theory of directed evolution, not one based on the selection of random variation by the local environment. The inheritance of acquired characters explained these trends as a consequence of the animals specializing for a particular way of life chosen by their ancestors.

The Lamarckians also insisted that some of the trends had no adaptive purpose. Hyatt in particular argued that the fossil invertebrates he studied showed patterns of progressive, Lamarckian evolution followed by an inexorable decline into racial senility and extinction. Whole groups of organisms developed increasingly bizarre non-adaptive characters until they eventually disappeared altogether. By the end of the century many paleontologists supported the theory of orthogenesis, in which variation was supposed to be pushed in some preordained direction by forces built into the constitution of the organisms. Modern paleontologists argue that these supposedly linear trends were actually a product of the imagination seeking to impose order upon a relatively small number of fossils. But at the time it was thought that the record had yielded up its secrets in sufficient quantities to allow confident assertions to be made about the character of the evolutionary mechanism.

By emphasizing the continuity of development within groups, paleontologists were forced to recognize that the actual origin of new groups represented a different kind of phenomenon. Evolution was not an absolutely continuous process because the crucial episodes in which life had advanced to an entirely new level of development could not be represented as extensions of the more predictable kind of evolution taking place within established groups. The origin of a new group often involved the development of new characters by an undistinguished member of an earlier class. Once the pioneers of a new level of development were given the chance to flourish, they radiated out into a whole range of orders and families to take advantage of all the various possible means of gaining a livelihood. But once established, these lines underwent predictable developments towards increasing specialization – and eventually to overspecialization and extinction.

Paleontologists became aware of the need to explain the radiation of the mammals at the beginning of the Cainozoic era in terms of the apparently sudden disappearance of the dinosaurs and other great reptiles at the end of the preceding era. Primitive mammals were known to have existed throughout the 'age of reptiles', but only after the dinosaurs had gone did they diversify into a wide range of specialized forms. Some paleontologists suggested that the reptiles had died out through racial senility in order to leave room for their successors, the mammals, who were waiting in the wings ready to seize control of the world.

By the early twentieth century, paleontologists were promoting a coherent model of evolution that differed significantly from that proposed by the early Darwinians. In a host of well illustrated books and in the displays erected in the great museums, the fossil record was used to support a theory in which evolution worked through a series of progressive episodes, each sending out waves of new and more highly organized creatures to displace their now-stagnant predecessors. Henry Fairfield Osborn, the influential Curator of Vertebrate Paleontology at the American Museum of Natural History, coined the term 'adaptive radiation' to denote the episodes of expansion, and insisted on the orthogenetic character of evolution within each specialized group when once established.

Although the new experimental science of genetics was now turning its back on Lamarckism and the idea of directed variation, the paleontologists could still influence popular understanding of evolution. During the early years of the new century, Osborn and his colleagues at the American Museum of Natural History created striking displays of the great dinosaurs whose remains had been unearthed in the west, solving the complex technical problems involved in reconstructing the skeletons of such exotic beasts. Casts of these dinosaurs were sent to other museums around the world; the industrialist Andrew Carnegie presented a cast of the gigantic *Diplodocus* to the Natural History Museum in London, where it was erected with great ceremony in 1905. E. Ray Lankester had become the director of the museum in 1898. Although trained as a morphologist, he had a long-standing interest in the fossil record and published a well illustrated book on *Extinct Animals* in 1905, which provided the information on prehistoric animals for Arthur Conan Doyle's popular novel *The Lost World* of 1912.

Paleontology had always had a geographical dimension, and this was exploited more thoroughly as the exploration of the fossil record became possible on a world-wide scale. On the *Beagle* voyage, Darwin had discovered evidence for the 'law of succession of types' by unearthing South American fossils related to the unique modern inhabitants of that continent. Some mammalian families must have migrated to that continent and evolved there in isolation from the rest of the world. By the late nineteenth century, paleontologists were becoming increasingly aware of this phenomenon and (in the absence of a theory of continental drift) were postulating temporary 'land-bridges' between the continents, which were raised and lowered by geological forces, providing migration routes in some epochs and not in others. The preservation of the marsupials in Australia was explained in terms of the breaking of a land-bridge connecting that continent to the rest of the world before the appearance of the placental mammals.

Drawing upon contemporary geological theories, the Canadian-born paleontologist William Diller Matthew (1871–1930) created a synthesis that integrated the temporal and geographical dimensions. Some geologists believed that the earth's climate alternated between periods of warm, moist conditions and intervals in which the northern regions had been subject to a much harsher climate. In his *Climate and Evolution* of 1914 Matthew argued that, since the last cold spell began in the Eocene, northern Asia had become the main centre for mammalian evolution. Unlike Osborn, Matthew was prepared to accept that evolution occurred primarily in response to environmental change, and – unusually for the time – he accepted the theory of natural selection. He argued against the view that the discovery of primitive members of a group in a certain location meant that this was the point of origin from which the group had spread out. On the contrary, the point of origin was where the highest members of the group were found. As soon as more advanced types evolved in response to a stimulating environment, they spread out around the world, driving earlier, more primitive forms into marginal locations where they were able to survive for some time before extinction.

Matthew drew upon a long-standing belief that the flora and fauna of Eurasia were dominant over those of the rest of the world and combined this with the paleontologist's theory of the periodic explosion of new types at certain points in the fossil record. Although his support for Darwinism was unusual, his concept of successive waves of migration radiating outwards from a centre of progressive evolution seems to have struck a chord in the minds of his contemporaries. The introduction of a geographical dimension at this level could be taken up even by those who saw evolution as something more than the selection of random variation. In the age of imperialism, an older model of history based on the rise and fall of empires could be projected back onto the fossil record to justify the world-wide expansion of modern Europeans.

Human Origins

We can illustrate this ideological dimension by comparing Matthew's theory with contemporary developments in ideas about human evolution. The late nineteenth century saw the emergence of a model of human origins based very much on Haeckel's linear progressionism, but this was replaced in the early years of the twentieth century by a theory more in tune with the new paleontology.

The question of humankind's origin from the apes was one of the most controversial aspects of Darwin's theory. Even some of his closest supporters found the prospect of a natural origin for humankind unacceptable. Both Lyell and Wallace opted for a theory in which a supernatural force had affected the later stages of human development. Clearly, if humans were not to be excluded from the theory, our ancestry must be linked to that of our closest relatives, the great apes. Huxley's *Man's Place in Nature* of 1863 had stressed the physical resemblances between humans and apes. But Huxley said nothing about the process that might have led one branch of the primate order to develop the upright posture and the enlarged brain that are the hallmarks of the human race. The objections voiced by Lyell and Wallace were eventually overcome, allowing the majority of late-nineteenth-century thinkers to accept an ape ancestry. But this was only made possible by treating the emergence of the human race as the last phase of the general progress of life towards higher levels of organization and consciousness.

Darwin's *Descent of Man* of 1871 addressed this issue directly. Unlike most of his contemporaries, Darwin realized that the human race was not the predictable end-product of a universal progressive trend. His theory required him to specify a reason why the human branch of the evolutionary tree had advanced so much further than those which led to the living apes. He postulated that the first step in the differentiation was our ancestors' adoption of an upright posture in order to move about on open plains instead of in trees. The brain had enlarged later in order to exploit the tool-making capacities of hands that were now freed from the task of locomotion. But hardly anyone took this seriously, and much of Darwin's own book was devoted to a discussion of the factors that would have led to mental progress.

As far as most nineteenth-century thinkers were concerned, it was obvious that the driving force of human evolution must have been a steady expansion of the brain, which merely continued the progressive thread that ran through the whole evolution of life. The possibility that the human race might have been an unplanned accident resulting from an adaptive modification was simply not acceptable. The most complete expression of this preconception was a linear model of human origins in which the white race was the last and highest rung in the ladder of progress. This had the advantage that 'lower' races could be dismissed as earlier steps in the advance that had survived into the present in parts of the world where there had been less stimulus for mental development.

Women, too, were dismissed as less highly evolved than men, thus fitting the prejudices of the male scientists who constructed the theories. The second half of Darwin's *Descent of Man* was devoted to sexual selection, because Darwin thought that this was the only process that could explain some of humankind's unique features such as our loss of body hair. If males compete for access to the females, selection will develop any character that encourages successful breeding – even if this character is of no benefit in terms of adaptation to the environment. In species where the males compete to gather 'harems' of females, sexual selection will develop the weapons used in male combat. The horns of male deer, for instance, have increased in size because over many generations those stags with large antlers have mated most successfully. In bird species where the males display to attract the females (the peacock is a classic example), extravagant plumage has developed because the best-endowed males have always mated more frequently. Wallace objected to this aspect of Darwin's theory on the grounds that 'female choice' attributed human characteristics to animals.

Applied to the human race, sexual selection allowed Darwin to account for the loss of body hair by assuming that this character had become sexually desirable among our distant ancestors. The mechanism also allowed him to 'explain' the different intellectual and emotional qualities of men and women. Men were naturally more active and intelligent, while women had evolved characters that fitted them for life in a world dominated by aggressive males. They were less intelligent, but more sympathetic, providing the glue that held the family together. The vast majority of evolutionists shared both the racist and the sexist assumptions that Darwin built into his theory.

To fill in the details of this ladder of progressive steps leading from ape to white man, it was necessary to define the intermediate stages. Black males and white women could be used to represent the next highest rungs of the ladder, but lower ones would have to be found in the fossil record. In his *Man's Place in Nature* Huxley had discussed an ancient human skull found at Neanderthal in Germany in 1857. He noted that its heavy brow-ridges gave it an ape-like appearance, but rejected it as a 'missing link' because of its large capacity. In the 1880s more Neanderthal-like remains were found, and many paleoanthropologists began to argue that they comprised a distinct species intermediate between apes and humans. Haeckel endorsed this view and regarded it as being confirmed by Eugene Dubois' discovery of the yet more ape-like *Pithecanthropus* (now known as *Homo erectus*) in Java during the 1890s. Haeckel's disciple Gustav

Schwalbe created a linear progressive sequence from the great apes to *Pithecanthropus* to the Neanderthals and finally to modern humans. All the known forms were fitted into a single sequence so that the Neanderthals became our own immediate ancestors.

In the early twentieth century this linear model of human evolution was rejected in favour of a theory in which there were several parallel lines of human development, only one of which has survived through into the present. Instead of being treated as our own ancestors, the Neanderthals of Europe and the more primitive 'Java man' were dismissed as distinct forms of pseudo-humanity, driven into marginal locations by the expansion of true humans with a more advanced level of intelligence. Central Asia was now regarded as the cradle of humanity, a view endorsed through into the 1930s by H. F. Osborn and other eminent paleontologists. From this source, a series of distinct radiations had emanated, each representing a new line of human evolution destined ultimately to displace the others. The parallel between this theory and Matthew's more general model of mammalian evolution is obvious enough – but it is also significant that throughout the previous century Asia had been regarded as the source from which successive waves of higher races had invaded Europe. The myth of a dominant 'Aryan' race had its origins in this approach, the original Aryans being pictured as invaders from the east who founded the main European cultures: Greece, Rome and the Teutons. Darwin had favoured Africa as the cradle of humankind, a view that was consistent with the link to the modern great apes. Now Africa was dismissed in order to revive the more traditional view that Asia was the centre of all human progress.

Far from extending Darwinian gradualism, the scientists who tried to reconstruct the past had returned to a theory of history that assumed that progress took place through a series of distinct episodes, representing the rise and fall of groups that evolved in parallel rather than in sequence. Each distinct form of humanity had its origin at a particular point in space and time, from which it expanded throughout the world before collapsing in the face of pressure from a more highly evolved type. The ideological dimensions of this model are evident from the racist implications drawn from it by the Oxford geologist W. J. Sollas in his influential book *Ancient Hunters* of 1911. Sollas made it clear that the 'primitive' races of today are the remnants of earlier human types marginalized in remote areas such as Australia or the Arctic. Their fate was a necessary consequence of progress[10]:

> What perhaps is most impressive in each of the cases we have discussed is this, that the dispossession by a new-comer of a race already in occupation of the soil has marked an upward step in the intellectual progress of mankind. It is not priority of occupation, but the power to utilise, which established a claim to the land. Hence it is a duty which every race owes to itself, and to the human family as well, to cultivate by every possible means its own strength; directly it falls behind in the regard it pays to this duty ... it incurs a penalty which Natural Selection, the stern but beneficent tyrant of the organic world, will assuredly exact, and that speedily, to the full.

Although Sollas referred to natural selection, he elsewhere ridiculed Darwin's theory as incapable of explaining how the higher types were actually produced. Natural selection merely justified the marginalization of primitive races by the later forms of humanity developed by some inherently progressive force. This form of 'social Darwinism' emphasized the destructive capacity of struggle, but drew even more heavily upon pre-Darwinian models of historical development, which had now become fashionable in the age of imperialism.

EVOLUTION AND THE ENVIRONMENT

For Darwin and his closest followers it was evident that the driving force of evolution was the process that adapted populations to their ever-changing environments. There were two areas in which this assumption could be explored. One was in the field of biogeography, where the migration of small populations to new areas could be seen as the chief cause of the branchings in the tree of life. But it was also possible to make direct studies of how a single population adapted to changes taking place in its local environment.

Paths of Migration

Matthew's theory drew upon a long tradition of applying evolutionism to the problems of biogeography begun by the first Darwinians. Even the evolutionists who accepted that species change in response to adaptive pressures realized that something more than simple adaptation was involved. The actual course of evolution was often determined by the factors that limited a species' access to new areas of the world. It was also necessary to explain how divergence occurs – how a single original species can be divided into a group of distinct but related forms, each of which can then undergo a unique process of further evolution. Geography was important because it would help the evolutionist to explain how populations were able to migrate and adapt to new locations. For the Darwinians, biogeography became a historical science. By looking at the present distribution of animals and plants, and by comparing this with what was known about geological changes in the past, the evolutionist could seek to explain how the populations of the earth's various regions had been built up through successive migrations.

Darwin had to think carefully about what is now called the problem of speciation: how a single species splits up into a number of 'daughter' species. Natural selection was a process taking place within the whole breeding population making up the species, and if a division were to occur, the single original population must somehow be split into several distinct subpopulations that did not interbreed. Darwin regarded the local varieties that exist within many species as incipient new species, on the assumption that further change would increase the separation between the varieties to such an extent that they would eventually be unable to breed together. For this to happen, it was essential that the local variations did not interbreed and thus blend their characters back together again. Since natural selection produced only slow, almost imperceptible changes, Darwin could not rely on the sudden appearance of a new form that did not interbreed with its parents.

To begin with Darwin had solved this problem by invoking geographical isolation. The Galapagos islands had given him the clue by showing that small populations derived from the same original species could evolve in different directions when separated by a geographical barrier such as the ocean. But by the time he wrote the *Origin of Species* he had begun to doubt that the physical separation of the original populations was essential. He now thought that speciation could take place across a continuous territory if the conditions at the extremes of a species' range were different enough to promote distinct ecological specializations. Varieties and eventually species would be formed despite any tendency for interbreeding with the main body of the population in the centre.

He was opposed on this point by the German naturalist Moritz Wagner, whose book *The Darwinian Theory and the Law of the Migration of Organisms* of 1868 insisted on the need for a period of geographical isolation. Modern biologists tend to agree with Wagner. Varieties can develop behavioural 'isolating mechanisms', such as differences in courtship, which prevent interbreeding even when it is genetically possible – but a brief period of physical isolation is needed to allow the populations to develop the differences that will prevent them blending back together again if they come into contact. Darwin's rejection of geographical isolation encouraged his opponents to postulate sudden transformations or saltations to explain how distinct new species were formed.

At the broader level of explaining the world-wide distribution of animals and plants, the Darwinians challenged the traditional system in which the earth's surface was divided into a number of distinct zoological and botanical provinces based on 'centres of creation'. They accepted that some regions had distinctive populations resulting from local evolution. But the notion of distinct provinces seemed to smack too strongly of the old pre-evolutionary approach. Darwin and his followers set out to explain the global distribution of animals and plants by postulating a complex sequence of migrations limited by geographical barriers. Since those barriers could be temporarily removed by geological agents (e.g. a lowering of the sea level during the ice age), or could occasionally be overcome by accidental means (e.g. birds blown across the ocean by storms), it would be possible to reconstruct the process by which the unique mix of species occupying any given territory had been built up through the periodic influx of newcomers.

In botany, J. D. Hooker interacted extensively with Darwin on this topic and defended the idea of land-bridges. A. R. Wallace became one of the leading authorities on zoological distribution. Wallace's *Geographical Distribution of Animals* of 1876 provided the model for the next generation of naturalists studying biogeography. Darwin himself had explained the strange inhabitants of isolated oceanic islands such as the Galapagos in terms of accidental migrations from the nearest mainland. But Wallace's main concern was the distribution of animals on large continental land masses. He accepted the ornithologist P. L. Sclater's view that there were six main zoological provinces. These could not be entirely distinct centres of evolution, but the existence of provinces implied that migrations were controlled by something more than accidental factors. Wallace shared Darwin's view that the continents were essentially permanent but realized that other geological changes might have affected the possibility of overland migrations. Although the basic position of the continents had remained unchanged since the Mesozoic, variations in the sea level might have opened up land passages between areas now separated by shallow seas. In the absence of a theory of continental drift, the Darwinians were forced to postulate migration to explain similarities in the populations of widely separated continents – they could not envisage an original population being split up by the physical separation of the land mass it occupied.

Wallace saw the great northern continents of Eurasia and North America as the chief focus of progressive evolution from which higher types had radiated out from time to time. Occasional crossings between the two chief areas were possible during periods when the Bering Strait became passable to land animals. Earlier, less advanced forms were driven south by the expansion of superior types in Asia, Africa and America, but occasionally a barrier appeared that held back the advance and allowed the earlier forms to take refuge in an isolated southern area. Ocean barriers opening up during the early phases of mammalian evolution had protected the marsupials in Australia and the lemurs and other unique animals of Madagascar. During his explorations in the 1850s Wallace had traced the line dividing the Asian and Australian faunas in South East Asia, still known as 'Wallace's line'. He appealed to Croll's theory of the ice ages to explain how, at certain points in time, the amount of water locked up as ice had caused a lowering of the sea level and hence the opening up of land connections. Wallace's line existed because there was one particular strait, between the islands of Bali and Lombok, that was so deep that it could never have been exposed as dry land in recent geological times.

Similar processes would explain anomalies such as the existence of closely related species in widely separated locations. Wallace argued against those naturalists who postulated temporary 'land-bridges' across even the deepest and most extensive oceans to explain these cases. Such bridges were both geologically implausible and unnecessary. The occurrence of tapirs in South America and South East Asia, for instance, could be explained quite easily without recourse to anything more than minor changes in the earth's conditions[11]:

> During Miocene and Pliocene times tapirs abounded over the whole of Europe and Asia, their remains having been found in the tertiary deposits of France, India, Burmah, and China. In both North and South America, fossil remains of tapirs only occur in caves and deposits of Post-Pliocene age, showing that they are comparatively recent immigrants into that continent. They perhaps entered by the route of Kamchatka and Alaska, where the climate, even now so much milder and more equitable than on the north-east coast of America, might have been warm enough in late Pliocene times to have allowed the migration of these animals. In Asia, they were driven southwards by the competition of numerous higher and more powerful forms, but have found a last resting-place in the swampy forests of the Malay region.

Here Wallace combined paleontological and modern evidence to throw light on the problem of migration, thus pioneering the technique later expanded by Matthew throughout the fossil record.

In botany too, Hooker, Thiselton-Dyer and others favoured the view that Eurasia was the principal centre from which more advanced types had radiated. It was known that many European plants and animals flourished when introduced into areas such as Australia, thus fuelling the imperialist assumption that the products of Europe – including the human products – were destined to rule the world. In the case of plants, however, the means of dispersal were so prolific that theorizing became impossibly vague. The fact that seeds could be carried for hundreds of miles by strong winds allowed the supporters of dispersal to explain any case of anomalous distribution by invoking such an accidental circumstance. By the turn of the century, some botanists were protesting against the excessive speculations of the dispersal theorists, arguing that in many cases those species with the highest dispersal power were not the most widely distributed. In the 1890s the British botanist H. B. Guppy, who specialized in the study of oceanic islands, began to argue that dispersal was merely a function of time: the oldest genera were the most widespread, whatever their powers of dispersal. Guppy postulated major saltations to explain the origin of the main groups and limited adaptive evolution to minor modifications within these groups following their dispersal.

Evolution and Adaptation

Whatever the means by which they were transported to their present location, many evolutionists agreed that species must undergo changes adapting them to the local environment. A great deal of effort was expended on studying the degree of adaptation and the processes by which adaptive change was brought about. Darwin himself investigated the means by which insectivorous plants catch and digest insects, which he interpreted as an adaptation to soils with low nitrogen content. He also studied climbing plants, suggesting that their ability was merely an extension of the power of movement possessed to some slight degree by all plants. Darwin was anxious to show that even quite complex adaptations could have been built up gradually by natural selection from rudimentary origins that were common to all animals and plants. Every stage in the process had to offer an advantage to the organism in the struggle for existence: natural selection cannot plan ahead and can only promote those characters that are useful in the short term. Complex adaptations have developed because in certain circumstances the same selection pressure has been maintained over a long period of time.

The study of animal coloration was a particularly active area. It had long been recognized that many species are protected from predators by colouring that provides camouflage. The Darwinians assumed that this kind of coloration had been developed by natural selection. Since in most species the individual animals cannot control their own colour, Lamarckism could not be invoked to explain the phenomenon; selection was the only alternative to divine creation. The Darwinians were also impressed with the discovery by Henry Walter Bates (1825–92) of a new kind of mimicry in South American insects, where an edible species copies the warning colours that have been developed by another species that is distasteful to birds. In the later decades of the century the professor of entomology at Oxford, E. B. Poulton (1856–1943), made extensive studies of animal coloration, which he explained in Darwinian terms. As far as Poulton was concerned, the fact that many animals possessed protective coloration proved the effectiveness of natural selection.

More direct efforts to measure the effects of selection upon a population were undertaken by the biometrical school of Darwinism. Inspired by Darwin's cousin, Francis Galton (1822–1911), the biometricians sought to provide hard evidence that a population subjected to selection pressure exhibited a measurable difference after a few generations. Galton had become convinced that Lamarckism was ineffective because the character of each individual is rigidly determined by inheritance. He developed statistical techniques to study the variability of wild populations and to measure changes in the range of variability over time. His disciple, Karl Pearson (1857–1936), built upon these techniques to provide detailed arguments defending the selection mechanism against its critics. Drawing upon detailed field studies by W. F. R. Weldon (1860–1906), the biometrical school was able to show that selection did indeed have a measurable effect even on the short term. Weldon studied crabs in Plymouth harbour that were being forced to live in water that was muddied by human activity. He also studied the coloration of snails' shells in different local environments. In each case it seemed clear that local adaptation was taking place, showing – as far as Pearson and Weldon were concerned – that natural selection was an effective evolutionary mechanism.

The biometrical defence of Darwinism was paralleled by the work of the German biologist August Weismann (1834–1914) on the cellular basis of heredity. After doing some early work inspired by the recapitulation theory, Weismann began to study the behaviour of the reproductive cells under the microscope. When the cell nucleus was stained to bring out its internal structure, minute rod-like bodies were observed and were called 'chromosomes' because they absorbed the colouring so well. By the 1880s Weismann was convinced that these chromosomes were the bearers of heredity. He argued that they consisted of a material substance, the 'germ plasm', which transmitted characters from parents to offspring via the nucleus of the egg and sperm cells. He proposed a theory in which the germ plasm was totally isolated from the adult body that transmits it to future generations. This meant that characters acquired by the adult body could not be incorporated into the germ plasm and could not be inherited. The Lamarckian effect did not work, and the only conceivable mechanism of evolution was the natural selection of variations arising spontaneously in the germ plasm.

Not everyone was convinced, however, and there were many efforts by anti-Darwinians to provide evidence in favour of alternative evolutionary mechanisms. In America the neo-Lamarckian zoologist Alpheus Packard (1839–1905) studied the blind animals that inhabit many caves. Everyone agreed that the loss of eyesight was a consequence of the animals taking up residence in an environment with no light, but the Lamarckians argued that the inherited effects of disuse provided a better explanation of the process than natural selection. The Darwinians had to assume that a useless organ was gradually reduced to a rudimentary state because natural selection would favour those individuals who did not waste their energy growing a useless structure. But to the Lamarckians it seemed much more natural to assume that the wasting away of an individual's eyes when there was no light would be passed on to the next generation, resulting in a rapid loss of eyesight in the whole population. Such indirect arguments are scorned by modern biologists, but at the turn of the century they were the mainstay of support for an active Lamarckian movement.

At the same time, there were some naturalists – and even more experimental biologists – who had doubts about the role of any adaptive process in evolution. The German zoologist Theodor Eimer (1843–98) attacked Poulton's approach to animal coloration. He scorned the claim that camouflage might be the product of adaptive evolution. Those who invoked this factor to explain the colour on the outside of a snail's shell had no way of explaining differences on the inside. In the case of insects apparently mimicking other species' warning coloration, Eimer argued that the similarities were due to an entirely different cause. He believed that evolution worked by orthogenesis, in which rigidly predetermined trends drove variation in a particular direction whatever the needs of the organism. If two different species showed the same coloration, this merely confirmed that they were both subject to the same orthogenetic trend.

The early supporters of Mendelian genetics also objected to the emphasis placed on adaptation by both the Darwinians and the Lamarckians. Even before he became one of the pioneering geneticists, William Bateson (1861–1926) had abandoned Darwinism and had begun to insist that evolution is driven by discontinuous variations or saltations. Unlike Eimer's orthogenetic trend, the saltation was an instantaneous reorganization of the reproductive process generating an entirely new character – but in both cases the end-product had no adaptive value. Bateson had been trained as an evolutionary morphologist and had published a theory to explain the origin of the vertebrates. In the 1890s he turned against this whole programme in disgust, realizing that there was no likelihood of fossil evidence coming to light. His *Materials for the Study of Variation* of 1895 argued strongly against adaptive evolution and in favour of saltations produced by forces internal to the organism. He dismissed the adaptive scenarios postulated by the Darwinians as 'just so stories' that had no scientific value. They were no more testable than the claims of Paley and the other advocates of natural theology[12].

Bateson now argued that the direct study of variation was the only way of trying to understand how evolution actually works. Like many others, he turned to an experimental approach as a means of transcending the limitations of traditional 'Darwinism'. Neither field studies nor the purely descriptive work of the morphologist satisfied the demands of a new generation of biologists seeking to put biology on a new, more scientific footing. The experimental study of heredity led Bateson to breeding studies and soon to the newly rediscovered laws of Mendelian heredity.

The sudden outburst of interest in the laws of heredity proposed in 1865 by Gregor Mendel was a by-product of the demand for an experimental science of life. By hybridizing different varieties of peas, Mendel had shown that some characters are inherited as units on an 'all or nothing' basis – they did not blend together as Darwin and almost everyone else assumed. This insight was ignored until the new generation of experimentalists began to see heredity as a process transmitting characters through the line of reproductive cells, independently of the adult body. In 1900 the new theory based on the inheritance of unit characters was announced by Hugo De Vries and Carl Correns – it was Bateson who coined the name 'genetics' a few years later.

Bateson resisted the claim that the gene was a material entity, but soon the more materialistic geneticists were arguing for a modified version of Weismann's germ plasm theory in which the gene was a segment of the chromosome responsible for transmitting a particular character from one generation to the next.

De Vries soon lost interest and began to promote his 'mutation theory', which stressed the role of saltations backed up by evidence that significant new characters were being produced in cultivated populations of the evening primrose. De Vries' 'mutations' were subsequently shown to be the result of hybridization, not the production of new genetic characters. At first, however, most geneticists joined De Vries and Bateson in assuming that new genetic characters appear suddenly by saltation. They also insisted that selection by the environment had no effect on the characters that appeared in this way. De Vries at least believed that there might be a struggle for existence between the mutated forms, but most geneticists refused to accept this compromise. Shortly before he took up the study of genetics, the American biologist Thomas Hunt Morgan wrote a book, *Evolution and Adaptation* (1903), arguing that evolution was governed solely by the production of saltations, and had nothing to do with adapting the species to its environment. The geneticist R. C. Punnett countered Poulton's arguments by trying to show that single mutations could produce the colours attributed to mimicry. It would be well into the 1920s before the majority of geneticists began to concede that natural selection might have a role to play by regulating the spread of mutated characters within the population.

THE ORIGINS OF ECOLOGY

As the geneticists of the early twentieth century turned their backs on field studies and the role of adaptation, a very different group of biologists were striking out in the opposite direction. They too were determined to 'modernize' biology, but they could not abandon fieldwork because the phenomena they wished to study could only be observed in the wild. The modern science of ecology emerged when naturalists trying to understand how a species' distribution is limited by environmental factors sought to develop more precise ways of studying relationships that had been largely taken for granted by previous generations.

The extent to which the history of science has been shaped by modern concerns is evident from the fact that only in recent years has the origin of ecology begun to attract much attention. As concern for the environment grows, this area may eventually come to rival the Darwinian revolution as a subject for scholarly analysis. There is little direct connection, because many early ecologists were not interested in evolution. The outsider might be tempted to assume that scientific ecology was inspired by the growth of environmentalism in the late nineteenth century. The Green movement has appropriated the term 'ecology' for its own purposes by pretending that anyone aware of the complexity of the interactions between species must be concerned to preserve the natural balance. Yet serious studies of the interactions between animals, plants and their physical environment were often initiated by scientists who hoped to modify the natural balance in order to allow sustainable exploitation. Whatever the modern relationships between scientific ecology and environmentalism, we must beware of assuming that similar relationships governed the attitudes of the biologists who first began the systematic study of ecology in the 1890s.

Darwin's theory threw the relationship between the organism and its environment into a new light by stressing the pressures that were exerted on any population by limitations of resources. The 'balance of Nature' was a shifting affair at the mercy of geographical factors that might change the physical environment or allow the invasion of rival species. The complex adaptation of the organism to its physical and organic environment was built up over time through the constant application of pressures that are never stable. It would be wrong to claim that Darwinism had no impact on the study of what would now be called ecological relationships, but the impact was indirect. There was little effort at first to study how the ecological balance could shift through time as the result of evolution. Most early ecologists simply assumed that the physical environment was stable and that existing species would establish natural relationships with one another in each area. The problem of working out how adaptations arose through evolution was too complex to be combined with detailed study of how those adaptations worked in the present.

Were the naturalists who studied adaptation doing real ecology before the modern name for the science was coined – or did something recognizably like modern ecology only come into existence when the term 'ecology' became popular in the 1890s? It was certainly possible to study what would now be called ecological relationships before the founding of scientific ecology. But the natural theologians' assumption that God had designed a harmonious order of Nature was hardly a suitable basis upon which to build a modern science. Some historians argue that modern ecology only became possible in the post-Darwinian age in which all natural relationships were seen to be fragile. Against this claim must be set the fact that many of the early ecologists were not, in any significant sense, evolutionists; for all practical purposes they were still dealing with an essentially static world view.

Recent historical studies stress the importance of scientific disciplines and research programmes. The emergence of a science of ecology depended not so much on changing assumptions about Nature as upon the creation of a community (or communities) of researchers who saw the study of natural relationships as their primary goal. The plural 'communities' is necessary because from the beginning there were a number of different schools of ecological thought. Animal and plant ecology long remained separate disciplines, and plant ecology itself existed in several different forms. We may look for generalizations linking all these groups, but we should also be aware of the different routes by which biologists came to realize that it was possible to specialize in what soon became known as 'ecology'.

The New Biology

Huxley's effort to create a scientific 'biology' based on morphology and physiology had broken down, partly because it was obvious that field studies were necessary to solve certain kinds of biological questions. In some areas it was necessary to assume that adaptation of the organism to its environment is crucial both for survival and for long-term evolution. Ecology was a second-generation response to the problem of creating a scientific biology. In the last decade of the nineteenth century there were a number of quite different moves to create new disciplines that would transcend the limitations of the old approach to the study of Nature. The experimental approach to the study of heredity led to the creation of the new science of genetics. Ecology was a parallel move designed not to transcend field studies, but to transform them in a way that would make them genuinely scientific. Evolutionism played only a minor role because it was associated with the old, rather amateurish techniques that led to unlimited speculation. Field studies would become scientific by adopting the techniques of measurement and quantification that had proved so effective in the 'hard sciences'. Ecology emerged not from Darwinism, but from an extension of physiology into the realm of the organism's relationship to its surroundings.

There was no single discipline of ecology created in response to a theoretical initiative, or to a particular philosophical position. Instead, several different groups of biologists tried, each in their own way, to modernize field studies in a manner that would allow them to compete for the ever-increasing supply of research funding being made available by governments and private institutions. Since these groups came from different backgrounds and worked in different institutional settings, they chose to explore different aspects of the environment and adopted different theoretical models. The term 'ecology' could be used as an umbrella to cover a wide range of research traditions that did not interact at the detailed level, but which could be seen to deal with one aspect or another of the relationship between living things and their environment. Many scientists could do 'ecology' while retaining their primary disciplinary loyalty elsewhere, in oceanography or forestry for instance. Ecology could exist at the research level even though it was at first difficult to introduce it into the universities and schools as an academic discipline. Far from being a coherent response to the growing problem of environmental degradation, ecology was a product of the new age of specialization ushered in by the rapid expansion of the scientific profession.

The term *Oecologie* was actually coined in Ernst Haeckel's *Generelle Morphologie* of 1866 to denote the study of the interactions between organisms and the external world. The term was derived from the Greek *oikos*, referring to the operations of the family household, and it seems that Haeckel intended his readers to visualize a kind of global organic economy in which all species played a part. His monistic philosophy promoted a sense of the unity of Nature among later generations of environmentalists, but it was several decades before serious work on what would now be called ecological topics began. By the 1890s self-conscious research schools were starting to emerge in several areas. In 1893 the eminent physiologist J. S. Burdon-Sanderson (1828–1905) told the British Association for the Advancement of Science[13] that 'oecology' was one of the three great divisions of biology, along with physiology and morphology, and was in some ways the most attractive of the three because it came closest to the spirit of what had once been called the 'philosophy of living nature'. The modern spelling 'ecology' was established at the International Botanical Congress in the same year.

One foundation upon which the new studies could be built was the Humboldtian search for the geographical factors that limited the distribution of particular species. Detailed studies of the distribution of animals and plants were made throughout the nineteenth century, often employing amateurs to do the actual surveying under the leadership of a few professionals. These surveys tended to assume that certain species were typical of a particular habitat, although there were also studies of the changing proportions of species in differing locations. By the end of the century C. Hart Merriam, director of the Bureau of the Biological Survey of the US Department of Agriculture, was providing extensive maps depicting the 'life zones' or habitats of various animals and birds running east–west across the North American continent. These zones were identified by the observers' fairly casual assessment of the dominant species in each locality. There was no quantitative work and later ecologists dismissed this whole technique as an example of exactly what they were trying to avoid. If this kind of field study was to have any value at all in the modern world, it would have to be transformed by the introduction of more rigorous techniques.

The new ecologists were particularly inclined to stress the interactions between the different species making up the population of each area. The old 'balance of Nature' theory had accepted the existence of such interactions, and Darwin had certainly stressed their importance as factors determining the selection pressures that caused evolution. In the course of his career, Darwin had in fact turned away from the idea that the physical environment is the crucial factor to which species must adapt. He became more concerned with specialization within a given environment, a process that he saw as a consequence of the struggle between the different inhabitants of that environment. While continuing to have some interest in biogeography, Darwin nevertheless deflected attention towards inter-species competition as the main driving force of natural selection – a force that continued to operate and produce change even in a perfectly stable environment.

In the *Origin of Species* Darwin gave an illustration of the unexpected complexity of the interactions between the organisms inhabiting a particular area. He suggested[14] that 'it is quite credible that the presence of a feline animal in large numbers in a district might determine through the intervention first of mice and then of bees, the frequency of certain flowers in that district!' The cats caught the mice, which otherwise would destroy the nests of the bees, which were crucial to the fertilization of certain plants. This example was meant to illustrate the unexpected links that might make up a chain of natural interactions, but it is significant that no experimental work was done to test the idea. Darwin worked extensively on the relationship between insects and the flowers they fertilize; his book *On the Various Contrivances by which British and Foreign Orchids are Fertilized by Insects* appeared in 1862, and the subject was generalized in Herman Müller's *The Fertilization of Flowers* (translated 1883). But these studies looked at the interactions between species only to explore the process by which evolution produced complex adaptive structures. They were a product of Darwin's particular view of evolution, not a true anticipation of ecology.

The switch to a more deliberate emphasis on the way in which all the inhabitants of an area depend upon one another came about at least in part because of developments in social evolutionism. Herbert Spencer, the philosopher of the evolutionary movement, argued that a society could be seen as a kind of biological organism in which the whole was more than the sum of the parts. Each specialized trade was like an organ within an animal's body, providing a function that benefited the whole and depending on the continued existence of the whole for its own survival. The possibility that natural communities could be visualized in similar terms occurred to many biologists in the later decades of the century, thus providing an incentive to study the interactions upon which the whole network depended.

An explicit statement of what is now called the concept of a biotic community was made in Karl Möbius' study of oyster beds (1877). The Illinois naturalist Stephen A. Forbes gave an address on 'The Lake as a Microcosm' in 1887, emphasizing that all the species within a lake were linked into a functioning community maintained by the balance between predators and prey[15]. Karl Semper's *Animal Life as Affected by the Natural Conditions of Existence* of 1881 developed the concept that would later be known as the food-chain and noted that the total numbers of predators must be much lower than the numbers of the prey organisms to maintain a stable system.

Semper's book was a pioneering study of animal ecology, but it was a forerunner rather than a true foundation-stone upon which others were to build. Terrestrial animals move around too much for the naturalist to think easily in terms of a stable community, and the main development of animal ecology did not take place until the twentieth century. Notable exceptions were in marine biology and limnology (the study of lakes), where serious efforts were made to study how communities functioned. The work of Möbius and Forbes falls into these traditions, while the pressure to study fish resources provided a direct incentive to investigate how populations could be affected by external disturbances such as human activity. The German marine scientist Victor Hensen (1835–1924) led an expedition to Greenland and the tropics in 1889, which showed that – contrary to everyone's expectations – the colder waters supported the greater quantity of life. Ernst Haeckel (who described some of the *Challenger* specimens) argued that Hensen's sampling techniques must have been faulty, but further studies gradually confirmed his results.

Hensen believed that the plankton represented the life-blood of the ocean and set out to study its growth in different areas by applying the methods of physiology. Variation in the nutrient content of the water was clearly the major factor involved. Here, as in other areas, we see ecology emerging from a deliberate revolt against the evolutionary morphology of earlier decades. Hensen established an important research school at Kiel in Germany, aided by state support provided for the study of fisheries stocks. The foundation of the International Council for the Exploration of the Sea in 1902 also acted as a stimulus. The most significant product of this school was the work of Karl Brandt (1854–1931), which applied the methods of agricultural chemistry to the study of plankton growth. Brandt switched the emphasis from geographical variation to the study of the annual cycle of plankton growth found in the northern oceans. By stressing the changing supply of chemical nutrients he and his followers were led to discover the plankton 'bloom' that occurs in the seas every spring. Brandt's emphasis on the annual variation of light and nutrients was modified by Alexander Nathansohn (1878–1940), who pointed out that the vertical circulation of water played a vital role in bringing nutrients to the surface.

By the early decades of the twentieth century, the Kiel school's theory of plankton growth had been widely accepted, but German oceanography collapsed after the First World War and the impetus passed to British and American scientists. Throughout this period, marine ecology remained isolated from other branches of the science. The separate funding provided for the study of fisheries allowed oceanographers to develop their own institutions and professional networks, which were not closely linked to those of the universities or agricultural research stations. These professional barriers ensured that there would be little interaction even when the disciplines were working on parallel problems.

Plant Ecology

The botanists pushed ahead to develop a number of rival schools of ecology at the turn of the century. Historians have been most active in studying the emergence of plant ecology in the United States, but it is agreed that much of the initial inspiration for the American work came from Europe, especially from Germany and Denmark. In part, these European studies came from an extension of Humboldt's approach to distribution. Oscar Drude (1852–1933) of the Dresden botanical gardens published a plant geography of Germany in 1896, which sought to illustrate how local factors such as hills and rivers combined with the overall climate of a region to determine the actual distribution of plants. Drude attempted to depict the relative abundance of each species in an area according to a ranking system with categories ranging from 'social' (where a single species formed an overall mass) down to 'scarce'.

Other plant ecologists saw physiology as their starting point, because it was the interaction between the plant as a living organism and its environment that determined whether or not a particular species could live in a certain area. Only by studying how each species coped with different levels of moisture, heat, light, etc., was it possible to explain why the species flourished in one area but not in another. There was already a substantial body of information on plant physiology, but in the 1890s a number of botanists realized that the discipline could be extended to include how the plant coped with different environmental factors. Some of the inspiration for this move came from botanists who had travelled to the tropics and other exotic locations. The establishment of a botanical laboratory at Buitenzorg in Java was particularly influential. As Andreas Schimper (1856–1901) wrote[16]:

> The greater prominence of physiology in geographical botany dates from the time when physiologists, who formerly worked in European laboratories only, began

to study the vegetation of foreign countries in its native land. Europe, with its temperate climate and its vegetation greatly modified by cultivation, is less calculated to stimulate such observations; in moist tropical forests, in the Sahara, and in the tundras, the close connexion between the character of the vegetation and the conditions of extreme climates is revealed by the most evident adaptations.

Many of the botanists who studied the adaptation of species to extreme environments became convinced that the effect of the climate upon an individual plant could be transmitted to its offspring, thus building up the species' level of adaptation through a Lamarckian process.

Perhaps the most influential of the European ecological botanists was the Dane, Eugenius Warming (1841–1924), whose *Plantesamfund* of 1895 was translated into German the following year and into English as *Oecology of Plants* in 1909. Warming was trained as a plant physiologist, spent some time in Brazil as a student, and finally developed his ecological approach as an alternative to both pure physiology and the sterile emphasis on classification of many field naturalists. He argued that the physical capabilities of the plant determined where it could or could not live, and studied the effects of various environmental factors, especially water, in determining the prevailing vegetation. But Warming also realized that the plants of a particular area formed a 'community' united by a variety of interactions including parasitism and symbiosis (the mutual interdependence of two distinct species)[17]:

The term 'community' implies a diversity but at the same time a certain organized uniformity in the units. The units are the many individual plants that occur in every community, whether this be a beech-forest, a meadow, or a heath. Uniformity is established when certain atmospheric, terrestrial, and any other factors discussed [above] are co-operating, and appears either because a certain, defined economy makes its impress on the community as a whole, or because a number of different growth-forms are combined to form a single aggregate which has a definite and constant guise.

The relationships that created a particular kind of community were real enough, but Warming resisted the temptation to regard the community as a kind of super-organism with a life of its own. These relationships were a natural consequence of evolution working to produce mutual interactions from which all the species involved would benefit.

Warming was influenced by a movement within German science reacting against the idealists of previous generations who had sought mystical unities in Nature. He was a reductionist who believed that all the relationships he investigated could be explained in materialistic terms. Perhaps the clearest indication of this is his reluctance to postulate a single kind of community that was the 'natural' occupant of any particular territory. He knew that wherever an unoccupied land surface was formed, by fire, landslip, the filling in of a bog, or human activity, plants moved in to colonize the area. There might be a succession of different communities, as when a bog is gradually converted to a meadow. But it would be wrong to suppose that the final community in such a sequence had some kind of privileged status that guaranteed its stability. Nature was a scene of constant struggle in which species were trying to extend their territory – a message emphasized by Darwin, but also by earlier writers such as de Candolle. As Warming put it[18]:

This struggle is caused by endeavour on the part of species to extend their area of distribution by the aid of such means of migration as they possess. 'Situation wanted' is the cry in all communities, whether these be human or vegetable. Millions upon millions of seeds, spores, and similar reproductive bodies are annually scattered abroad in order that species may settle in new stations; yet millions upon millions perish because they are sown in places where physical conditions or nature of the soil check their development or where other species are stronger.

The consequence of this pressure was that in some circumstances, at least, a community that was destroyed might never be re-established, because other species would invade the territory and prevent the original inhabitants reclaiming it. When humans cut down the forest, it might be replaced by a permanent grassland because the soil itself had been modified in a way that prevented the trees from reseeding themselves.

Warming's materialism influenced one of the pioneering American plant ecologists, Henry C. Cowles (1869–1939) of the University of Chicago. Cowles developed an ecological approach within an academic framework that stressed rigorous pure research. He studied the changing vegetation around the shores of Lake Michigan, arguing that local factors such as the lake interfered with the natural plant covering characteristic of the region. As one moved away from the edge of the water, the plants changed systematically until eventually the natural woodland of the surrounding region was encountered. Cowles postulated that the sequence thus observed in space was equivalent to that observed in time when a section of lake was filled in and converted to dry land. The fact that the 'natural' climax community for the region could be held at bay by geographical factors such as the presence of a lake showed that there was no super-organism designed to flourish in a particular region.

The concept of the community as a super-organism was promoted most effectively by another school of ecological botany based on the study of the grasslands of the midwestern United States. This school had a different institutional framework based on applied research supported by agricultural stations, and had its roots in the European tradition of plant geography. Charles Edwin

Bessey (1845–1915) came to Lincoln, Nebraska, in 1884 as professor of botany at the state university and state botanist. He was a plant physiologist who had great success in promoting the study of botany as a science, but he was also concerned with conservation and wanted to study the grasslands of the prairie before they were completely destroyed by the plough. His most important student was Frederic E. Clements (1874–1926), who joined with Roscoe Pound (1870–1964) to begin the study of grassland ecology in the area. They read Drude's plant geography and decided to do a similar study, but soon found that the European techniques did not work on the almost featureless prairie. Bessey was sceptical of the whole approach as little more than revived natural history, and, in order to convince him that scientific ecology was possible, Clements and Pound looked for a method of quantifying their studies. They realized that the old technique of depending upon the trained observer's overall impression to characterize the natural vegetation of an area was unreliable. The only way to get precision was to mark off an area and count every individual plant within it.

Thus was born the technique of intense study of typical 'quadrats' or squares marked off to define a sample of the vegetation for the surrounding area. Quadrats of a large scale had been used before to determine the precise range of various species, but Clements and Pound marked off small areas, often only a metre square, so that every plant of every species could be identified. In this way a precise sampling of the vegetation could be obtained, and, by setting up quadrats over a wide area, a systematic assessment of distribution could be worked out. By clearing a quadrat of all plants and then checking it periodically, the succession of plants leading to a mature sustainable 'climax' vegetation could be observed. Pound and Clements' *Phytogeography of Nebraska* of 1898 became a standard text for American botany, and Clements' *Research Methods in Ecology* of 1905 was the first textbook to describe the new methodology.

Pound eventually abandoned botany for law, but Clements became professor first at Nebraska and then at Minnesota, before moving to the Carnegie Institution of Washington. Although originating in the universities, the new grassland ecology did not become a formal academic discipline, and flourished best among botanists with a practical concern for the problems experienced by the farmers who were interfering with the natural vegetation of the area.

Clements was an influential writer who developed a philosophy of ecology that differed fundamentally from the reductionism of Warming and Cowles. Following the model of the social organism proposed by Spencer he adopted the view that the plant community was a genuine super-organism with a life of its own. The community was something more than a collection of species working together for mutual advantage – it obeyed laws that could only be understood at a level transcending that of the individual organisms. This is most apparent in his views on the development of a community through time. Clements was convinced that there was a single type of climax vegetation that formed the mature covering for any particular area, depending on the physical conditions. If anything disturbed that natural covering, the vegetation would follow a pattern of development that led through a series of recognizable stages to the mature climax once again[19]:

> Succession must then be regarded as the development or life-history of the climax formation. It is the basic organic process of vegetation, which results in the adult or final form of this complex organism. All the stages which precede the climax are stages of growth. They have the same essential relation to the final stable structure of the organism that seedling and growing plant have to the adult individual.

Clements had little interest in evolution. He thought that the prairie grassland climax had been stable since the last ice age and could be restored at any time even after interference by humankind. Yet his language when discussing the development of the community is strongly reminiscent of the developmentalism typical of Haeckel and the progressionist evolutionists of the Darwinian era. Here the old belief that Nature tended to develop inevitably towards a predictable goal resurfaced in one of the newest branches of science, which had been designed to replace the old approach to natural history.

Clements' holistic, almost vitalistic viewpoint was not linked to any romantic 'back to Nature' environmentalism. On the contrary, it flourished within a discipline that had emerged to help the farmers who were struggling to exploit the prairies in a way that destroyed the traditional climax. Unlike the continental European ecologists, Clements was working within a system that emphasized practical rather than pure research and saw science as a way of controlling the economy as a whole. His appeal to the concept of a super-organism allowed him to present ecology as a science that would show us how to manage the natural productivity of an entire region. The grassland ecologists studied the development leading towards the natural climax as a way of helping farmers to understand the tensions created by their interference with the system. They offered the essentially optimistic message that nature was strong enough to resist destruction by human agencies – the climax could easily be restored if humankind chose to set aside areas for conservation.

This optimism would be undermined by the Dust-Bowl of the 1930s, which showed that the destruction of the soil made possible by farming would have a permanent effect on the land, just as a follower of Warming's more materialistic approach might have predicted. Nevertheless, Clements' philosophy represents a notable extension of nineteenth-century developmental attitudes into the early decades of the new century. It suggests that 'old-fashioned' vitalistic metaphors were quite capable of flourishing in an age that demanded that science have practical applications.

In Britain there was a tradition of plant geography supported by amateur activity. In the 1890s the brothers William and Robert Smith (1866–1928 and 1873–1900), influenced by the sociological approach of Patrick Geddes and by reading Warming's work, instituted botanical surveys of Scotland and of Yorkshire. But it was Arthur G. Tansley (1871–1955) who took over the leadership of the new approach, proclaiming ecology as the guide to future work in botany in a 1904 paper given to the British Association. Tansley had experience in the tropics, having travelled to Ceylon (Sri Lanka), the Malay peninsula and Egypt in the years 1900–1. Over the next few decades he developed the study of plant communities in Britain, using methods similar to those of Clements. Although impressed with Clements' views, Tansley eventually challenged the super-organism interpretation of the community (see chapter 11). Lacking the sources of funding available in the United States, he had to arouse public support in order to establish ecology as an independent discipline. In 1904 Tansley and William Smith set up the Committee for the Survey and Study of British Vegetation, later known as the British Vegetation Committee. In 1911 Tansley organized an International Phytogeographical Excursion around the British Isles, attended by Clements, Cowles, Drude and others. Tansley's pioneering *Types of British Vegetation* of 1912 was a product of this expedition. In 1913 the British Ecological Society was founded, the first society in the world devoted to the field.

The emergence of ecology as a science had no clear-cut links to the environmental movement that was advocating a reduction in the destructive activities required by modern civilization. Ecologists studied the natural environment, but often saw this only as a means of helping the human race to manage its interference more effectively. The relationship between the new science and earlier developments in biology also turns out to be remarkably complex. At one level, ecology was a reaction against the old natural history, a determined effort to put field studies on a new, more scientific footing. Various groups exploited the rhetoric of modernization in their efforts to create new research programmes that could exploit the opportunities now being offered by the flood of government and private money available for scientific research. Yet there was no unity within ecology, just as there was no unity in the life sciences generally. Many schools of thought flourished, each within its own professional environment, while others withered away, unable to find a niche within which they could develop.

In these circumstances, the tensions apparent within the philosophies of life available in the nineteenth century were able to perpetuate themselves into the twentieth. There was no dramatic switch to a materialist paradigm in science; materialistic and holistic/vitalistic philosophies both survived the transition to a more experimental approach to the study of life, at least temporarily. Although some biologists had become aware that Nature was not a system designed to progress inevitably towards higher levels of development, others retained the kind of progressionism that had been used to make the general idea of evolution palatable in the wake of Darwin's *Origin of Species*. Those who ignored the role of adaptation were able to flourish at the same time as those for whom adaptation was the primary consideration, because the range of problems that could attract research funding allowed both approaches to find suitable niches for themselves. A new institutional framework for science was emerging, but far from imposing a conceptual unity, the network of academic, government and industrial research allowed a massive fragmentation of theoretical perspective. Each research school developed ideas that seemed consistent with its own set of problems and techniques. In these circumstances, many of the different philosophical perspectives that had flourished in the nineteenth century were able to update themselves and remain viable in the new age of professional specialization.

The Earth Sciences

By the early decades of the twentieth century, the relationship between science and society had begun to take on the form familiar to us today. In the era of 'big science', the massive involvement of government and industry in the financing of research has pushed the private individual onto the margins. There are no more Darwins and Humboldts creating major scientific reputations on the basis of their own resources. The industrialists who make the money now hire scientists to do the necessary research, and the successful scientist must increasingly be as much a manager as a theoretical innovator. Pure research survives in the universities, often supported for the sake of national pride and prestige – although this function can easily be threatened when governments adopt a more pragmatic approach and join the industrialists in the search for short-term gains. Whatever the source of support, however, the modern scientist has become acutely conscious of the need to maintain funding within the intensely competitive world of professional rivalries.

By the end of the nineteenth century it had become clear that science had played a vital role in promoting the technological power that had enabled Europeans to dominate the world. America had emerged as a new source of influence in western culture and was determined to press its advantages. The west was certain that its dominant role was the natural consequence of its cultural – perhaps its racial – superiority. Western people and western society were inherently progressive, and this progress was destined to continue indefinitely. Technological superiority was a consequence of moral worth, justifying the extension of European power on a world-wide scale. There were critics within, including a small, anti-materialistic lobby concerned with the health of the environment. Others were worried about the growth of national rivalries in the age of imperialism. The expectation of an earlier generation that industrial expansion would make nationalism redundant had not been justified. The threat of war seemed to grow with the expansion of military technology. It was still possible to argue, however, that national competition was a spur to progress, replacing the individualism that had fuelled the first phase of industrial expansion.

The First World War cast serious doubts on the optimistic philosophy of progress. Perhaps the European powers were morally no better than the savages they had conquered in the name of civilization. For many intellectuals, the moral superiority of Europe had vanished in a cloud of poison gas drifting over the trenches of the Western Front. Such feelings would fuel the development of the anti-materialistic viewpoint that questioned the right of humankind to tear the earth apart for its own material benefit. But for those in control of society, confidence in humankind's ability to conquer the world was dented but not destroyed. If Nature could not be trusted to progress by itself, then human beings would have to use their intelligence to ensure further development. Fears were expressed about the possibility of degeneration within the white race, but governments could surely take steps to ensure that the gains made in the previous century would not be lost. There was a renewed emphasis on the need for the scientific management of society and the economy, an emphasis that would be renewed after the depression of the 1930s once again brought the superiority of western values into doubt.

There were social experiments designed to find workable alternatives to the capitalist system. In Russia the Bolsheviks turned to state control for the whole economy, but Stalin's drive to industrialize the country ensured that the exploitative dimension of the traditional world view would not be effectively challenged. In Germany, the Nazis too ended up by putting the whole economy on a military footing – yet their ideology had its roots in respect for a traditional lifestyle that was more in tune with Nature (chapter 10). Britain, France and America retained the traditional system and continued the exploitation of those parts of the world colonized during the age of imperialism before the war.

Following the Second World War, the colonies were freed, or freed themselves, although economic shackles remained. There was no let-up in the drive to increase the level of industrialization, in either the east or the west. What became known as the Third World suffered the consequences of exploitation driven by the demand for cash crops and minerals, coupled with the problems of debt and rapid population increase. The harmful consequences of industrialization for both the developed and the undeveloped nations were becoming ever more obvious, and in the 1960s a massive challenge to the philosophy of control and exploitation at last emerged throughout the west. For many, this challenge entailed a rejection of science itself. There had been scientists who opposed the philosophy of materialism, but such attitudes had now been marginalized and science was perceived as a monolithic source of the exploitative attitude that characterized the military–industrial complex.

Most scientists still work within that complex and have a vested interest in its survival. But the strength of the environmental movement has ensured that an increasing number of scientists have turned back to the kind of holistic viewpoint that had once formed a viable alternative to materialism. Public concern over the state of the environment has at last ensured that governments divert at least some funding into projects designed to expose the dangers of unlimited exploitation.

SCIENCE IN THE MODERN WORLD

These changing attitudes have affected the ways in which science functions in society. Wars, both 'hot' and 'cold', have had a direct impact by channelling money into those areas of research with military applications. The long-term consequences of this have not always been negative from the viewpoint of the environmental sciences – meteorology and oceanography have both benefited directly from military spending, and modern geology has drawn upon some techniques that began as military projects. The demands of various industries have provided employment for a vast range of scientists and have dictated their areas of research. The massive expansion of the oil industry has had an obvious effect on the science of geology. The emergence of funding groups dedicated to environmentalism has added a new dimension to the scene, creating sources of opportunity for research and providing a framework within which some scientists can articulate a non-materialist philosophy of Nature.

The complexity of science's role in modern society has stood in the way of efforts to provide a unified theoretical perspective in many fields. External influences have played a role by providing support for different theoretical positions, especially in the life sciences. When biological concepts are applied to the human race, they have obvious social implications, and there is no shortage of evidence suggesting that the scientists themselves make use of these implications when promoting their ideas. But even within the scientific community there are social pressures that affect the ways in which the various disciplines develop. Pure researchers working within a university may adopt a very different approach to that favoured by applied scientists, even when addressing similar questions. When there are large numbers of different funding bodies offering to support research, the opportunities for carving out new economic niches for research programmes are multiplied.

Expansion and Fragmentation

A number of new disciplines, including genetics and ecology, were created in the early years of the twentieth century. This multiplication has continued through to the present, with the emergence of an ever-widening array of journals, societies, etc., devoted to ever more specialized research areas. Sociologists of science argue that there is a limit to the number of individuals who can participate fruitfully in a coherent research discipline – beyond a hundred or so it becomes impossible to keep up with what everyone is doing, and there is a natural tendency for the discipline to fragment into groups that may go off in their own directions. In the age of big science, there are powerful social pressures acting to prevent the creation of truly wide-ranging theoretical paradigms.

The changes in the social structure of science have involved reorganizations as well as fragmentation. In some cases, research topics that were once isolated have been joined together to create new disciplines. The emergence of a science of 'geophysics' in the early twentieth century is a case in point, where various studies once isolated in different parts of geology and physics were brought together to form a coherent area of study. In the course of the twentieth century, geophysics began to play an increasingly powerful role as compared to the traditional research programmes of geology. The modern term 'earth sciences' has been introduced to replace 'geology' partly in order to symbolize the changing emphasis that has come about as a consequence of this social reorganization within the area.

Our modern willingness to talk about the 'environmental sciences' as though they represent an obvious grouping of related studies is itself a product of our growing concern with the problems of pollution and over-exhaustion of resources. A group of disciplines with different professional structures have been brought together because all are seen to have a bearing on problems arising from human activity.

The amount of unity generated at the level of actual research should not be overestimated. The problems that are associated with the environment are due to humans, and they encompass a range of scientific disciplines whose very size and complexity ensure that different techniques and theoretical perspectives will flourish within the supposed unity. From the seventeenth century through into the age of Humboldtian science there had been a growing trend towards specialization, but the sharp boundaries we take for granted today had not yet appeared. The interests of individual scientists often created interactions between various areas of study even after the appearance of specialized journals and societies. But the story of the twentieth century is one of professional fragmentation and increasing theoretical incoherence. Environmentalist groups may lament our inability to visualize the earth as a whole, but the recent history of the 'environmental sciences' shows that the regrouping implied by this all-embracing label runs against the grain of modern scientific development.

Historians are particularly interested in the creation of new disciplines, and in the changes that take place within traditional fields when confronted by new initiatives. Those who believe in 'scientific revolutions' have tried to seize upon the creation of the modern theory of plate tectonics as an example of such a revolution taking place in the earth sciences. But if the creation of new disciplines is involved, there are serious difficulties lying in the way of any attempt to treat the new theory as a 'revolution' within an established body of theoretical knowledge.

By recognizing the disciplinary complexity of modern science, we are forced to confront the multiplicity of theoretical perspectives that can be brought to bear on the same question. At the turn of the century there was little unity within either the earth or the life sciences. The classic nineteenth-century theories – the cooling-earth theory and evolutionism – had proved incapable of sustaining coherent research programmes. In both geology and biology there were a host of competing subdisciplines, each with its own characteristic theoretical approach. At the practical level, scientists have no difficulty with this situation; on the contrary, they are very good at burying themselves in their own specialization and not worrying about what is going on in neighbouring areas. At the level of public relations, however, it is extremely embarrassing to have scientists promoting very different answers to what everyone else sees as the same question. Since science is supposed to be searching for the truth about Nature, it must be presumed that all its practitioners will eventually converge on the same interpretation.

The new rhetoric of science as a source of technical expertise demands a degree of conformity, at least in public pronouncements on fundamental issues. This desire for a superficial appearance of unity probably underlies the marginalization of holistic and vitalistic theories in twentieth-century biology. The conflict between material- ism and vitalism that had flourished in the previous century was resolved when the materialists became strong enough to present their philosophy as the prevailing orthodoxy, dismissing all support for the opposing position as pseudo- science. Holism has certainly not thereby been eliminated as a philosophy, however, and in an age of environmental consciousness it may be able to make a comeback even within science (chapter 10).

At a more technical level, the history of science has been influenced by the scientists' own desire to present a facade of unity. Many areas of science have attracted little attention from professional historians; if their histories have been written at all, it is by the scientists themselves (often after retirement). The result is a collection of highly technical histories that take the current theoretical perspective of the discipline for granted and are of very little interest to anyone except a specialist. The sheer number of disciplines created by the vast expansion of modern science has created a problem for any historian trying to provide an overall survey. There are so many areas to study, each with its own mass of highly technical literature, that there is no realistic prospect of giving the outsider a well-balanced account of what has been going on. A comprehensive history of twentieth-century science is impossible because no one would have the patience to wade through the mass of technical detail.

Historians have chosen instead to concentrate on a few selected topics where there are important conceptual revolutions that can be highlighted to attract the interest of outsiders. They concentrate on innovations that have resolved outstanding problems and permitted a degree of unity to emerge from an earlier state of chaos. Two obvious examples considered below are the theory of plate tectonics – which produced the unity of the modern earth sciences – and the 'modern synthesis' of Darwinism and genetics (chapter 10). The resulting studies tend to endorse the claim that science is advancing towards a consensus on important theoretical issues. They also provide case studies to illustrate the mechanism of a scientific revolution.

In the last decade or so historians have become suspicious of this artificial concentration on unifying theories. Precisely because we have become more aware of the disciplinary complexity of the scientific community, we have realized that the appearance of unity given by the new theories may sometimes be only skin deep. Several related areas often pay lip-service to an overarching theoretical perspective, but at the practical level the scientists may still be doing very different things. It has also become apparent to English-speaking historians that the disciplinary boundaries taken for granted in Britain and America do not necessarily correspond to those in place elsewhere. For this reason the theoretical unities perceived by scientists in different countries may not correspond. This is not to deny that cross-disciplinary initiatives do occur, but it is important that the historian exercise caution when assessing the scientists' own rhetoric of unity. We should look at what they do, rather than at what they say in public. It is unlikely that historians will lose interest in the theories that are presented as the great success stories of twentieth-century science. But they have begun to take a more realistic view of the social interactions within the scientific community that allow theories to attain the status of paradigms.

The Age of Big Science

The vast expansion of science in the twentieth century was made possible because both industries and governments now recognized the practical advantages to be obtained from supporting scientific research. Geology has always benefited from the interest of the mining industry, and the world-wide search for petroleum oil has merely carried on this trend. In some cases new industries have directly contributed to the emergence of scientific disciplines, as when the advent of air travel focused attention on the need for a sophisticated programme of research in meteorology. Industries have usually concentrated on applied science, although some of the most powerful industrialists have sought to spread resources more widely by setting up research trusts and institutions to finance research that – if not exactly pure – was freed from the need to show an immediate profit. The Carnegie Institution of Washington, founded in 1902, supported research in genetics, geophysics and many other areas. Indeed, these great foundations could dramatically alter the direction of research by deciding which areas were suitable for funding. For motives often more subtle than mere profit, big business still influenced the development of science to fit its vision of what counted as a significant contribution to knowledge.

In general, though, research at the most fundamental level has benefited from – and been controlled through – the massive involvement of governments. Here again America has led the way through the establishment of a vast number of research organizations devoted to all aspects of science. The sheer multiplicity of organizations has often led to confusion and conflict of interest – as early as 1903 President Theodore Roosevelt set up an abortive commission to study the problem of duplication. Roosevelt's concept of a scientifically managed economy was discredited in the boom years following the First World War, but the New Deal that followed the depression of the early 1930s was based on a renewed interest in state control. The Tennessee Valley Authority linked the scientific resources of the Department of Agriculture and the Geological Survey to establish a unified approach to the problem of land use, with a strong emphasis on conservation. The National Science Foundation, founded after the Second World War to co-ordinate government funding of research, directed money into areas that offered, or could claim to offer, information on issues of social concern.

The two world wars themselves focused attention on science as the hostile governments sought new and more frightful weapons. Oceanography benefited from the demands of submarine warfare, which generated new techniques for studying the sea bed. Meteorology too gained support as the armed forces demanded better weather predictions and offered unlimited technical resources to do the job. Wartime research was often very short-lived, but the new developments provided foundations upon which scientists could continue to build in times of peace.

Military and industrial pressure has merely accelerated an increase in the level of technical sophistication, which has had an immense impact on the character of modern science. New devices were invented to study hitherto unknown phenomena, and in some cases the results helped to create entirely new disciplines or to direct the internal development of existing fields. The science of seismology began in the early twentieth century when it became possible to study the propagation of shock waves in the earth's surface. Such studies helped to reveal the inner structure of the earth, and at the same time gave new insights into the nature of earthquakes and new techniques for locating oil. The development of acoustic techniques for underwater probing was boosted by anti-submarine research, but generated new information on the structure of the sea bed, which helped to create the modern theory of the earth's structure. The space age has affected most of the environmental sciences by offering techniques of remote sensing that have revolutionized our ability to monitor what is happening across the whole surface of the earth.

Sciences have always transcended national boundaries, and the twentieth century has seen a steady increase in the formalization of international relationships. The world wars inevitably interfered with the international organizations that were becoming more important in many sciences. Co-operation in the study of the environment was an obvious casualty, although the global communications networks were restored fairly rapidly once the hostilities were over. Today, many sciences are studied by researchers in a wide range of countries who co-operate freely together. But national differences remain, and disciplines that have become clearly defined in one country may have a subordinate status in another. The rapid expansion of science in America has made it easier for new disciplines to establish themselves there. In Europe, financial restrictions and a highly centralized academic framework have often limited the ability of new fields to emerge as autonomous entities. The high visibility of international organizations should not blind us to the fact that the scientists who participate in them often have to work within very different institutional structures at home.

The End of Exploration

By the early years of the twentieth century there was little of the earth's land surface left unexplored. In Britain, the

Royal Geographical Society promoted Captain R. F. Scott's ill-fated expedition to the south pole in 1912, but insisted that such expeditions were now scientific surveys rather than simple efforts to open up unknown territory. When Scott and his men died on their way back from the pole, they were still carrying thirty-five pounds of geological specimens with them. The British tended to sneer at the achievement of the Norwegian explorer Roald Amundsen, who got to the pole first and survived. Apart from the fact that he ate his sledge-dogs, his expedition was seen as a mere 'dash to the pole' with no scientific value. Amundsen had actually trained himself in the study of the earth's magnetic field so that he could investigate the north magnetic pole during his successful navigation of the North West Passage (1903–6). When the American Robert E. Peary made a dash for the north pole in 1909, his observations were so poor that serious doubts have remained as to whether he did, in fact, get there. But Peary was hailed as a hero by the privately funded National Geographic Society, which was anxious to gain from the publicity associated with the first man to the pole.

As the age of discovery at last came to a close, emphasis necessarily switched to the ever more detailed study of little-known areas. There were still challenges to attract those determined to prove themselves against a hostile environment – Mount Everest was assaulted by a series of much-publicized expeditions until it was finally conquered by Edmund Hillary and Tenzing Norgay in 1953 – but the image of the explorer as an adventurer striking out into the unknown has gradually faded into the background. Modern expeditions have a wide range of specialists who study every aspect of the environment in the region they have selected. The most significant area of exploration in the mid twentieth century was the sea, and here scientific oceanography and geophysics took over as the leading disciplines.

If the purpose of a geographical society was not to encourage the filling-in of blank spaces on the map, what then was to be the role of geography in twentieth-century science? Geography is a classic example of a subject that can disappear as a separate entity, each of its functions siphoned off by a new area of specialization. Much of what is called 'physical geology' could be dealt with quite happily by geomorphologists, geologists, climatologists and others. The exploration of humankind's relationship to the physical world could be taken up by anthropologists and various branches of the social and economic sciences. In the United States geography has indeed almost disappeared as a distinct academic discipline. European geographers resisted the potential dissipation by a deliberate act of will. This began when the Prussian government founded a professorship in geography at the University of Berlin in 1874. Soon university departments had been created throughout Germany, and other countries followed this lead in later decades. Geography thus survived as an independent field, although its emphasis has gradually transferred from the physical to the human dimension as independent sciences have taken over the study of so many aspects of the natural world.

Geography has retained its integrity by projecting itself as a science devoted to the interaction of humankind with the physical environment, serving in effect as a bridge between the natural and the social sciences. To some extent, the link thus preserved through the emphasis on geographical diversity has run counter to an overall trend that has led the human sciences to separate themselves off into distinct fields which no longer interact directly with the natural sciences. Sociology and anthropology, once integral parts of the nineteenth century's evolutionary world view, have become independent disciplines that fiercely resist any suggestion that they should take their cue from biology. But the social sciences are dismissed as 'soft' areas by scornful exponents of the physical or biological sciences. The kind of unity that Humboldt had once called for in the relationship between the study of humankind and the environment has been hard to preserve in the modern age of specialization. Geography is an exception – but how many 'hard' scientists regard geography as a science? There are many in the environmental movement who would regard the specialization of the sciences as a key factor in the disintegration of western society's ability to see the earth as a unified system within which the human race must function as an integral component.

At the more popular level there had always been a strong interest in the exotic, and the development of photography allowed everyone to participate, at least vicariously, in the exploration of even the remotest parts of the globe. In America the National Geographic Society has continued to exploit this interest throughout the twentieth century. The editor of its magazine through the first half of the century, Gilbert Grosvenor, exerted strict control over the Society's image and built up its reputation by creating the impression that the public was directly involved in science. Major expeditions were sponsored, allowing members to feel that they were contributing to research. The resulting stories in the magazine could thus be presented as first-hand reports from the frontiers of real science, despite their popular appeal. The Society has been criticized for its paternalistic attitude to non-white peoples, although in more recent decades it has certainly helped to focus attention onto environmental issues. Even here, though, the Society has influenced the public perception of issues by stressing the work of some scientists at the expense of others. Scientists in turn have been influenced by the possibility of attracting research funds if they can generate projects that will catch the Society's attention.

A GLOBAL PHYSICS

The most active efforts to create a global science of the physical environment have arisen within those areas which have sought to understand the processes at work within the earth as a whole. Geophysics emerged as a distinct discipline in the late nineteenth century, sidestepping the geologists' concern for local description to concentrate on the forces that determine our planet's structure and behaviour. The best-known aspect of geophysics is the study of the earth's crust, and here the new discipline would eventually challenge geology for supremacy. But geophysics was never seen purely as a rival to geology: from the start it was designed to synthesize knowledge of all the physical processes operating within the earth's crust, its oceans and its atmosphere. Major figures such as Alfred Wegener (1880–1930) moved freely between meteorology and the earth sciences, much to the confusion of those trained within the tradition of geology. Historians' attention has continued to focus on the earth sciences, because here geophysics has upset the conceptual foundations established by earlier geologists. But this should not blind us to the significance of the developments taking place elsewhere.

Geophysics was by its very nature concerned not with minute details of crustal structure, but with the physical properties of the earth as a whole. The subjects that make up the particular field of geophysics are not, in themselves, new. But they were originally studied in isolation, as subordinate aspects of old-style geology or as applications of general physics. In the late nineteenth century it was at last recognized that these isolated elements might be grouped together to form a coherent discipline that could stand as an alternative to geology. A journal, *Terrestrial Magnetism*, was founded in 1896 to publish material on this particular aspect of the earth's physics, and was expanded to become the *Journal of Geophysical Research* in 1949. The American Geophysical Union was formed in 1919, originally as a committee of the National Academy of Sciences. Although snubbed by the geologists, geophysics developed as an independent science and seized the initiative in the years after the Second World War, when the vast explosion of new techniques allowed it to become the centre for theoretical innovation in the study of geological processes. The International Geophysical Year (actually July 1957–December 1959) symbolized the triumph of the younger discipline.

Geophysics and oceanography went hand in hand because many of the most important developments in the study of the earth's crust centred on the ocean bed. The late-nineteenth-century expeditions to explore the oceans had generated considerable interest on an international scale and had shown that there was still much work to be done in this area. In 1902 representatives of Denmark, Sweden, Norway, the United Kingdom, Holland and Russia met in Copenhagen to found the International Council for the Exploration of the Seas. Textbooks of oceanography began to appear, along with maps designed to reveal the underlying structure of the world's ocean basins. The First World War encouraged research on echo sounding for submarine detection, and after the war private institutions continued to apply similar techniques to mapping the ocean floor. In America the Scripps Institution of Oceanography was founded at San Diego in 1922, and 1930 saw the founding of the Woods Hole Oceanographical Institution in Massachusetts. The Second World War again saw a massive investment by the military, confirming a transition from ocean biology to physics as the chief focus of attention. By the 1960s it was possible to argue that, from the perspective of the geophysicist rather than the geologist, the oceans were better known than the continents.

Weather Patterns

There were also close links between geophysics and the rise of meteorology. International co-operation in the collection of weather data had begun in the mid nineteenth century and had led to the creation of the International Meteorological Organization in 1873 (reconstituted as the World Meteorological Organization in 1950). The advent of air travel focused attention on the problem of weather forecasting and promoted interest in what Wladimir Köppen (1846–1940) called 'aerology', the study of the air above the level where its movement was limited by interaction with the ground. Aerological observatories were established in many countries, and improved balloons provided the opportunity for studying the conditions at high altitudes. In 1905 the rapid circulation of the upper atmosphere now known as the 'jet stream' was discovered. Later on the development of minute radio transmitters for use on balloons greatly improved the scope of observation.

The first major school of meteorological theory was founded at Bergen in Norway by Vilhelm Bjerknes (1862–1951) and achieved international prominence during the 1920s and 1930s. Bjerknes was trained as a mathematical physicist, but in the early years of the century he realized that this area was being eclipsed by the exciting new developments in atomic physics. He began to exploit his mathematical skills to create models for the circulation of the atmosphere, thus linking up with the strong development of geophysics that was taking place throughout Europe. The growing interest in aviation made it clear that meteorology would soon become a topic of considerable public interest, offering new opportunities for scientists to gain both funding and international influence. Bjerknes became professor at a newly founded institute for geophysics at Leipzig in 1912, where he hoped to develop an exact physics of the atmosphere. During the war he returned home to Norway and worked at the geophysics institute of the Bergen museum. He now began to see the need for practical, short-term weather forecasting and saw how the improved communications offered by the development of radio would aid the collection of data and the dissemination of predictions.

In the years immediately following the war Bjerknes turned away from detailed mathematical modelling to concentrate on the basic structure of weather patterns as revealed by his increasingly dense network of observation posts. He recognized that much of the northern hemisphere's weather was dominated by 'cyclones' involving the interactions between masses of warm and cold air moving westwards. Soon the boundaries between the air masses were being called 'fronts' – a metaphor borrowed explicitly from the military terminology of the war years. The problem of the weather was now to be conceptualized using a model based on the image of two antagonistic powers engaged in a global conflict[1]:

> We have before us a struggle between a warm and a cold air current. The warm is victorious to the east of the centre. Here it rises up over the cold, and approaches in this way a step towards its goal, the pole. The cold air, which is pressed hard, escapes to the west, in order to make a sharp turn towards the south, and attacks the warm air in the flank: it penetrates under it as a cold west wind.

Cyclones were seen as disturbances running westwards along a permanent line of confrontation between polar and tropical air circling the whole hemisphere, the 'polar front'. Although there were exaggerated claims about the predictive power of this model, it made an impact on the whole science of meteorology and still influences the way our weather forecasts are presented today.

International aviation began soon after the end of the First World War and by the 1920s the public was gripped by a fever of enthusiasm for the prospect of transatlantic air travel by either airship or aeroplane. Bjerknes was determined to exploit this excitement in order to gain his Bergen school a dominant role in meteorological science. He lectured widely, was appointed to international commissions on aerial navigation, and proposed a global network of observatories that would be staffed by people trained at Bergen. Transatlantic aviation did not develop as quickly as at first anticipated, but it became a commercial reality in the 1930s and ensured that Bjerknes' proposed network was at last established. The Second World War focused yet more attention onto meteorology – the weather was a crucial factor in the timing of D-Day for the Allied invasion of Europe in 1944. Meteorology benefited directly from the invention of radar, and in later decades from the space programme, both begun as military projects but offering the atmospheric scientists new tools for remote sensing of weather patterns.

In the 1950s the Bergen school's models were transcended by new ideas pioneered by Carl-Gustaf Rossby (1898–1957). Although trained by Bjerknes, Rossby moved to the United States and built up a programme in meteorology at the Massachusetts Institute of Technology, drawing support from the Guggenheim Fund for the Promotion of Aeronautics. He showed that the circulation of the upper atmosphere would have to be taken into account in predicting the weather on the ground. At high levels there were several different layers of westerly winds that interacted to give what became known as Rossby waves, which in turn could be linked to the well known cyclones. The theoretical sophistication of these new models, coupled with a vast increase in the amount of information now available from observation networks, at last began to offer the prospect of weather forecasting based on exact mathematical predictions. Here the introduction of computers provided the speed of calculation needed to process the data quickly enough, pioneering efforts being made in the 1950s by a team of mathematicians led by John von Neumann (1903–57). The Global Atmospheric Research Programme, begun in 1966, has carried on these studies at a theoretical level.

Ice Ages Again

Although practical issues naturally drew meteorologists to the study of the earth's atmosphere in its present state, there was a long-standing question centred on major climatic variations in the recent geological past. By the beginning of the century it had been firmly established that there had been a sequence of ice ages, and most geologists accepted that there had been four major glaciations separated by warmer interludes (chapter 6). James Croll had tried to account for this phenomenon in terms of variations in the sun's heat resulting from eccentricities in the earth's orbit. Many geologists believed that the timing predicted by Croll's theory did not fit the facts, since the last ice age seemed to have ended much more recently than his calculations would allow.

Early-twentieth-century climatologists were still interested in the problem of past variations in the climate, two leading figures being Wladimir Köppen and his son-in-law Alfred Wegener. A new attack on the possibility of an astronomical cause for the ice ages was initiated by the Yugoslav engineer Milutin Milankovitch (1879–1958), who wanted to produce a mathematical model of how the sun's heat affected the upper atmosphere. Köppen informed Milankovitch that – contrary to Croll's assumption – it was a sequence of cold summers, not winters, that would trigger off an ice age, because the ice built up in each winter would not be completely melted. Milankovitch now checked his astronomical calculations to determine when such sequences of cold summers would have occurred in the past.

Köppen and Wegener were sufficiently impressed with the results to publicize Milankovitch's theory in their *Die Klimate der geologischen Vorzeit* (*Climates of the Geological Past*) of 1924. Further works on the theme from Köppen and from Milankovitch himself followed during the 1930s. Although some Europeans were sympathetic to the idea, many Americans opposed it. There was some fit between Milankovitch's cycles and the accepted sequence of glaciations, but the timings were not close enough to convince many specialists. It was also argued that Milankovitch concentrated solely on variations in the sun's heat, making no allowances for the circulation of the atmosphere as it was now understood by meteorologists.

In the 1950s there was a general reaction against the astronomical theory, due partly to the new radiocarbon dating technique, which seemed to show that there had been a warm spell in the middle of a period of minimum solar heating. But different dating techniques were beginning to throw doubts on the whole of the accepted pattern of glaciations. Cores drilled from the ocean bed offered new ways of checking the earth's past temperatures by noting variations in the populations of the minute sea creatures known as Foraminifera. By the 1960s the old system of four main ice ages postulated by Penck and Brückner had been broken down by the emergence of evidence for a much more complex sequence of glaciations. By the 1970s some geologists were arguing that the evidence could fit with a modified version of the Milankovitch cycle, although the topic remains controversial.

DRIFTING CONTINENTS

Alfred Wegener is remembered not for his work on climates, but as the first major champion of the theory of continental drift. He was an exponent of a German tradition in geophysics that drew no sharp line between the study of the earth and of its atmosphere. From the perspective of this tradition, it was quite natural for the same person to do research on paleoclimatology and on the structure of the earth itself. The fact that continental drift has subsequently emerged as a consequence of the mechanisms postulated by the modern theory of plate tectonics has drawn attention to one aspect of Wegener's career – and has highlighted a single factor within the complex debates that took place within early-twentieth-century geophysics. There are now several historical studies outlining Wegener's theory of continental drift and its reception by other scientists. The relationship between his theory and the modern conception of the earth's structure has been minutely analysed. Much attention has also focused on the reasons why Wegener's ideas were greeted with such scepticism at the time, even though we now know (with hindsight) that he was partly right. The proposal of new ideas about the earth's history has been treated as a test case for evaluating various theories about how science is supposed to proceed.

Whatever the value of this kind of analysis, there can be little doubt that the focusing of so much historical effort upon the debate over continental drift has seriously distorted our understanding of what happened in the earth sciences during the first half of the twentieth century. The fact that certain aspects of Wegener's theory turned out to be right (for reasons he could not anticipate) has led historians to write about the drift debate as though it were the centre of everyone's attention at the time. Yet a history of the earth sciences written in 1950 would have treated continental drift not as its main theme but as a maverick theory that had gained some support but had never been able to impress the majority of geologists or geophysicists. Hindsight has led us to focus on this one theory because we know that it turned out to contain a kernel of truth – but at the same time it has led us to neglect other debates that would have seemed far more important to those who actually participated in them. There is little that a survey of the period can do to redress the balance, since information on many topics is not available outside the original technical literature.

Geology in Crisis

The geology of the early and mid nineteenth century had concentrated on stratigraphy, and this area certainly did not disappear. The basic divisions of the geological record were consolidated at the second International Geological Congress of 1881 and revised by an International Commission of Stratigraphical Classification in 1901. Geologists still use the traditional system to establish the relative age of formations, and the twentieth century has seen notable developments in the precision with which fossils, especially microfossils, are used to identify strata. But these are refinements of a system that was already established by the 1850s. By the end of the nineteenth century, the theoretical interests of geologists had already switched to the structure of mountains and continents, with efforts being made to date and explain the various earth movements involved (chapter 6).

The earth sciences were in a state of unusual tension at the turn of the century. Existing theories were known to be inadequate, and continental drift, far from solving the problem, seemed to add yet another controversial alternative. A consensus was not reached until the 1960s, when the theory of plate tectonics emerged. This offered a completely new model of the earth's crust, and the scientists involved were well aware that they were participating in a theoretical revolution of major proportions. But historians are by no means certain that they should support the view that geology underwent a 'scientific revolution' in the post-war years. A conceptual transformation certainly took place in theories of the earth's structure, but should this be regarded as a revolution within the existing science of geology? An alternative view is that the revolution took place when the relatively new science of geophysics staged a takeover bid, displacing the traditional discipline of geology from its original position as the main science of the earth. The new ideas came from within geophysics, and as a result the earth sciences were reconstituted along a new axis, with the traditional disciplines being forced into a subordinate role. On this interpretation, the theoretical confusion of the early twentieth century represents the death-throes of the once-powerful discipline of geology, and the birth-pangs of the modern earth sciences.[2]

The weakness of the old theories was already becoming apparent in the early years of the twentieth century. Even the information generated by orthodox research techniques was proving difficult to assimilate. The complex nature of the deformations affecting strata was becoming ever more apparent. The Alps had already been interpreted as a series of folds and overthrusts due to lateral compression, but extensive flooding during the drilling of the Simplon tunnel (1898–1915) revealed that the internal structure of the mountains was still not properly understood. More detailed studies were made during the early decades of the new century, confirming the extent to which whole sections of crust had been thrust one on top of another, as opposed to simple folding.

Far from understanding the causes that produced these horizontal compressions, however, geologists were finding them even more difficult to explain. The traditional assumption was that the earth itself was gradually shrinking as it cooled, thus wrinkling its crust like the skin of a drying apple. The resulting pressures not only produced mountain ranges, but also the more general arching of the crust that depressed some areas to become sea bed. This theory, advocated throughout the late nineteenth century by Eduard Suess, was based firmly on the assumption that the earth had been formed as a molten mass that had gradually cooled down. Estimates of the age of the earth by Lord Kelvin and others were based on the assumed rate of cooling, and had given figures that were (by modern standards) far too low. Now the whole basis of this conception of earth history was challenged from an unexpected direction. Kelvin and the other physicists of the mid nineteenth century thought that they knew all the forces governing the operations of the material universe. The discovery of radioactivity by Henri Becquerel (1852–1908) in 1896 revealed a new source of energy that could not have been included in earlier calculations. It soon became apparent that radioactive elements gave off heat energy over a long timescale, and that such elements were present in at least small quantities throughout the earth's crust.

This point was already recognized by one of the earliest

investigators of radioactivity, Ernest Rutherford (1871–1937), who later recorded an embarrassing incident in 1904 when he lectured to an audience including the now very elderly Kelvin[3]:

> I came into the room, which was half dark, and presently spotted Lord Kelvin in the audience and realized that I was in for trouble at the last part of the speech dealing with the age of the earth. To my relief, Kelvin fell fast asleep, but as I came to the important point, I saw the old bird sit up, open an eye and cock a baleful glance at me! Then a sudden inspiration came, and I said Lord Kelvin had limited the age of the earth, *provided no new source of heat was discovered*. That prophetic utterance refers to what we are now considering tonight, radium! Behold! the old boy beamed upon me.

In fact, Kelvin continued to defend his old position, but by 1906 Rutherford's idea was being developed in detail by R. J. Strutt, later Lord Rayleigh (1875–1947). Strutt's student, Arthur Holmes (1890–1965), summed up the new approach in his book *The Age of the Earth* in 1913. The immense timescale once demanded by Charles Lyell and the uniformitarians was now vindicated, and the physicists were beginning to develop techniques that would use the proportions of radioactive elements and their decay products to estimate the actual ages of the rocks in which they were contained. These techniques, forerunners of the modern potassium–argon method, provided the first really secure dates for the earth's history and suggested that the planet itself was several billions of years old. The modern estimate is approximately 4.5 billion years.

By showing that the earth had cooled much more slowly than Kelvin anticipated, the physicists effectively undermined the logic of the 'wrinkled apple' analogy. If the earth was not cooling, then it was not contracting, and there could be no horizontal forces to produce the crumpling and arching postulated by Suess. The claim that continental areas could be depressed to form deep sea bed had, in any case, been challenged by evidence suggesting that the continents and the ocean floors were composed of totally different rocks. Gravity measurements confirmed that the overall density of the crust was not the same at all points on the earth's surface. The lighter sial (silicates of aluminium) formed the continents, while the denser sima (silicates of magnesium) formed a uniform layer that was exposed on the ocean floor but was covered by the continental sial. The continents were, in effect, rafts of lighter rock floating on the denser material beneath. On this model the continents were permanent features of the earth's surface – they could not be depressed by horizontal forces to provide new basins for the waters to drain into.

Wegener's Initiative

It was in the full knowledge that these latest developments had undermined the late-nineteenth-century synthesis that Alfred Wegener went one step further and declared that the continental rafts might actually move horizontally across the face of the earth. There is a fairly broad consensus that Wegener's background as a geophysicist may have given him the intellectual freedom necessary to conceive a hypothesis so much at variance with the assumptions of traditional geological theory. But beyond this, historians are divided over the degree of significance to be assigned to Wegener's proposal. Since the modern earth sciences have accepted the basic idea of drift and incorporated it into the theory of plate tectonics, there is a natural tendency to present Wegener as a bold pioneer who saw the evidence for drift at a time when his contemporaries were still blinded by old-fashioned preconceptions. Against this must be set the fact that Wegener did not anticipate the major theoretical revolution that generated the modern view of the forces that actually produce continental motion. He could not have foreseen the developments in oceanography that precipitated a revolution in ideas on the structure of the crust. It is thus possible to argue that Wegener's theory was not really an important anticipation of later developments; it was a symptom of the confused state of geological thinking at the time that just happened to anticipate a single aspect of the modern theory.

Wegener certainly appreciated the significance of some lines of evidence still accepted by modern earth scientists, but he was unable to come up with a reasonable explanation of why drift occurred. He accepted this limitation, but insisted that the evidence for the fact that drift had occurred was strong enough to be convincing. Geophysicists should accept the fact, while waiting for the more fundamental developments that would provide an explanation. The majority of his contemporaries were not willing to give him the benefit of the doubt.

Wegener was certainly not the first scientist to suggest that the continents might have moved in the course of the earth's history. In 1910 the American geologist F. B. Taylor (1860–1938) tried to explain the pattern of mountains formed during the Tertiary era by claiming that the continents had crept southwards from the Arctic. Greenland was the remnant from which North America and Eurasia had pulled away as they moved south. But Taylor's ideas attracted little attention, and it was left for Wegener to develop the first comprehensive argument for drift as it is understood today.

Wegener began his career as a meteorologist. He studied the climate of Greenland between 1906 and 1908, returning to a lectureship in meteorology at Marburg. He became increasingly interested in geophysics, leading to his work on paleoclimatology and his theory of continental drift. He conceived his theory in 1910 when he noticed the striking 'fit' between the east coast of South America and the west coast of Africa on a map. He soon began to quarry the literature for evidence that would fit the case for a separation of the continents in the course of the earth's history, and by 1912 was confident enough to lecture on the topic. His book *Die Entstehung der Kontinente und Ozeane* appeared in 1915 and was enlarged in a series of subsequent editions. The third edition of 1922 was translated into English as *The Origin of Continents and Oceans* (1924). Although the theory was controversial, Wegener enjoyed a successful career as a geophysicist and obtained a professorial chair at Graz in 1924. He died during another expedition to Greenland in 1930.

Wegener stressed the weakness of the contraction theory and the need to regard the continents as rafts of lighter rock floating on a denser medium beneath. He then went on to survey the evidence suggesting that these rafts had moved across the face of the earth in the course of geological time. He proposed that all the land masses had once been collected together in a single supercontinent, 'Pangaea', which had begun to break up in the Mesozoic era, the fragments steadily moving further apart until they took up their modern positions. The most powerful geological evidence for the break-up of the supercontinent was the 'fit' of South America and Africa. This was based on far more than mere geographical juggling with coastlines; Wegener was able to stress that the actual geological formations on both sides of the Atlantic also showed remarkable parallels. He argued[4]:

> It is just as if we were to refit the torn pieces of a newspaper by matching their edges and then check whether the lines of print run smoothly across. If they do, there is nothing left but to conclude that the pieces were in fact joined this way. If only one line was available for the test, we would still have found a high probability for the accuracy of fit, but if we have n lines, this probability is raised to the nth power.

The dating of the split was based on evidence from paleontology showing that up to the Mesozoic the inhabitants of the two continents were very similar. After this, evolution seems to have diverged to generate distinct South American and African species, something that could only be explained if the once-continuous area had been separated by an oceanic barrier.

Another line of evidence came from Wegener's interest in paleoclimatology. There were many indications in the geological record that parts of South America, Africa, India and Australia had experienced glaciation in the Permian era. This was difficult to explain had these areas been in their present locations, but made sense if they had once been concentrated around the south pole before drifting northwards. The tropical conditions enjoyed by parts of North America and Europe at the same time would then be explained by the fact that these land masses had occupied an equatorial position. The mountains running down the western edges of both North and South America could be explained as a crumpling effect produced on the leading edge of land masses being pushed in a westerly direction, while the Himalayas were due to the collision of India with the rest of Asia in comparatively recent geological times.

Much less convincingly, Wegener also used the similarity of glacial moraines on both sides of the North Atlantic to suggest that Europe and North America had still been linked in the most recent ice ages. This would imply a very rapid rate of drift since the glaciers disappeared, a point that Wegener thought was confirmed by measurements suggesting that Greenland and Europe were now separating at a rate of over ten metres per year.

The weakest part of Wegener's argument lay in his suggestions as to the causes that might have moved the continents. He assumed that the underlying layer of dense sima was static, and that something must be responsible for pushing the rafts of lighter sial across the surface, overcoming the immense frictional resistance. He argued that the earth's crust might have a consistency similar to pitch; it seemed rigid when struck a sharp blow, but would gradually yield when subjected to a steady pressure. As to the sources of this pressure, he had two suggestions: one involved the postulation of a westwards tidal force produced by the moon; the other a centrifugal effect tending to move continents away from the poles towards the equator. As though aware that these were hardly convincing, he argued that the Newton of drift theory had yet to come. Scientists had accepted that the planets orbited the sun long before Newton explained this fact in terms of the sun's gravity; similarly they should accept drift even though it was not yet clear how it occurred.

The Reaction to Wegener

The drift hypothesis attracted few adherents, and criticism of it increased rather than subsided over the following decades. There were, however, significant differences in the reactions of the various national scientific communities. In Germany, drift was taken as an interesting suggestion that would need much further research to prove itself. Only a few geologists openly rejected the idea. Reactions in the English-speaking world soon became much more hostile. In 1922 the British Association hosted a discussion of the theory, which was lively, but produced few outright rejections. In 1926, however, the American Association of Petroleum Geologists held a seminar in which most speakers rejected Wegener's arguments altogether. One of the most persistent critics, Bailey Willis, later wrote an article[5] entitled 'Continental drift: Eine Märchen [a fairytale]'. Through into the 1950s, drift would be seen as a wild idea supported by a few enthusiasts, but with no real standing in the scientific community as a whole.

There were many technical objections to the evidence that Wegener had brought to bear on the question. His geological parallels between South America and Africa were challenged, as was the supposed 'jigsaw' fit of the continental outlines. Paleontologists still preferred the idea of a now-sunken land-bridge to explain the similarities between the inhabitants of the continents up to the Mesozoic. English-speaking geologists were particularly hostile to Wegener's methodology, arguing that he had simply gathered a highly selective collection of facts from the literature, ignoring all the difficulties. As the British geologist Philip Lake wrote[6]:

> Wegener himself does not assist his reader to form an impartial judgment. Whatever his own attitude may have been originally, in his book he is not seeking truth; he is advocating a cause, and is blind to every fact and argument that tells against it.

There was a widespread feeling that a real scientist should base his or her hypothesis upon massive fieldwork – simply combing the literature for evidence did not count. This criticism did not arise in Germany, where Wegener's technique was seen as a legitimate way of testing a general hypothesis (Suess and other eminent theorists had not engaged in extensive fieldwork).

Another methodological objection centred on whether or not the theory was compatible with the prevailing enthusiasm for 'uniformitarianism'. Charles Lyell's insistence that the geologist should explain everything in terms of observable causes was now part of the rhetoric (if not the actual practice) of most British and American geologists. Wegener certainly argued that drift was still taking place today, but his theory supposed that the continents had suddenly begun to break up in the Mesozoic and were now gradually dispersing. This postulation of a mysterious 'starting point' followed by a directional process leading towards the present state of the earth was identified by some geologists with the world view of the now-discredited catastrophism.

Perhaps the most substantial objection of all, however, was the incompatibility between the theory of continental movement and prevailing views on the structure of the earth's crust. Even accepting that the continents were like rafts of lighter material resting on a denser substratum, what force could possibly push them across the face of the earth, overcoming the immense friction that must exist between the two layers? As long as the internal structure of the earth was seen as essentially static, continental drift remained implausible. The British geophysicist Harold Jeffreys (1891–1990) brought out this objection in his textbook *The Earth* first published in 1924, pointing out that the mechanisms suggested by Wegener were many orders of magnitude too small to overcome the friction between the continents and the underlying crust.

A few geologists stood out against the general hostility. The Harvard geologist R. A. Daly (1871–1957) proposed a new mechanism based on the possibility that the continents might be sliding down from a polar 'bulge' in the earth under the force of gravity. Most enthusiastic of all was the South African Alexander Du Toit (1878–1948), who was particularly impressed with the geological similarities between his native continent and South America. He improved the theory by abandoning Wegener's insistence on a late and very rapid separation of North America and Europe. He also postulated two original continents, Laurasia and Gondwanaland, instead of the monolithic Pangaea. Du Toit's book *Our Wandering Continents* of 1937 summed up all the arguments for drift and made it clear just how revolutionary the new idea was[7]:

> While in the future some closer approximation in viewpoint will become possible, it must frankly be recognized that the principles advocated by the supporters of Continental Drift form generally the antithesis of those currently held. *The differences between the two doctrines are indeed fundamental and the acceptance of the one must largely exclude the other.*

Small wonder, perhaps, that most of his contemporaries dismissed Du Toit as another bigoted advocate striving to revive long-discredited arguments through the use of flowery and over-enthusiastic language.

The most innovative support for drift came from the British geophysicist Arthur Holmes, who proposed a mechanism anticipating some components of the modern explanation. As a leading student of the geophysical effects of radioactivity, Holmes was well aware of the weakness of the old contraction theory. At first he supposed that the heat given out by radioactive elements would merely slow down the cooling of the earth. But by 1925 he had become so impressed by the evidence for vast upwellings of molten magma at certain points in the earth's history that he challenged the whole idea of a cooling earth.

Holmes now supposed that radioactivity produced so much heat that the real problem was how the earth was able to get rid of it fast enough to prevent a general heating-up of the interior. The occasional discharge of molten rock onto the surface was one obvious mechanism, but in 1927 Holmes began to argue that there might actually be convection currents within the crust. Hot, softened rock rose gradually to the surface over a concentration of radioactive heating, then spread out in a horizontal current before sinking back to the depths when cool. He soon saw that such currents would provide a new mechanism for drift: instead of the continents being forced over a static underlying crust, they would be carried along by movements within the crust itself. He supposed that convection currents would be most likely to arise under a continent, which provided a blanket trapping heat in the crust beneath, thus explaining the general tendency for large continental areas to be broken up through time.

Holmes did not anticipate the modern concept of 'sea-floor spreading' (see below), but his model of convection currents in the crust represents a pioneering version of the mechanism accepted today. His theory attracted few adherents, however, and produced no general wave of enthusiasm for drift. Historians seeking to use this episode to throw light on the general process of scientific innovation ask why the geologists of the 1920s and 1930s were so opposed to drift, even after the appearance of a new mechanism. It is understandable that those trained in traditional geology should have been unwilling to accept the new idea – but geophysicists such as Jeffreys also opposed Wegener because they could see no plausible mechanism. Yet if the lack of mechanism was crucial, why was Holmes' suggestion not greeted as a breakthrough? One possible answer is that the effects postulated by Holmes were invisible, given the techniques available to geologists at the time. There was no way of telling whether or not the proposed convection currents existed. The breakthrough would come when an explosion of research into the study of the ocean bed provided a new model for the structure of the crust in the 1950s. Without the onslaught of new evidence and new techniques, the old idea of a static interior for the earth was too strongly embedded in scientists' thinking.

PLATE TECTONICS

This breakthrough came in the years following the Second World War, and owed a great deal to scientific research that had actually been stimulated by the war. New evidence for drift came to light from paleomagnetism (the study of the past state of the earth's magnetic field as preserved in some rocks), and from the extensive exploration of the ocean bed. The oceanographic research eventually proved that new areas of crust are actually being created along the mid-ocean ridges, as hot rock works its way to the surface. A new model of convection currents in the crust was worked out and backed up by an impressive array of evidence. Continental drift now became not merely plausible, but an inevitable consequence of the new 'mobilism' – a commitment to the idea that the whole of the earth's crust is in a state of flux. Because it divides the earth's surface into a series of plates bounded by the creation and disappearance of crust at the edges, the new model is often called the theory of 'plate tectonics'. By the 1970s the new paradigm was well in place, and most earth scientists were confident that their field had undergone a major revolution.

Paleomagnetism

The original breakthroughs in the study of paleomagnetism came from a small group of British physicists and geophysicists in the 1950s. Their work was stimulated by a controversy over the fundamental cause of the earth's magnetic field. The most popular view, promoted by Edward C. Bullard (1907–80), was that the field arose from dynamo-like effects produced by convection currents in the earth's core. But Patrick M. S. Blackett (1897–1974), who had won the Nobel prize in 1948 for work in astrophysics, proposed a radical new theory in which magnetism was generated by any rotating body. Blackett tried to test his theory by searching for a field produced by rotating a large mass of gold borrowed from the Bank of England.

Controversy also centred on the evidence for the past state of the earth's field, since, if Blackett was right, the field would have stayed the same throughout the period in which the earth had rotated on its present axis. Evidence for the past direction of the earth's field could be found 'frozen' in some rocks; as they were formed the rocks had preserved a slight 'remanent magnetization' recording the direction of the field at the time. These minute fields could only be detected by very sensitive magnetometers, but such instruments were now becoming available. Blackett had worked on the detection of magnetic mines during the war, and he now turned his skills to the production of a magnetometer that would detect the fields for which the geophysicists were searching.

Although Blackett's fundamental theory of magnetization was rejected, the rivalry between his group and supporters of the conventional view of the earth's field stimulated a substantial programme of research into remanent magnetization. It was found that rocks formed at different periods often recorded directions of the earth's field differing substantially from the present. One possible explanation of this was that the earth's magnetic poles had moved in the course of time, the continents remaining in fixed positions. If this were so, then rocks formed at the same period should all show the same direction of magnetization. Studies in India and other parts of the world soon showed that this was not the case; the remanent magnetization of rocks from the same period gave apparently different directions for the field in different locations. This could be explained if the continents had not been in their modern positions with respect to the poles at the time. Wegener had, in fact, suggested that India had moved northward and collided with the rest of Asia. The study of remanent magnetization now seemed to confirm this and other continental displacements. By the early 1960s the British geophysicists were arguing for a revival of Wegener's theory.

The evidence for changing directions of the magnetic field recorded by the rocks was linked to a puzzling anomaly: in many cases the actual direction of the field seemed to be reversed. One explanation was that some rocks are able spontaneously to reverse the direction of their remanent magnetization, but a few geophysicists were already beginning to suspect that the earth's magnetic field itself reversed from time to time. On a timescale of millions of years, there would be occasional reversal events in which the north and south magnetic poles would change place. Remanent magnetization would record these events, and if the rocks could be dated accurately enough a complete scale of reversals could be plotted against the geological record. Careful tests were needed to rule out the possibility of spontaneous reversal of the remanent magnetization within the rocks, but in the course of the 1960s the weight of evidence began to point towards reversals of the earth's whole magnetic field.

By the late 1950s several European geophysicists had begun to explore the possibility of constructing a magnetic-reversal scale that would divide the stratigraphical column into periods of normal and reversed polarity, with the timing of the reversals precisely fixed by comparison with geological dating methods. The Dutch geologist Martin G. Rutten published the outline of a scale in 1959, but was unable to follow up the proposal in part because of the lack of precise dating techniques. By this time, however, American scientists at Berkeley, California, had begun to refine the radiometric technique for dating rocks, using especially the potassium–argon method (this compares the proportion of a radioactive isotope of potassium and its decay product, argon; the greater the proportion of potassium transformed into argon, the older the rock). In the early 1960s a group at Berkeley led by Richard Doell (*b*. 1923), Alan Cox (*b*. 1926) and G. Brent Dalrymple (*b*. 1937) began to combine the study of magnetic reversals with potassium–argon datings, producing a crude timescale of reversals for the Pleistocene era in 1963. Australian scientists produced their own scale based on the dating of Hawaiian lava flows, and at first there were some discrepancies. It was soon realized that immense care had to be taken when sampling the rocks – a difference of a few metres in the point of collection might represent a move from one lava flow to another of an entirely different date.

By 1965 Doell and Dalrymple were working on a series of rock samples from New Mexico and identified the most recent reversal event, which they named the 'Jaramillo event' after Jaramillo Creek where the crucial samples had been gathered. It was dated at 0.9 million years before the present. Doell later recalled how precarious their evidence seemed when first analysed[8]:

Alan [Cox] was a little doubtful about the Jaramillo event at the time. I can remember that night, having the bet for a martini about that. He thought we were going too far out, to base this just upon a few rocks. 'But, hell,' we argued with him, 'we had no more rocks than that when we hypothesized the whole thing [the polarity-reversal scale] in the beginning.'

The evidence did hang together, and Doell and Dalrymple subsequently telegraphed Cox, who was on a field trip in the Yukon, to claim their Martini. Their paper outlining the now completed scale for the Pleistocene was published in the journal *Science* in 1966. The Jaramillo event was soon to play a crucial role in establishing the evidence for the theory of sea-floor spreading.

Sea-Floor Spreading

The interview with Richard Doell quoted above tells us that we have moved into the most recent period of the history of science, where history becomes virtually inseparable from what is still going on. We are now sufficiently conscious of the importance of new developments in science for histories to be written while the participants are still alive to reflect on what they have done. Once a consensus has been reached on a major innovation, those who pioneered the new idea may write accounts for publication, or may be interviewed by eager historians anxious to question them about what really happened. Shelves of tape recording are now piling up in the archives along with the more traditional paper records (this is just as well, considering that the telephone has dramatically reduced the number of casual interactions committed to paper). The histories of recent events that have been based on this kind of evidence provide an intimate glimpse into the world of research, showing us the uncertainties, the rivalries and the manipulations that take place within the scientific community. There is a great deal at stake in the modern world of big science, and those who play the game – both successfully and unsuccessfully – are well aware of this. Their recollections provide a vital source of information for the historian of modern science, although it is information that reflects personal feelings and must thus be evaluated with caution.

The same level of immediacy is conveyed by accounts of the developments in oceanography that generated the new theory of the earth's structure. In the 1930s efforts were made to explore the crust beneath the oceans by adapting the seismological techniques already used on dry land. The shock waves emanating from explosions were reflected from the subsurface features and could be detected by a seismograph (the instrument used for detecting the shocks produced by natural earthquakes). The massive investment in oceanographic research produced by the military demands of the Second World War and the subsequent Cold War generated an immense amount of new information about the structure of the sea floor. Before the 1950s the oceans had been unknown territory as far as geophysicists were concerned, but by the end of the 1960s they were better known than the land. The study of remanent magnetization in underwater rocks received a boost from the military's efforts to detect submarines. By the 1950s a magnetic detector was available that could be towed behind a ship to give a record of the magnetization on the sea floor beneath. The US Navy was anxious to have maps of the natural magnetization of the ocean bed so that the presence of enemy submarines could be identified more easily.

This research overturned almost everyone's expectations of what the sea floor would be like – including the predictions of Wegener and his followers. Many areas were remarkably uniform, and were composed of basalt only six kilometres thick. The continents, by contrast, were of granite over forty kilometres thick. Far from being ancient, as predicted by the theory of a static earth, the oceanic rocks turned out to be comparatively young. It had long been known that the uniformity of the sea bed was broken by occasional ridges and deep trenches, but now the world-wide pattern of these features was revealed. The mid-Atlantic ridge – a line of underwater mountains running the length of the North and South

Atlantic – was matched by similar features in other oceans. Dredging showed that the ridges were the youngest rocks of all, being formed of recently produced volcanic material. Measurements of heat flow confirmed their volcanic origin by detecting a much higher heat output from the ridges than from the surrounding material. Holmes' prediction that the 'hot spots' would be underneath the continents was falsified. The ridges were also a focus for shallow earthquakes. The ocean trenches, by contrast, were the source of deep earthquakes, although they too were associated with extensive volcanic activity on nearby land masses.

Magnetic surveys showed that there were patterns of anomalies in the rocks surrounding the mid-ocean ridges. On either side of the ridges there were parallel stripes of rocks bearing alternately normal and reversed remanent magnetization. To begin with, no one linked this phenomenon with the timescale of normal and reversed states of the earth's magnetic field being produced by the researchers into paleomagnetism. In the end, however, a particularly clear set of magnetic anomalies detected around the Juan de Fuca ridge (off the west coast of North America) by the research vessel *Eltanin* in 1965 was to prove crucial in establishing a new theory to explain the structure of the ocean bed.

The search for a new theory was well under way by 1960, since none of the existing views of the earth's crust made sense in terms of the new data. The geophysicists were able to use the evidence that the ocean floors were essentially different from the continents as the basis for a direct challenge to the authority of the traditional, land-based geologists. If there was to be a new theory in the earth sciences, it would be based on the evidence provided by marine geophysics; the problems identified by the study of the continents would have to be solved as a by-product of a theory whose structure would be determined by the new disciplines. The latest evidence for continental drift provided by paleomagnetism would form part of the new theory. An alternative favoured by some geophysicists was that the earth is actually expanding as crust is created at the mid-ocean ridges. Eventually, however, attention would focus on a model of sea-floor spreading linked to the idea of convection currents in the crust. Although differing significantly from Holmes' earlier suggestion, this again provided a mechanism that would explain the movement of the continents.

A leading figure in the articulation of the new theory was the American geophysicist Harry Hess (1906–69). Already in the 1930s Hess had become interested in the idea of convection currents in the crust as an explanation for some of the evidence even then beginning to emerge about the ocean floors. During the war he commanded an American naval vessel in the Pacific and had used the sonar equipment for surveying. In the 1950s Hess suggested that the mid-ocean ridges might be the sites at which new material was brought to the surface by the convection currents resulting from radioactive heating.

By 1960 Hess was ready to generalize this idea as the basis for a global synthesis in which the oceans were the chief centre of activity. Ridges were where new crust was produced, and trenches marked the sites at which old crust was drawn downwards into the depths, thus completing the cycle of the convection current. The ocean floors were young because they were constantly being produced and destroyed; only continents – too light to be drawn down by the currents – would preserve records of the distant geological past. The continents would move across the surface, carried by the horizontal motion of the convection current. Hess was able to incorporate the old arguments for drift into his theory, but the mechanism was now firmly based on the ocean basins. His theory was published in 1962, but aware of its highly speculative nature he presented it as 'geopoetry' – a global vision that needed much more work before it could be refined into a real theory. The term 'sea-floor spreading' was coined by Robert Dietz in 1961.

Hess gained few immediate converts, but his suggestions were seen as sufficiently important to become a major focus of attention for researchers at the various oceanographical institutes. The Canadian geologist J. Tuzo Wilson (*b*. 1908) linked the theory to traditional concerns over mountain-building expressed by land-based researchers. Wilson argued that if oceanic islands were produced by the upwelling at mid-ocean ridges and then carried away by the convection current in the sea bed, then the further an island was from the ridge, the older it should be. This was confirmed by the available geological evidence. In 1963 Wilson published a popular article on continental drift in *Scientific American*, bringing the new ideas to a wider audience and confirming that drift had at last become a major force in the thinking of earth scientists. Edward Bullard, a key figure in the British paleomagnetism group, provided a new map confirming the 'fit' of the continents and supported drift at a 1963 meeting of the Geological Society of London. At this stage there was still a great deal of opposition, but in the same year a dramatic new line of evidence confirmed sea-floor spreading and began to tip the balance in favour of the new approach.

The new suggestion came from Fred Vine (*b*. 1939) and Drummond Matthews (*b*. 1931) of Cambridge University, both of whom had been impressed by Hess' interpretation of mid-ocean ridges. As a graduate student, Vine was trying to make sense of the patterns of magnetic anomalies surrounding the ridges, and he realized that, if Hess' theory was correct, the new rock upwelling at the ridge would be imprinted with remanent magnetization corresponding to the direction of the earth's field at the time. After a reversal of the field, the later rock would be imprinted with the opposite direction of magnetization. The result would be a pattern of horizontal stripes parallel to the ridge, each side being a mirror image of the other because there was an equivalent spreading of the sea floor in both directions. The same idea was conceived independently by the Canadian

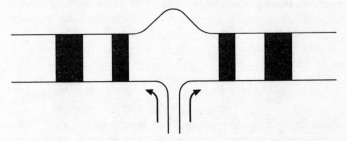

A cross-section through a mid-ocean ridge shows new crust being formed at the ridge as material rises from below, older crust being pushed to either side in the process. As the crust solidifies, it preserves the direction of the earth's magnetic field at the time. The dark bands represent crust formed during periods of reversed polarity – note that these are symmetrical on either side of the ridge.

geophysicist Lawrence Morley, who submitted it as a letter to the prestigious journal *Nature*. The editor told him that there was no room to print it, and when it was sent instead to the *Journal of Geophysical Research* Morley was told[9] that 'such speculation makes interesting talk at cocktail parties, but it is not the sort of thing that ought to be published under serious scientific aegis'. Morley had to give up, but Vine and Matthews were able to get an article into *Nature* in 1963, thus putting the suggestion onto the table for further discussion.

There was much opposition, and most of the workers at the Lamont Geological Observatory were sceptical. It was Lamont's survey vessel *Eltanin* that was producing the best surveys of magnetic anomalies, including results from the Juan de Fuca ridge in which a pattern of parallel stripes was obvious. On one sweep across the ridge, known as *Eltanin* 19, the mirror image characteristic of the stripes on either side of the ridge was especially clear. In 1965 the news of the Jaramillo event broke, thus completing the record of reversals in the earth's field. When Vine heard of this at a meeting of the Geological Society of America, he saw how the patterns could now be made to fit much more neatly into his theory[10]:

> The crucial thing at that meeting was that I met Brent Dalrymple for the first time. He told me in private discussions between sessions, 'We think we've sharpened up the polarity-reversal scale a bit, but in particular, we've defined a new event – the Jaramillo event.' *I realized immediately that with that new time scale, the Juan de Fuca ridge could be interpreted in terms of a constant spreading rate.* And that was fantastic, because we realized that the record was more cleanly written than we had anticipated. Now we had evidence of constant spreading; that was very important.

Even the Lamont group now realized that the evidence for sea-floor spreading was too powerful to be ignored, although one critic dismissed *Eltanin* 19 as too perfect a match to be true. Within a year or two, the majority of earth scientists began to accept that the new model represented the best answer to the questions they had been asking.

It was also in 1965 that J. Tuzo Wilson proposed the concept of the 'transform fault' to clarify the structure of mid-ocean ridges. Although the active section of the ridge (where new material was being produced) corresponded to a linear feature marking the edge of a convection cell, sections of the ridge were sometimes displaced from one another by transform faults to give an overall zig-zag effect, which would also be reflected in a corresponding displacement of the magnetic anomaly stripes. The notorious San Andreas fault in California is a transform fault linked to the Juan de Fuca ridge. The earth's surface could now be divided into sections or plates, each corresponding to a convection cell, and each bounded by active zones in which new crust was being either created (mid-ocean ridges) or destroyed by subduction beneath the existing crust (trenches).

Wilson's idea was developed into the full version of plate

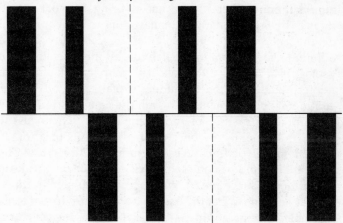

From above we can see parallel bands of crust preserving normal and reversed polarity on both sides of the ridge (vertical dotted line). The horizontal line represents a transform fault, where the ridge and the whole associated pattern of remanent magnetization is displaced at right angles to the ridge itself.

tectonics by Jason Morgan, Dan McKenzie and Xavier Le Pinchon in the mid 1960s. They realized that many of the geographical features of Wilson's plates were a direct consequence of the earth's spherical shape, which imposes geometrical constraints on the plates that are not apparent on a two-dimensional map. Le Pinchon produced a simplified scheme in which the earth's surface was divided into six major plates, the boundaries of which could be explained in terms of the convection-current hypothesis.

It was soon shown that deep earthquakes were produced where one section of crust was being forced beneath another, the same process also being responsible for the volcanic activity in areas such as the Andes of South America. In the case of both North and South America, the mountains running down the western edge of the continent result from the fact that the continental 'raft' is on the leading edge of a plate, facing the oncoming material from other plates that is being forced beneath them. The Alps and the Himalayas have been produced by the collision of continental areas, each driven by a different plate system. In the light of this interpretation, the geologists of the late 1960s and 1970s undertook a massive reinterpretation of the traditional explanations offered for continental phenomena. Although the data on stratigraphy and on the structure of mountain ranges remained the same, they could now be explained in terms of a totally revised model of the forces operating in the crust. The geographical and paleontological evidence for the break-up of continents suggested by Wegener was explored in much greater detail now that the basic concept of drift had become plausible.

Geologists were well aware that their discipline had undergone a conceptual revolution based on the creation of a picture of the earth's interior that was totally at variance with traditional ideas. The fact that the strongest evidence for the new theory had come from the new discipline of ocean-based geophysics forced a restructuring of the profession, bringing the exponents of the new fields into positions of influence and forcing the traditionalists onto the sidelines. The increasing use of the term 'earth sciences' rather than the older 'geology' seems to reflect a general sense that the conceptual revolution has been accompanied by major realignments within the scientific community. Whatever continuity there might be between Wegener's theory and plate tectonics is outweighed by the gulf separating even his form of geophysics (let alone the nineteenth-century tradition of geology) and the modern emphasis on the ocean floor as the key to the understanding of the earth's crust. Without the creation of the new research tools for exploring the ocean, and the emergence of a younger generation of geophysicists who saw the new data as the chief source of conceptual problems, the revolution would have been unthinkable.

Past and Present

The only level at which the modern earth sciences claim an intellectual link back to the nineteenth century is in the area of methodology. Charles Lyell is still a hero figure because the theory of plate tectonics is consistent with the principles of uniformitarianism. The past is explained in terms of processes that can still be observed going on in the world today. Lyell himself had speculated that a race of intelligent beings living in the sea would have a very different view of the earth's structure, a view that seems amply borne out by the role of oceanography in the modern earth sciences. Better techniques for observing the whole of the earth's crust have simply altered our conception of the observable processes available for the geologist to use when trying to reconstruct the past.

It must be remembered, though, that Lyell extended his uniformitarian principles so far that he claimed it was unscientific to expect that the geologist might see far enough into the past to find evidence of a period when the earth was significantly different from the present. Despite our reliance on uniformitarian principles in interpreting the recent geological past, modern earth scientists nevertheless accept that we have access to evidence telling us about the creation and very early state of the earth. There was a time when the continental 'rafts' and the underlying plates were still in the process of being formed. When we look billions rather than millions of years into the past, we do indeed see a state when things were quite different from the present.

The geologists' reliance on uniformitarian principles was challenged in the 1950s by the Russian–American scholar Immanuel Velikovsky, whose *Worlds in Collision* took the records of catastrophes left by many ancient cultures as the basis for a rival view of the earth's history. Velikovsky also used the arguments of early-nineteenth-century catastrophist geologists to bolster his case for massive upheavals.

He proposed a theory of interplanetary collisions that most astronomers regarded as quite fantastic. Velikovsky was dismissed as a crank by the geologists, some of whom organized a clumsy attempt to suppress the publication of his book. As a result, he became a hero to those members of the counter-culture who saw science as an expression of ruthless materialism. Velikovsky's books are still in print today, serving as a potent reminder that not everyone accepts the consensus reached by the professional scientists. It should also be remembered that many creationists dismiss the orthodox view of the earth's history and argue that the sedimentary rocks were deposited during Noah's flood – a view that the earth sciences have not taken seriously since the late seventeenth century.

Velikovsky's attack came from outside the scientific community, but the 1970s saw a revival of a more serious form of catastrophism based on the claim that there was evidence of major asteroid impacts causing world-wide devastation at certain key discontinuities in the geological record. This thesis has been linked to the fear of a 'nuclear winter' expressed by many proponents of disarmament, since the dust thrown into the atmosphere by an asteroid impact would produce consequences similar to those of a nuclear war – including the extinction of many forms of life. The evidence for the supposed impacts has been disputed by many earth scientists. The general preference for uniformitarian explanations has led most geologists to favour the view that the 'discontinuities' marking the boundaries between the geological periods are more apparent than real. The best-known example of such a discontinuity is the Cretaceous–Tertiary boundary, marked by the disappearance of the dinosaurs and many other forms of life. According to the orthodox view, there was no 'mass extinction' at a single point in time, only the completion of a trend that had been building up over a vast period. But if the exponents of the new catastrophism are right, then the discontinuities represent genuine mass extinctions caused by effects that lie outside the scope of the earth sciences.

Not surprisingly, the new theory is favoured more by cosmologists than by geologists and paleontologists. Nevertheless, it is important that we conclude by noting the continued existence of theoretical debates over the nature of the earth's past. The new catastrophism is different from the old because it accepts the uniformity of internal processes and appeals to external factors (asteroids from outer space) to justify its emphasis on discontinuity. But despite the triumph of plate tectonics, arguments continue over some of the most fundamental issues that have disturbed the geologists of the last two centuries. The emergence of a new paradigm does not mean that geology has lost its ability to stir up passionate debate. As soon as one issue is settled to the satisfaction of the majority, a new one rears its head. Many non-scientists remain suspicious of the whole theoretical framework upon which the modern view of the earth's history is built. Now, as ever, scientists – and their opponents – are as much concerned with evaluating new ideas as with the mere collection of factual data.

Darwinism Triumphant

The sciences dealing with various aspects of the earth's biosphere have undergone conceptual revolutions even more fundamental than those in the earth sciences. Developments here have also been more controversial, partly because topics such as the Darwinian theory of evolution have always stirred people's emotions through their impact upon how we perceive ourselves as human beings. But in addition to the ongoing debates over the threat of materialism, the environmental dimension of the life sciences has become controversial through the growing public concern over the state of our planet as an abode for life. A century ago the question of whether or not we are related to the apes would have been debated heatedly at all levels of western culture. Now only certain religious groups object to the link, and their concerns have been marginalized by the growth of a more general fear that we may wipe out the creatures that represent our closest living relatives – along with many others. The genetic relationships postulated by evolutionism are taken for granted by those whose concern is the imbalance between humankind and the rest of the living world. Far from being 'Nature's crowning glory', the human race is increasingly being presented as a threat to the very system that produced it.

If the basic idea of evolution has been taken for granted by most intellectuals, our understanding of how the process works has undergone a major revolution. Darwin's theory of adaptation by natural selection had fallen into the background around 1900, but it was revived in the 1920s as it became clear that the new science of genetics had undermined most of the rival mechanisms. By the 1940s the 'modern synthesis' of Darwinism and genetics had begun to dominate all of the sciences to which evolutionism could be applied. The new emphasis on adaptation and reproduction helped to focus attention onto the behaviour of animals in the wild state, and the science of ethology became the latest addition to the ever-widening range of biological disciplines. The study of animal behaviour naturally interacted to some extent with the growing emphasis on ecology. Having established itself as an independent discipline in the early years of the twentieth century, ecology struggled to retain its place until catapulted into the foreground by the growing environmental concerns of the post-war years (chapter 11). Now the image of gorillas or chimpanzees living naturally in the wild has come to symbolize the fragile nature of the physical and biological environments that make such natural behaviour possible.

The historian seeking to chart these developments must thread his or her way through a minefield of interlocking issues and problems, some of them with explosive moral or ideological implications. Even within the sciences themselves there are hidden complications and rivalries, which have often been concealed by the scientists' own tendency to ignore or disparage anything that does not fit their public image. Here, as in the earth sciences, the emergence of new disciplines, and the resulting inter-disciplinary rivalries, has often determined the direction of theoretical innovation. In biology the relationship between the laboratory and field disciplines has often been one of tension and recrimination. The early geneticists openly repudiated the field naturalists' work on adaptation and geographical isolation, creating instead an abstract model of evolution that had little relevance for the real world. The synthesis of Darwinism and genetics was important both as a theoretical initiative and as a means of bringing the laboratory and field dimensions of biology back into communication. Even in the post-synthesis era, there are still major differences between the ways in which the various disciplines interpret their common theoretical foundations.

Although the increasing technicality of the sciences has pushed them beyond the comprehension of even the informed layperson, the implications of some theoretical perspectives have remained controversial. Opposition to the whole idea of evolution has been largely confined to religious fundamentalists, whose fortunes have by no means declined in the twentieth century, at least in America. But the Darwinian theory of natural selection has continued to draw fire from a wide range of groups who see its 'trial-and-error' mechanism as the negation of all hope that living things might play a creative role in Nature. Various strands of holistic thought have continued to emphasize that a living organism is something more than the sum of its parts. This viewpoint has been effectively marginalized in those areas of biology which are influenced by the experimental method, but a similar anti-reductionist approach played an important role in ecology until the post-war era. As concern for the state of the environment has grown, more scientists may be tempted back to the image of Nature as a system of fruitful interactions.

Opposition to the claim that the universe is nothing more than a machine driven by the blind laws of physics is but one facet of an increasingly strident challenge to the exploitative ethos that has triumphed within most areas of science. This is not a purely moral issue, however; it has a political and ideological dimension that creates immense problems for the historian. Many areas of science offer us practical means of controlling Nature, but, even where direct control is not involved, science can be used as a source of images and metaphors that shape our attitudes and beliefs. The image of Nature as a machine that can be tinkered with at our pleasure is a potent symbol of the exploitative view of humankind's relationship to the rest of the universe. Through books and museums, various aspects of the life sciences have been used to shape public opinion in a way that may be undetectable by the man or woman in the street who is the target of the imagery. The areas of race and gender offer obvious examples where twentieth-century biology has been exploited for ideological purposes. The environmental movement seeks its own counter-images to challenge those projected by the military–industrial establishment.

Modern historians of science are constantly extending their ability to detect the indirect ways in which science has been exploited for ideological purposes. The scientists themselves are often unwilling to admit that their work may be influenced by external factors, but in many cases they may be simply unaware of what is going on – a classic case of not being able to see the wood for the trees. There is often a good deal of tension between historians seeking the overall picture and scientists convinced that their own particular research has been conducted according to purely objective standards. This tension can also develop in the reverse direction: the opponents of materialism do not like to be told that their past heroes may have had feet of clay, or that ideas now praised for their liberating influence have been associated with less wholesome activities in the past. Yet the historian can show that anti-Darwinists, for instance, have been quite capable of endorsing the application of their own theories to justify racial discrimination. The truth is that all sides in the ideological debate wish to reconstruct history to vindicate their own position. Historians can hardly claim to be free of all prejudices, but they are at least aware of what is going on and can try to expose the more obvious cases of manipulation. If they are damned from all sides for exposing the weakness of everyone's favourite images, they can be sure that they are doing their job.

SCIENCE AND IDEOLOGY

Science has played an integral role in the process that brought western culture to a position of world domination in the early decades of the twentieth century. Not surprisingly, it has continued to play a similar role even as the west's position has become more uncertain. The fact that Europe and America could no longer command instant obedience from subordinate cultures did not alter the need for symbols that demonstrate an assumed moral superiority. In the seventeenth century, Protestant Christianity seized upon the image of a world machine constructed by God for humankind's benefit to justify the claim that Europeans were entitled to exploit Nature's resources on a global scale. By the late nineteenth century, this claim was increasingly based on the idea that the white race was the pinnacle of evolutionary progress – with the same virtues of industry and initiative being rewarded by Nature rather than by God.

The situation has been made more complex by the rise of alternative ideological positions even within western culture. The emphasis on material progress has been challenged by reformers and revolutionaries determined to show that Nature is really on their side, not on that of the commercial establishment. Marxists have called for the redistribution of wealth within society, but have tended to assume that humankind should still exploit Nature for its own benefit. Anti-materialist and environmentalist groups have developed a more radical challenge, which questions the value of exploitation and the search for material riches. Perhaps the individual should transcend the petty world of commercialism, seeking the meaning of his or her own life in artistic or spiritual endeavour. Perhaps we should all abandon the world of science and technology, and retreat into an idyllic past of yeoman husbandry. Perhaps the west's fascination with dominating the material world is but a symptom of a deeper malaise caused by turning our backs on 'feminine' values that were once the foundation for a more organic view of Nature.

Some of these alternatives represent positions with a long pedigree, while others are relative newcomers. All seek to exploit history for their own ends, rewriting the story of the past to give an interpretation that highlights their own concerns. Just as orthodox scientists may write a history of their discipline stressing its contributions to factual knowledge and our material well-being, so the opponents seek to justify their rejection of materialism by emphasizing the heroic struggles of their forebears against the growing dogmatism fostered by the military–industrial complex. Some of these critiques relate directly to the history of science by showing us how the scientists' theories have reflected particular value-systems. Others challenge the values of western civilization as a whole and seek the origins of modern problems in the very foundations of the Judeo-Christian world view.

The professional historian is inevitably drawn into the resulting debates, often as a critic seeking to expose the distortions of a particularly biased revision of the past. But historians are human beings too; each has his or her own axe to grind, and will target appropriate trees in the forest of interpretations on offer. In any highly controversial area, it is almost impossible to write unbiased history. The outsider may wonder why anyone should bother with such a morass of conflicting interpretations. But the range of critical perspectives does at least force all of us to be on our guard. An informed reader is much less likely to be tricked into accepting a particular view as self-evident simply by being ignorant of any alternative. As new perspectives such as feminism are injected into the debate, our awareness of the possible distortions based on the manipulation of scientific concepts expands. There is no such thing as a completely objective account of how science has developed, but this does not mean that we should despair. By recognizing the impossibility of such a goal, we put ourselves in a position from which we can make a critical evaluation of the many different attempts that have been made to influence our beliefs.

Scientific theories can acquire many different meanings depending upon the interests of whoever tries to exploit them. Those writers who approach the past with a single-minded intensity shaped by a particular modern concern may easily be led into the trap of oversimplification. Certain theories or disciplines are targeted as the source of a harmful influence, and their opponents are automatically assumed to have been motivated by a superior morality. But deeper analysis often shows that those who were the 'good guys' on one issue were capable of being very nasty when they turned their minds to something else. Darwinism has been blamed for many modern evils, and its opponents have tended to stress the more humane viewpoint of those who adopted anti-materialistic theories such as Lamarckism. Environmentalists have assumed that anyone from the past who shared their perspective must have been on the right side on a whole range of related issues. Those who have made such assumptions may get a nasty shock when the more critical historian points out that many Lamarckians were vicious racists, or that there was an environmentalist component in Nazi ideology.

The simple fact is that critical history is a two-edged sword. It can be and is being used to expose the ways in which science has been used to support positions that many of us now find objectionable. Those who reject materialism are only too glad to welcome new evidence of the ways in which scientists built social values into the theories that they presented as objective knowledge of the world. They cannot then object when equally careful analysis shows that their own heroes sometimes let the side down. Once we have opened Pandora's box by admitting that scientific theories do reflect the interests of those who promote them, we must be prepared to admit that bias exists on all sides. At the same time we must acknowledge that in past circumstances people had perspectives that differed considerably from our own. An anti-materialist from the early twentieth century may not have shared the hatred of racism that is taken for granted by most modern intellectuals. Perhaps racism was then so prevalent that all biological theories were bound to reflect this attitude, whatever their implications in other directions. The assumption that a certain theoretical perspective guarantees ideological purity on a whole range of issues cannot be justified. The political and moral labels attached to theories are manufactured by their supporters – and the label affixed by an earlier generation may not suit those who seek to exploit the same theory in the very different climate that prevails today.

Images of Power

The concept of progressive evolution was used to justify the white race's drive towards world domination (chapter 8). The triumphs of industrial capitalism were presented as the last phase in the ascent of life, driven by the same forces as those governing the development of animal life towards the human form. There were notable shifts of emphasis in the decades around 1900, but the rhetoric of imperial expansion proved surprisingly durable. The scramble for colonies following the opening up of the African interior fuelled national rivalries that would eventually explode in the First World War. Many intellectuals saw the war as a demonstration of the west's moral bankruptcy, the product of a spiritual decay that had accompanied the industrial revolution. But the intellectuals did not control the funding agencies that financed science, nor did they have access to the mass media that shaped public opinion. Except in Russia the established power structure adapted itself to the new environment and ensured that the image of progress would continue to influence the way most people thought about the world. If unrestrained free enterprise no longer seemed enough to guarantee progress, strong control by central governments would be needed to preserve the west's dominant status. Confidence in the inevitability of progress may have waned, but it was still thought possible for the west to hang onto its gains as long as society resisted the threats that might lead to degeneration.

The natural-history museums that occupied a prominent location in many great cities still stood as potent symbols of western power. By the early twentieth century the great age of exploration was over, and the whole panorama of life on earth could be displayed in order to demonstrate the power of industrial civilization. One of the last species of large mammal to be brought to the knowledge of scientists was the okapi (a relative of the giraffe from central Africa). Enough material for a reconstruction of the animal's appearance reached the Natural History Museum in London in 1901, and the museum's director, E. Ray Lankester, gained a considerable amount of publicity by describing it. Now that the whole biosphere was known (the public did not care about the vast number of insect species still being discovered and described by specialists), the emphasis could shift to a more visually powerful representation of the natural world that was dominated by science.

Perhaps the clearest evidence of this use of spectacular displays to convey a hidden message about the status of western civilization can be seen in the dioramas created by Carl Akeley (1864–1926) in the American Museum of Natural History in New York. Akeley hunted in Africa with both the camera and the gun; he brought back the skins of his prey to be stuffed and mounted in realistic settings, which seemed to convey the onlooker to the very heart of the wilderness. One of the most spectacular displays includes a male silverback gorilla, now seen as the last step but one before the advent of humankind. The dioramas transformed the museum from an abstract display of the booty of conquest into a time machine capable of transporting the visitor back to an age in which there was still a wild world out there to conquer.

There was precious little wilderness left, especially in America, and the museum was intended to help preserve the spirit of manliness that had allowed the world to be conquered in previous centuries. By serving as a memorial to Theodore Roosevelt – whose equestrian statue stands outside – the museum would help to prevent the urban visitor from losing the virtues that had once been stimulated by the challenge of the wild. Roosevelt's words, inscribed on the museum's walls, identify the spirit of social responsibility that rests with those who still appreciate the mystery of nature[1]: 'There are no words that can tell the hidden spirit of the wilderness, that can reveal its mystery . . . The nation behaves well if it treats its natural resources as assets which it must turn over to the next generation increased and not impaired in value.' The problem was: could the environment be protected by science without at the same time destroying its spiritual value?

If prominent members of the establishment such as Roosevelt were concerned about the increasingly unnatural lives led by urbanized humanity, there were more radical thinkers prepared to take this line of thought much further. One of the most controversial links suggested by some historians identifies the 'religion of Nature' promoted by writers such as Ernst Haeckel with a number of right-wing policies, including those adopted by the Nazi party in Germany[2]. Haeckel developed an almost mystical evolutionary philosophy based on the assumption that there was a spiritual dimension built into the material universe (chapter 8). This 'monist' philosophy was taken up by many thinkers who distrusted the mechanical world view of orthodox science, including some who were directly involved with the attempt to revive occult ways of thought. The 'Monist League' was active in early-twentieth-century Germany and promoted a number of anti-democratic values. Haeckel had certainly been a racist who saw the 'lower' races as relics of earlier stages in the evolution of humankind. He was concerned about the degeneration caused by the urban lifestyle increasingly adopted by the 'highest' race, and urged a return to a way of life more in tune with Nature.

Historians have attempted to discern a link between the network of values promoted by Haeckel's anti-mechanist philosophy and a number of scientific and cultural movements. Supporters of this approach formed the last bastion of defence for vitalism in physiology and for Lamarckism in evolutionary biology. There are hints of a link to the studies of animal behaviour performed by Konrad Lorenz. Outside science, the racist element of Haeckel's system provided a natural foundation for right-wing views, while his concern for a more natural way of life fuelled the demands of radical environmentalists calling for a halt to industrialization. Both the racism and, to a lesser extent, the radical environmentalism were taken up by the Nazis, and some historians see Haeckel's monism as the source for much of the early-twentieth-century's right-wing thought. Others urge a more cautious approach when it comes to suggesting direct links between monism and Nazism. The Nazis did not like Haeckel's evolutionism, because the concept of a descent from the apes threatened their image of the innate superiority of the Aryan race. The Monist League did not encourage militarism, although Haeckel threw his influence behind the German war effort in 1914. Many of the values ascribed to the Monists were, in any case, pervasive through much of early-twentieth-century thought.

We cannot afford to ignore the role played by radical anti-materialist philosophies both within certain areas of science and in the promotion of extremist social policies. But at the same time we must be careful not to be carried away by the hunt for villains upon whom we can place the blame for the misuse of scientific theories. The fact is that the established social leadership of the industrialized world was pervaded by racism and the worship of individual and national enterprise. The historian must explore the ways in which science was exploited both by the establishment and by the more radical social philosophies. We are obviously concerned when it is suggested that a certain scientist may have put his work at the service of a repressive regime such as that of Nazi Germany or Stalin's Russia. But we must be equally concerned to expose the ways in which western culture as a whole has exploited scientific imagery to support its position in the world. The opponents of materialism often blame Darwinism for all our society's ills, creating a grotesquely oversimplified view of history in which a single theory lies at the heart of imperialism and militarism. Historians can reverse this trend by revealing the link between anti-materialist thought and other, equally destructive social philosophies. What we really need is a comprehensive effort to reveal the ideological underpinnings of many allegedly objective scientific theories.

Nature and Culture

If the hidden message of evolutionism still spoke of dominance, there was an underlying concern about the future that had been lacking in the previous century. The real problem now was to consolidate and preserve existing conquests by preventing the slide into cultural decay that could so easily follow when there were no major challenges left. Many were shocked by the cultural degeneracy that had flourished in the *fin de siècle* atmosphere of the 1890s, the cynicism of Oscar Wilde and the eroticism of world-weary artists such as Aubrey Beardsley. The greatest need of the new century was to prevent the loss of a vitality that had once been taken for granted. Radicals and conservatives agreed on this, but distrusted each other's recipes for salvation. Technocrats such as Roosevelt wanted a carefully managed world whose resources would not be wasted, and assumed that the white race would take on the responsibility of stewardship. But where the technocrats wanted yet more scientific management to solve the problems, the radicals thought that industrialization had already gone too far and must be reversed.

Of particular concern to many social thinkers was the alleged threat to the white race posed by the unchecked breeding of the least-fit members of the human species. The eugenics movement flourished in the early twentieth century, dedicated to the claim that governments must step in to limit the reproduction of the feeble-minded and degenerate types flourishing in the slums. Founded by Darwin's cousin, Francis Galton, the eugenics movement was based on the assumption that the individual's character is fixed absolutely by inheritance: no amount of improved education or social services can help those who are born with limited intelligence. This approach was stimulated by the development of biological theories in which heredity became a process by which absolutely fixed characters are transmitted from one generation to the next. After 1900 the new science of Mendelian genetics was thought to provide a firm foundation for such beliefs. Although this episode in the development of biological theories lies outside the scope of the present work, the link between eugenics and the fear of degeneration cannot go unnoticed. The claim that heredity imposes constraints on individual behaviour provides the foundation stone upon which a host of rival theories has been built, each claiming to show how a particular biological imperative has been imposed on us by 'Nature'.

In America eugenics was linked to the fear of degeneracy that might be produced by the influx of immigrants derived from inferior racial groups. Here evolution theory was more directly involved, as a host of biologists and physical anthropologists sought to consolidate the notion of a hierarchy of racial types established in the nineteenth century. Evolution was now seen not as the ascent of a single ladder, but as a process by which multiple parallel lines advanced to different levels of the same hierarchical scale. The leading paleontologist of the American Museum of Natural History, Henry Fairfield Osborn, built his career around the reconstruction of such patterns and did not hesitate to extend them to the human race. He saw the races as having a vast antiquity, so vast that to all intents and purposes they were distinct species. Each had its own unique character, and of course the white race had advanced further up the scale of development than any other. Evolution theory was thus exploited as a means of condemning the 'inferior' races to their lowly status. One did not need to be a geneticist to believe that the individual's character is firmly fixed by his or her racial heritage. As a paleontologist of the old school, Osborn would have nothing to do with the upstart science of genetics – yet he threw his weight behind the eugenics movement and its calls for immigration restriction.

Osborn was also opposed to the Darwinian theory of natural selection. Like many paleontologists of the early twentieth century he thought that the evolution of each group was controlled by built-in biological trends that world drive all members of the group in the same direction whatever the environment to which they were exposed. But those who pioneered this view of evolution at the turn of the century still accepted that the Darwinian law of struggle might be applied at the racial rather than the individual level. The claim that groups of human beings naturally come into conflict with one another to determine which will survive to undergo further development had been extended to the supposedly distinct lines of racial evolution within humankind. It could even be applied at the national level, and there is some evidence that the ruthlessness of the German war machine in 1914 had been stimulated by half-understood talk about the 'struggle for existence'. The American biologist Vernon Kellogg (1867–1937) visited Belgium and talked extensively with the German invaders. His book *Headquarters Nights* offered a chilling exposure of 'social Darwinism' at the international level[3].

Anti-Darwinians such as Osborn were worried that the theory of natural selection could be misused in this way. In the 1920s Osborn also became involved in defending the general theory of evolution against the attacks being launched by fundamentalist Protestants. If the late nineteenth century had taken evolution for granted, the religious enthusiasts of rural America did not like the consequences. For them, evolution was the mainstay of a materialism that led to wars, crime and all the degeneracy of the big cities. Many southern states passed legislation forbidding the teaching of evolution in their schools. In 1925 this movement came to a head when a Tennessee schoolteacher, John Thomas Scopes, was brought to trial after deliberately violating such a law. The resulting 'monkey trial' brought public figures and crowds of reporters to the small town of Dayton, Tennessee (the film *Inherit the Wind* is a dramatization of the subsequent events). Although the creationists were ridiculed in the big-city newspapers, the prosecution was successful in that it kept the teaching of evolution out of American schools for the next three decades. The modern controversies over the demand for equal time for 'creation science' are a continuation of the same conflict, with a revived Darwinian theory still finding itself unable to shake off the materialist label.

Osborn wanted to present evolution as a theory that did not depend on Darwinism so he could challenge the link with materialism. Unlike most evolutionists he also tried to distance the theory from the claim that humans had evolved from apes. He understood how the image of an ape ancestry could be exploited by opponents seeking to discredit the whole theory by emphasizing its brutalizing implications for humankind. Osborn argued that the human line goes so far back into the past that the apes cannot have been our ancestors – they are a parallel and degenerate branch of primate evolution.

This in turn allowed Osborn to get rid of the claim that the human race evolved in Africa – a source of potential embarrassment in a culture used to thinking of all things African as inherently inferior. He argued that the more challenging environment of central Asia provided the stimulus for the development of humankind, while the lush jungles of the tropics produced the degenerate apes – and the equally degenerate branches of the human family that had been unfortunate enough to migrate there later on. Note how Osborn, in most respects an avowed anti-Darwinist, appeals to the struggle for existence to make his case because it allows him to stress the role of effort and initiative[4]:

> The high plateau country of central Asia was partly open, partly forested, partly well-watered, partly arid and semi-desert. Game was plentiful and plant food scarce. The struggle for existence was severe and evoked all the inventive and resourceful faculties of man and encouraged him to the fashioning and use of wooden and then of stone weapons for the chase . . . [Thus] while the anthropoid apes were luxuriating in the forested lowlands of Asia and Europe, the Dawn Men were evolving in the invigorating atmosphere of the relatively dry uplands.

The possibility of an Asian origin for the human race was widely endorsed in the 1920s. The hope of locating fossils to support this view was one of the motivations behind a series of much-publicized expeditions to Mongolia, the first of which was led by Roy Chapman Andrews in 1921.

Evidence of early hominids (members of the human family) in Africa was ignored at first. As early as 1924 the South African anatomist Raymond Dart (1893–1988) discovered the skull of a creature with a relatively small brain but which, he inferred, had already walked upright. He named it *Australopithecus africanus* and insisted that it represented the ancestral form of humankind. Most authorities rejected Dart's claim, partly because they were suspicious of African fossils and partly because of a widespread assumption that progress towards modern humanity would have been led by the expansion of the brain. It was thought that our ancestors would only have begun to walk upright after they had become intelligent enough to realize the advantages of this. Dart shared the view that a plains environment would have served as the best stimulus for human development, but he realized that Africa could provide such conditions too. He saw the earliest hominids as aggressive predators upon small game, and inclined to violence among themselves. According to Dart[5], *Australopithecus* was 'a cave-dwelling, plains-frequenting, stream-searching, bird-nest-rifling and bone-cracking ape who employed destructive implements in the chase and preparation of his carnivorous diet'.

Dart's scenario for human evolution in Africa was eventually taken more seriously when the paleontologist Robert Broom (1866–1951) uncovered more *Australopithecus* remains in the late 1930s. The advent of modern Darwinism threw more emphasis on the need to define adaptive differences between the apes and the earliest hominids, and by the 1950s it was generally accepted that the adoption of an upright posture constituted the crucial breakthrough leading towards humanity.

Although fascination with Asia was slow to decline, most authorities refused to accept Osborn's claim that the apes were completely unrelated to humankind. On the contrary, many scientists saw the link with the apes as a powerful tool in their efforts to pinpoint the animal origin of many human characteristics. As the study of animal behaviour became a science in its own right, primatologists such as Robert Yerkes (1876–1956) studied apes with a view to reinforcing prejudices about what was 'natural' behaviour for human beings (see below). It is small wonder that such studies tended to emphasize the existence of family groups dominated by males – the traditional structure that the supporters of western dominance wanted to preserve in their own society. There was a darker side to the link, however, proclaimed most openly in Robert Ardrey's 'anthropology of aggression' in the 1960s. Inspired in part by Raymond Dart's image of *Australopithecus* as a vicious small-game hunter, Ardrey used books such as his *Territorial Imperative* of 1966 to develop the theme that aggression is built into human nature by our evolutionary origins.

The creationists are not the only group to challenge this use of animal parallels to justify the claim that certain kinds of behaviour are 'natural' in human beings. The early years of the twentieth century saw the social sciences emancipate themselves from the evolutionary straitjacket that had been imposed upon them. Anthropologists such as Franz Boas (1858–1942) now directly challenged the claim that biological inheritance determines human behaviour. Boas rejected the whole idea of a racial hierarchy, and insisted that cultural diversity was not the result of innate racial differences imposed by biology. He was vilified by many opponents within the biological sciences, concerned that his approach would undermine the claim that the white race and western culture were intrinsically superior to their rivals. The effort to establish anthropology as a distinct discipline encountered many setbacks, but it was ultimately successful. Today the social scientists are among the most vociferous critics of the view that biology determines human behaviour. They argue that, in humankind, Nature has produced a creature with capacities so subtle that cultural development can proceed in many different – but equally valuable – directions. People's behaviour *can* be shaped by an improved environment and by better education. Anthropologists thus rejected Ardrey's theory of innate aggression, and continue to oppose the most recent manifestation of the biologists' desire to speak on human affairs: sociobiology (see below).

THE EVOLUTIONARY SYNTHESIS

The science of ecology had to struggle hard to establish its independence, and its ultimate success may have owed something to changes taking place within other disciplines. In the early decades of the century a wide range of biologists would have simply denied any close dependence of the organism upon its environment. They believed that evolution was a process driven by trends quite capable of producing features that were useless or even harmful to the organisms concerned. Extinction would only occur when there was a massive imbalance between the organism and the environment within which it must live. This situation began to change when Darwinism at last began to surface as the most promising theory of the evolutionary mechanism in the 1930s and 1940s. As theories of non-adaptive evolution crumbled, so biologists began to focus on the relationship between the organism and its environment. Darwinism also proclaimed evolution to be a process taking place within populations, again helping to produce a climate of opinion favourable to ecological studies. It would be wrong to pretend that evolutionists and ecologists have always worked hand in hand, but ecology could not have risen to prominence if biology had continued to be dominated by theories of non-adaptive evolution.

The belated completion of the 'Darwinian revolution' has involved conceptual changes of immense significance. The non-Darwinian theories that flourished at the turn of the century had certain implications that were extremely comforting to anyone still worried about the prospect of reducing the human race to the level of animals. Evolution was still seen as an essentially predictable process; the development of Nature was not a chapter of accidents, and the modern species – including of course the human race – were the inevitable outcome of laws built into the very nature of things. Initiative and intelligence were important qualities of living things, which enabled them to progress steadily towards higher states. The synthesis of Darwinism and genetics changed all that. Evolution became an essentially unpredictable sequence of events governed by random mutations and the hazards of an ever-changing local environment.

The materialistic implications of Darwin's ideas, long concealed by the popularity of alternative mechanisms, had at last become apparent. The creationists were not the only group to see the resulting world view as a dangerous nightmare. Many non-scientists had expressed support for the alternative theories, and they continued to lead an underground struggle against the materialism that was dominating the scientific community. Social scientists too objected when overconfident Darwinians sought to use their theory as the basis for predictions about human behaviour. All too often the opponents of Darwinism have blamed it for a host of modern evils, and have assumed that all supporters of their own position must share their hatred of intolerance and dogmatism. It cannot be denied that the idea of natural selection was used to promote 'social Darwinist' attitudes. But many anti-Darwinists also used their theories to support the concept of a racial hierarchy, and some even appealed to the beneficial effect of struggle. Neither side was free from the prevailing ideology of the times, and modern efforts to claim that one scientific theory or another guarantees moral superiority are invariably based on an oversimplification of history.

The general public may have been only dimly aware of what was going on within biology. For most ordinary people, evolution means the rise and fall of the dinosaurs and other prehistoric beasts displayed at their local museum. New ideas on the mechanism of change do not translate very easily into popular images. It is important that we recognize the ongoing work of paleontologists whose main concern is the reconstruction of the history of life on earth. There have been important developments in this area, some of them produced by the discovery of new fossils that help to fill in gaps in the existing record. At the same time, though, it would be wrong to visualize the attempt to reconstruct life's ancestry as a project that is totally unaffected by changes in the current theory of how evolution works. The new ideas on human origins that surfaced in the 1950s, for instance, were in part inspired by fossils unearthed in Africa – but those new ideas would have been unthinkable without a new model of evolution that threw doubts on the assumption of inevitable progress that had underpinned the non-Darwinian theories of earlier decades.

Parallel Evolution

The opposition to Darwin's theory of natural selection that had built up in the later decades of the nineteenth century did not die down immediately. Although the new science of genetics would ultimately revive interest in selectionism, this movement only began in the 1920s and did not percolate into some disciplines until the 1940s. In the meantime, many paleontologists and field naturalists continued to work with anti-Darwinian theories, including Lamarck's inheritance of acquired characteristics and mechanisms of non-adaptive evolution such as orthogenesis. Increasing tension built up as it became apparent that these theories were not supported by the findings of genetics, but there was a widespread hostility to the new experimental science of heredity, which allowed many biologists in traditional disciplines to ignore the latest developments. Outside science there was vociferous support for Lamarckism from literary figures and writers on moral questions, all of whom saw Darwinism as an expression of the materialistic view of Nature, which they abhorred. Because historians' attention has focused on the impact of genetics, it is all too easy to fall into the trap of assuming that support for Lamarckism was confined to non-scientists. In fact, the anti-Darwinian ideas continued to flourish in those areas of biology that remained aloof from genetics.

Paleontology was now committed to the reconstruction of the history of life on earth and the search for 'missing links' to throw light on those evolutionary episodes hidden by gaps in the known fossil record. By the very nature of the evidence they used, paleontologists were inclined to see evolution as a linear process: they tended to link the known fossils together into neat patterns expressing hypothetical evolutionary trends that continued over vast periods of time. Because they studied the structure of extinct animals as represented by their bones or shells, they had little interest in the pressures imposed by the need for organisms to adapt to a constantly changing environment. Although there was now a geographical element in the study of the history of life, made possible by the exploration of once-remote corners of the world (chapter 8), most paleontologists were still inclined to see evolution as a matter of trends brought about by factors controlled by forces internal to the organism.

An episode that illustrates the prevailing attitude to the history of life has recently been brought to light by Stephen Jay Gould[6]. In 1909 the American geologist Charles Doolittle Walcott (1850–1927) began to collect fossils from the Burgess Shale of British Columbia. This was rock from the Cambrian era – the earliest in which fossils are found in any quantity – containing impressions of the soft parts of the animals that had once lived there. Most fossils consist of the mineralized remains of only hard structures such as bones and shells, so the Burgess Shale offers a glimpse into details of the past that are normally hidden from the paleontologist. Modern studies suggest that the collection of creatures thus revealed includes many whose structure is so bizarre that they cannot be included within any of the main groups or phyla that make up the animal kingdom as we now know it. It is as though the first explosion of multi-celled organisms produced a vast range of 'experimental' forms, most of which subsequently died out.

Yet Walcott saw none of this in the specimens he collected. He was a busy administrator, rising to become head of the Geological Survey and secretary of the Smithsonian Institution, and he had little time for detailed work on his finds. But when he did describe them, he shoehorned all of the creatures into known phyla, turning a blind eye to their bizarre characteristics. Walcott was blinkered by the prevailing view that evolution must represent the steady development of types whose primitive foundations were laid down in the earliest times. He simply could not admit that Nature might first have experimented with forms quite different from those which are now so familiar to us. Significantly, he was a religious man who saw the idea of progress in evolution as a means of countering fundamentalist opposition. He also led the campaign against the anthropologist Franz Boas, whose relativistic view of human culture threatened the idea of progress. Only with the emergence of a more Darwinian perspective in recent decades has it become possible for biologists to admit the open-ended and unpredictable nature of the earliest phases of animal evolution. The whole episode confirms the theory-laden character of scientists' conclusions, even when they are attempting to describe specimens whose features ought – according to the conventional view of observation – to be perfectly obvious.

Most early-twentieth-century paleontologists believed that the fossil record would reveal a history of overall progress, but they accepted that the progress was episodic. There were certain key points when new and more advanced types of organization appeared; these then proliferated at the expense of the previously dominant classes, after which the various components of the new order settled down to a rather predictable history of development, maturity, degeneration and extinction. This emphasis on the predictable nature of evolutionary change was most clearly developed in the theory of orthogenesis, which supposed that groups of related species became endowed with built-in trends that would drive their future evolution in a fixed direction, whatever the environmental challenges to which they were exposed. The evolution of each group consisted of parallel lines advancing in the same direction, ending ultimately with the overdevelopment of their most prominent characters and extinction. On such a model, the dinosaurs were already in decline before they were challenged by the explosion of mammalian evolution that created the world of modern life.

One of the chief exponents of the theory of orthogenesis was Henry Fairfield Osborn. During the early years of the century, Osborn focused his attention on the explosion of new forms that characterized the rise of a new class such as the mammals, and coined the term 'adaptive radiation' to denote such episodes. But much of his detailed work concerned the evolution of particular mammalian groups, and here he was convinced that the fossil record confirmed the existence of rigid trends governing the whole pattern of each group's rise and fall.

In 1929 Osborn produced a monograph on the Titano-theres, an extinct group of mammals whose remains had been unearthed in the American west. Their horns showed a pattern of regular growth through geological time, leading to an eventual overdevelopment that produced structures so cumbersome that they may have contributed to the group's extinction. There was no sign, he claimed, of the kind of open-ended, irregular evolution that would be expected if the process were governed by the species' response to an ever-changing environment or by random variation[7]:

> Within certain phyla a tendency or predetermina-tion to evolve in breadth or length orthogenetically appears to be established, flowing in one direction like a tide, on the surface of which occur individual fluctuations and variations, like waves and ripples. The overdevelopment of certain proportions . . . seems to be a manifestation of this orthogenetic impulse, which appears to carry certain skulls to inadaptive extremes.

The struggle for existence came in only at the end, to eliminate those species whose horns had become so large that they were at a serious disadvantage.

Since orthogenetic trends were supposed to affect all the members of a group, evolution often consisted of parallel lines moving independently in the same direction.

Although Osborn emphasized the divergence that took place as soon as each new class was formed, he minimized the extent to which the subsequent development of each line was subject to branching. Similar species might not be closely related by descent, as Darwin had claimed, because each had arrived at its present state by following the same pattern of development independently over a long period of time. It was by applying this model of evolution to human origins that Osborn was able to develop the claim (discussed above) that humans and apes represent parallel lines of primate evolution. He could also argue that the human races are quite distinct, each having developed separately through parallel evolution.

Osborn believed that most evolution was essentially predetermined, but he had to confess that he had no idea how the trends that he postulated became fixed into the reproductive system of the organisms. When criticized by the geneticists on this score, he had to argue that paleontology reveals trends operating over a timescale so protracted that there may be no hope of laboratory verification. In effect, Osborn was resisting the geneticists' claim that their experiments offered the only way of uncovering the reproductive processes which when added together over many generations, constitute evolution. Paleontology had its own independent source of evidence, and it could not be dictated to by new-fangled laboratory sciences. In fact, Osborn's whole approach reflected an anti-Darwinian mode of thought more typical of the nineteenth century, based on the search for orderly patterns in Nature that would transcend the haphazard effects that could be expected if evolution were subject to migration and local adaptation.

It would be wrong to pretend that Darwinism lacked all support among paleontologists. W. D. Matthew continued to endorse the selection theory as part of his claim that central Asia was the chief centre of mammalian evolution (chapter 8). Osborn's subordinate at the American Museum of Natural History, William King Gregory (1876–1970), placed far greater emphasis on the functional adaptation of fossil organisms to their mode of locomotion. By stressing the importance of adaptation he was led to distrust Osborn's theory of non-adaptive trends and to favour a selectionist explanation of how the adaptation was produced. He also opposed Osborn's efforts to widen the evolutionary gulf between humans and apes. Matthew and Gregory show that early-twentieth-century paleontology was not totally dominated by the theory of orthogenesis, but they were important exceptions to the general orthodoxy promoted by Osborn.

The Defence of Lamarckism

When dealing with evolutionary trends that were obviously adaptive, many paleontologists still preferred to invoke Lamarckism rather than natural selection. Field naturalists also assumed that the inheritance of acquired characteristics was responsible for allowing populations to adapt to changes in their environment. There were important differences of emphasis, however. The field naturalists were more concerned with geographical factors and the process by which populations isolated by migration acquired the status of distinct species. In many respects the field workers were adopting a 'Darwinian' perspective, even though they still saw Lamarckism as the actual mechanism of adaptation. The paleontologists, in contrast, saw Lamarckism as a mechanism that would produce linear trends driving a species in a predictable direction. Once a new mode of behaviour had been chosen by one generation, the whole future of the species was mapped out in advance because the descendants had to continue specializing in the same direction.

Lamarckism had flourished in the late nineteenth century because it seemed to offer a way of avoiding the materialism of the selection theory (chapter 8). Natural selection was a kind of fatalistic lottery, a genetic Russian roulette in which the organism was completely at the mercy of its inheritance. If it inherited a harmful character, it was doomed, however hard it might struggle to improve itself. Lamarckism would allow living things to play an active role in evolution. The individuals making up a species could choose a new behaviour pattern to cope with any changes in their environment; the new habit would encourage them to exercise their bodies in new ways, and the resulting acquired characters would accumulate through inheritance to direct the evolution of the species. Evolution was a matter of choice, not of brutal necessity.

In the 1890s the American psychologist James Mark Baldwin (1861–1934) pointed out that there was no reason why the selection theory could not be adapted to make room for behavioural modification as a directing factor. If the animals chose a new lifestyle, then natural selection will favour those individuals who happened to possess physical characters better suited to the new habit. This mechanism, known as organic selection or the 'Baldwin effect', was endorsed by Osborn and gained some popularity in the following years. The enthusiasm it generated reveals the extent to which the Lamarckians' argument on the question of individual choice was influencing scientists' thinking. Unfortunately Baldwin was disgraced in a sex scandal in 1909 and his subsequent influence on the debate was slight.

Outside science there was still a great deal of support for Lamarckism based on the arguments put forward by Samuel Butler. Literary figures such as the playwright George Bernard Shaw continued to lambast Darwinian materialism and to proclaim Lamarckism as the foundation for a more humanistic model of evolution. In the preface to his *Man and Superman* of 1906 and again in *Back to Methuselah*, Shaw poured out his contempt for Darwinism[8]:

... you can be a thorough-going Neo-Darwinian without imagination, metaphysics, poetry, conscience, or decency. For 'Natural Selection' has no moral significance: it deals with that part of evolution which has no purpose, no intelligence, and might more appropriately be called accidental selection, or better still, Unnatural Selection, since nothing is more unnatural than an accident. If it could be proved that the whole universe had been produced by such Selection, only fools and rascals could bear to live.

He preferred Lamarckism, or what he called 'creative evolution', borrowing a term coined by the French philosopher Henri Bergson for the idea that evolution is driven by a non-material creative impulse or *élan vital*.

In more sober terms, the South African soldier and statesman Jan Christiaan Smuts (1870–1950) wrote his *Holism and Evolution* of 1926 to proclaim the link between Lamarckism and the holistic philosophy. If the organism was a system whose parts co-operated to generate something more than a mere machine, then it seemed obvious that characters acquired by the adult must somehow be incorporated into the genetic material for transmission to future generations. As late as the 1960s the same theme was still being developed by the author Arthur Koestler (1905–83), who called upon both the Baldwin effect and Lamarckism to defend the view that living things can direct their own evolution along purposeful channels.

The problem confronting all these anti-Darwinians was that the latest developments in experimental biology were making it increasingly difficult to believe that acquired characters really could be inherited. The breeding experiments used by the geneticists seemed to confirm that characters are transmitted as fixed units from one generation to the next. The theory elaborated to explain the Mendelian effects assumed that there was no way in which characters acquired by the adult organism could affect the genes transmitted to the next generation. Some paleontologists like Osborn tried to shrug off the implications of the new experimental discipline, but the Lamarckians increasingly found the lack of hard evidence for their mechanism an embarrassment. There were a number of efforts to remedy this deficiency, one of the most notorious being that centred on the work of the Austrian breeder Paul Kammerer (1880–1926). This episode has acquired a unique status in the history of biology owing to the attention focused on it by Arthur Koestler's book[9] *The Case of the Midwife Toad*.

In the years immediately prior to the First World War, Kammerer performed experiments on several different species in search of evidence favourable to Lamarckism. Those which generated most attention arose from his efforts to get the midwife toad, *Alytes obstetricans*, to breed in water. Unlike all other toads, this species normally mates on dry land, and the males have lost the typical pad on their forelimb used to grasp the female. Kammerer claimed that, when forced to mate in water (no one else could get them to do this!), the males acquired the forelimb pad just like any other toad species – and the character was transmitted to the next generation. After the disruption of the war, Kammerer tried to get British and American scientists interested in his work. He was well aware of the broader implications of Lamarckism, and spoke optimistically of using social reform as a way of improving the biological character of the human species. All races, he claimed, would be able to benefit from the improvements predicted by the Lamarckian theory. A visit to America in 1923 generated newspaper headlines about breeding a race of supermen.

The geneticists were already suspicious of Kammerer's experiments and were appalled at the publicity he was now gaining for the Lamarckian cause. Led by William Bateson they tried to discredit Kammerer through a series of articles casting doubts on the validity of his experimental claims. When an American scientist finally examined one of the crucial specimens of the midwife toad (now preserved in a jar), he found that the mating pad had been marked with Indian ink. Kammerer insisted that this must have been done by an assistant trying to prevent the original mark fading due to the action of the preservative, but shortly afterwards he shot himself. The geneticists took this as clear evidence of guilt. Koestler tried to revive interest in the experiments, arguing that Kammerer had been hounded to his death by a vicious campaign organized by a scientific community unwilling to permit any challenge to its authority. The geneticists, needless to say, do not accept this interpretation, and hard evidence for Lamarckism has not emerged despite several later efforts.

One of the few biologists to speak out on Kammerer's behalf in Britain was the embryologist E. W. MacBride (1866–1940). As one of the last remaining champions of the theory that individual development recapitulates the evolutionary ancestry of the species, MacBride preferred the Lamarckian view that acquired characters merely extend the normal growth process. To Koestler, MacBride was the 'Irishman with a heart of gold' who stood up to the geneticists in defence of a less materialistic view of life. In fact, MacBride was a vicious racist, a supporter of eugenics who urged that the Irish component in the British population should be sterilized. Later on he wrote letters to *The Times* defending Hitler's policies in Germany[10].

Koestler missed this side of MacBride's character because it never occurred to him that a Lamarckian could hold anything but the most humane views on all moral issues. In fact, Lamarckism has been linked to the idea of a racial hierarchy throughout its history. There is no guarantee that someone who adopts a holistic position on the question of evolution will thereby be prevented from taking up exploitative ideas on other issues. The connections between scientific theories and moral or social issues are manufactured to suit a particular set of circumstances. A well publicized link may tempt us to assume that there is a logical connection between a particular theory and a particular set of moral values – as with the often-repeated claim that Darwinism is the source of all society's ills. The historian's job is to show that in the real world things are not so simple.

The same point is evident from the Lysenko affair in Russia. T. D. Lysenko (1898–1976) was an agricultural scientist who gained the ear of Stalin during his drive to modernize the Soviet Union during the 1930s and 1940s. He developed a Lamarckian theory that, he claimed, would revolutionize food production by adapting plant species to Russia's harsh environment. Wheat could be 'vernalized' by freezing the seeds to adapt it to the short growing season. As part of his campaign to dominate Russian biology, Lysenko argued that both Darwinism and genetics were theories based on bourgeois values. Lamarckism was more in tune with Marxist thought, and anyone who doubted it was branded as ideologically suspect. Many geneticists lost their positions, and some were purged. The result was a disaster for Soviet agriculture, which missed out on the geneticists' efforts to breed better crop varieties, and for Soviet science, which was isolated from the international community for decades. Only after Stalin's death did genetics begin to revive in Russia.

The Lysenko affair certainly confirms that Lamarckians can behave as badly as anyone else. But the whole episode continues to raise questions over the link between science and ideology. To the western geneticists the situation was simple: Lysenko abandoned scientific objectivity by preferring one theory over another on ideological grounds, and he paid the penalty by being led up a blind alley. But recent historical studies suggest that the situation is not quite so clear-cut. Lysenko was certainly wrong, but it is by no means so obvious that the geneticists were completely right. The genetics of the 1930s presented the organism as a mosaic of characters, each rigidly predetermined by inheritance. It ignored the fact that the environment can affect the development of the organism (even if those effects cannot be inherited). Genetics thus upheld the hereditarian attitudes typical of the eugenics movement. Other continental biologists shared Lysenko's suspicions of the geneticists' rigid views (see below).

Population Genetics

As an experimental science dedicated to the study of heredity under controlled conditions, genetics lies outside the scope of the present survey. But the impact of this relatively modern discipline on debates concerning evolution theory and the relationship between the organism and its environment cannot be ignored. In its purest form, as developed especially in America, the 'classical' genetics of the 1930s treated the organism as little more than a collection of characters each transmitted as a unit from one generation to the next. Identification of the gene as a segment in the chromosome within the cell nucleus gave the theory a distinctly materialistic slant. Even before the discovery of the double-helix structure of DNA in 1953, it was assumed that the gene somehow stored the information necessary for reproducing a particular character in a material template, that template being transmitted from parent to offspring in a way that prevented any interaction with the adult body. This was Weismann's theory of the absolute isolation of the germ plasm from the body (chapter 8) translated into modern terms. The potential impact of such a theory was immense. At one stroke it undermined the plausibility of many anti-Darwinian mechanisms of evolution, including Lamarckism. Once the new theory could be applied to reproduction taking place within large populations, the way was paved for the creation of a new form of Darwinism far more powerful than anything that Darwin himself could have anticipated.

Historians have concentrated on the emergence of population genetics and the synthesis with the insights of the field naturalists who had preserved a Darwinian emphasis on migration and local adaptation. Yet this represents only the more visible manifestation of what was actually a major conceptual revolution. Darwin had challenged the idea that evolution is somehow programmed by internal factors and had focused on the interaction between the organism and its environment as the only directing agent. But he himself had failed to see that the full flowering of his new approach would depend upon a rejection of the old belief that variation (the appearance of new characters) is a function of the process by which the organism develops towards maturity. For Darwin, heredity and individual development were aspects of a single biological phenomenon, and this made it difficult to believe that anything affecting the individual (such as the acquisition of a new character by exercise) could be left completely outside the process of heredity. Genetics severed the link between the transmission of characters (heredity) and their development in the individual (embryological growth). The Mendelians studied only transmission, and they developed a theory in which the mechanism of transmission was divorced from the process of individual development.

The result was that a whole complex of ideas once taken for granted was rendered unacceptable. Neither Lamarckism nor the theory of orthogenesis by 'extension of growth' was plausible, since there was no way in which the new characters could feed back to influence the structure of the gene. The recapitulation theory fell by the wayside too, because the production of new characters could not occur by adding-on stages to individual development. New characters could be produced only by the creation of new genes, i.e. the changing of an old gene's code so that it now produced something new, a 'mutation'. Mutations occurred within the structure of the gene in a manner that could not be seen as a response to the environment. Evolution became nothing more than the process by which new characters produced by random mutation were incorporated into the population as a whole. If this point were accepted, the whole foundation of evolution theory would be transformed, because what happened to *individuals* was not necessarily reflected in what happened to *populations*. The only way in which the environment could affect the population was by determining which individuals would live and reproduce and which would die, not by altering the individuals themselves. The re-emergence of the selection theory was a by-product of the elimination of all other means by which the flow of new genetic characters could be controlled.

The potential for a synthesis with Darwinism was not recognized at first because the early geneticists went so far as to claim that the environment could have no effect, not even indirect, upon evolution (chapter 8). Both William Bateson and Thomas Hunt Morgan had begun their careers with an interest in embryology. Both became convinced that evolution proceeded through the sudden creation of entirely new characters by saltation, and neither would admit that the environment has any influence over the kind of new characters that appear. Mendelism seemed to fit the old theory of saltative evolution because the character differences it studied were discontinuous; the organism either had a certain character or it did not – which made it much easier to believe that the character had been produced as a unit by some internal rearrangement of the germinal material. As long as the geneticists' breeding experiments were confined to small experimental populations isolated under artificial conditions, it was possible for them to create a model of evolution that ignored the role of the environment. Even when studying natural effects such as mimicry in insects, geneticists such as R. C. Punnett claimed that the parallel characters were produced as units, not by the gradual selection of minute adaptive differences. Genetics at first merely contributed to the eclipse of Darwinism in the years around 1900.

The first signs of a changing attitude appeared around 1920, partly in response to new evidence on the nature of mutations, and partly as some geneticists began to think more carefully about what would happen in the wild. T.

H. Morgan became the founder of classical genetics by developing techniques that allowed the transmission of characters in breeding experiments to be linked to the behaviour of the chromosomes in the cell nucleus during fertilization. To do this he studied the appearance of new characters by mutation in what was rapidly becoming the geneticists' favourite organism, the fruit fly *Drosophila melanogaster*. Contrary to his early opinions, he was forced to recognize that mutations often produce not entirely new characters, but merely modifications of existing ones. There was no evidence that the mutations were of any benefit to the organism; indeed, they seemed to provide a flow of essentially random, i.e. undirected, variation. It soon became clear that large mutations are almost invariably harmful or lethal, and only the smaller ones have any chance of reproducing in the population.

By the time he wrote his *Critique of the Theory of Evolution* in 1916, Morgan was prepared to admit that natural evolution must to some extent be governed by whether or not the new characters produced by mutation have any adaptive value. Harmful characters will not be transmitted in a natural environment because the individuals carrying them will not survive to reproduce. There was thus a role for natural selection – but not, he insisted, the kind of ruthless selection postulated by Darwin. Morgan wrote[11]:

> Such a view gives us a somewhat different picture of the process of evolution from the old idea of a ferocious struggle for existence between the individuals of a species with the survival of the fittest and the annihilation of the less fit. New and advantageous characters survive by incorporating themselves into the race, improving it and opening it to new opportunities. In other words, the emphasis may be placed less on the competition between the individuals of the species (because the destruction of the less fit does not *in itself* lead to anything that is new) than on the appearance of new

characters and modifications of old characters that become incorporated into the species, for on these depends the evolution of the race.

There was no need for struggle because harmful characters were eliminated as soon as they were formed. There was no mass of unfit individuals to be eliminated, although useful characters would spread into the population because they would be reproduced faster than normal.

Over the next decade the majority of geneticists began to admit that evolution must proceed through the spread of minor character differences conferring adaptive benefits. Since their theory ruled out Lamarckism, selection (in the form of differential reproduction governed by adaptation) was the only way that the flow of new characters produced essentially at random by mutation could be controlled and directed. But as Morgan's remarks quoted above reveal, there was still a considerble gulf between such a position and a truly Darwinian view of selection acting upon a wide range of variation within the natural population. To apply the Mendelian laws of heredity to a large population already containing a fund of natural variation, new techniques were needed. The biometricians led by Karl Pearson had pioneered the use of statistics to study the variability of wild populations (chapter 8), but had refused to accept that Mendel's laws were generally applicable. Where Bateson stressed discontinuous evolution and discontinuous heredity, Pearson emphasized the existence of a continuous range of variation in all populations, and saw evolution as the gradual shifting of that range under the influence of natural selection.

Two things were needed to bring these approaches together. First, it had to be recognized that the pairs of alternate characters studied by Mendel represented only the simplest possible situation. Most characters in most species are affected by several different genes, whose effects are intermingled in such a way that *although each gene is subject to Mendel's laws* the overall effect is for the character to exhibit a continuous range of variation. A science of population genetics would seek to identify the ways in which the genes interact and duplicate themselves, in effect explaining the phenomenon studied by biometry in Mendelian terms. The second requirement for such a science to be created was that the fierce hostility between the champions of the two approaches, Bateson and Pearson, should be overcome.

The first person to tackle the problem was Pearson's student, R. A. Fisher (1890–1962). So sensitive were the issues that Fisher's first paper was rejected by the Royal Society of London and finally appeared in the transactions of the Royal Society of Edinburgh in 1918. He realized that there was a flaw in Morgan's version of selection: if a harmful gene is recessive, it will not disappear from the population immediately after being formed by mutation. The 'dominant' version or allele of the gene will completely mask the 'recessive' version when the organism inherits one of each; thus a population may contain some individuals who can transmit the recessive gene even though they do not manifest its character. Once produced by mutation, many genes will maintain themselves in the population even though they are not favoured by natural selection.

A natural population will thus contain a wide variety of genes for all characters, producing the range of variability that Darwin and the biometricians had studied. The 'gene pool' will maintain a fund of variability independently of the production of entirely new characters by mutation. If the environment changes so that a once-useless gene becomes beneficial, the individuals carrying that gene will reproduce more and the gene's representation in the gene pool will increase. The reverse will, of course, happen if a gene becomes positively harmful in a new environment. Selection will thus be able to modify the range of variation over many generations, gradually altering the overall character of the population. Fisher challenged the saltationist explanation of mimicry, showing that selection would gradually produce colour schemes that conferred some advantage. By the time he wrote his *Genetical Theory of Natural Selection* of 1930, Fisher had developed a sophisticated theory based on the assumption that evolution took place most readily in large populations. Selection was a mechanistic process gradually altering gene frequencies to match changes in the environment. Fisher's calculations assumed that the reproductive advantage conferred by any single gene would be very slight, thus requiring many generations for selection to be effective.

This last point was challenged by J. B. S. Haldane (1892–1964), who drew upon practical observations suggesting that in some circumstances selection could act much more rapidly. In a series of papers leading up to his *The Causes of Evolution* of 1932, Haldane explored the consequences of natural selection acting on genetic variability. He certainly agreed with Fisher's main point[12]:

> Quantitative work shows clearly that natural selection is a reality, and that, among other things, it selects Mendelian genes, which are known to be distributed at random through wild populations, and to follow the laws of chance in their distribution to offspring. In other words, they are an agency producing variation of the kind which Darwin postulated as the raw material upon which selection acts.

Haldane noted the case of industrial melanism in the peppered moth now called *Biston betularia*. In parts of Britain where the natural environment had been blackened by industrial pollution, the melanic or darkened form of this moth had begun to predominate in the population at the expense of the normal grey form. The normal form was well camouflaged against predators on unpolluted tree trunks, but stuck out a mile in smoke-darkened areas. Haldane showed that, for selection to have altered the local populations in less than a century, the melanic form must have its rate of reproduction increased by fifty per cent. Although based on a rare discontinuous character difference, this was far greater than anything postulated by Fisher.

Between them, Fisher and Haldane had created a science of population genetics that took it for granted that natural selection was the only process capable of producing changes in the genetic make-up of a wild population. Even so, their approach had little direct impact on the work of many field naturalists because their mathematical models were extremely complex and were based on the assumption that evolution takes place when a large population is exposed to a uniform change in its environment. As the case of industrial melanism indicated, environmental changes are often *not* uniform, and natural barriers often mean that there is no regular interbreeding between the members of the same species occupying different areas. What was needed was a geographical perspective, acknowledging that populations can be broken up into partially or even totally isolated groups, and providing techniques for understanding how evolution would proceed in these circumstances.

The American biologist Sewall Wright (1889–1988) had become interested in selection through animal breeding, and he knew that new characters are often produced most effectively in small, isolated populations where there is a great deal of inbreeding. By 1920 he had developed powerful mathematical techniques to show that these small populations often elicit apparently new characters because the inbreeding allows the production of unusual combinations of genes. The organism is not just a mosaic of genetic characters; in some cases a new combination of genes can produce an interaction with consequences that could not have been predicted. If a natural population is broken up into small groups, partially (but not completely) isolated to minimize interbreeding, new gene interaction systems may be produced in the subpopulations. If it subsequently turns out that one such new character is useful to the species as a whole, it will spread from subpopulation to subpopulation until eventually the whole species is affected.

Important developments also took place in the Soviet Union during the period before Lysenko eliminated genetics there. The entomologist Sergei S. Chetverikov (1880–1959) was convinced that wild populations contain much hidden genetic variation. When genetically pure strains of *Drosophila* were brought to Russia by Herman J. Muller (one of T. H. Morgan's colleagues), Chetverikov interbred them with wild populations to confirm this supposition. His students developed statistical techniques for dealing with small populations and showed that these could be expected to produce greater variability (thus paralleling Wright's work). The Russians anticipated the concept of the population as a pool of genetic variability and realized that small subpopulations may produce unusual genetic combinations. Because they worked with natural populations from the start, they were able to develop a form of population genetics that addressed the problems being studied by field naturalists. Chetverikov was arrested and sent to internal exile in 1929, thus breaking up his group. But the influence of his work was spread abroad, partly by N. W. Timoféeff-Ressovsky, who moved to Germany in 1925, and more powerfully by Theodosius Dobzhansky (1900–75), who – although not a member of the group – took information on its techniques with him when he went to work with T. H. Morgan in America in 1927.

The New Darwinism

The impact of this renewed support for selectionism was muted among field naturalists, many of whom could not understand the complex mathematical models used by the population geneticists. It took some time for them to realize that Darwinism now offered a more plausible explanation of adaptive evolution than Lamarckism. But the field naturalists were developing a Darwinian perspective in their own, very different way. They were becoming steadily more aware of the need to understand evolution in terms of how populations respond to the environment and to the new situations created by migration. The old approach to natural history based on classification had encouraged the view that species are collections of essentially identical individuals all modelled on the same type. No one had any interest in the natural variability of the population, and the extent of geographical variation within each species was minimized. By the late nineteenth century, however, a new level of sophistication in field studies had emerged. Variability was not only acknowledged, it was actively studied, along with the geographical factors that seemed capable of producing significant modifications in a species.

There has been some debate among scientists and historians over the relative significance of mathematical population genetics and field studies in the creation of the evolutionary synthesis. Genetics certainly represents the more obvious conceptual breakthrough, and had received more attention. But Ernst Mayr, one of the founders of the new Darwinism, has argued that the field workers provided their own input into the synthesis – they did not merely follow in the footsteps of the geneticists[13]. Long before they were certain that natural selection was the cause, the field naturalists were adopting the perspective that had led Darwin himself to recognize that evolution was caused by the response of each population to its local environment. Once they were encouraged to understand the implications of population genetics, they would realize that there was no need to retain their old Lamarckian assumptions. The evolutionary synthesis came about when the findings of population genetics were translated into terms that could be used by a field naturalist with no mathematical background. But at the same time, population genetics itself would benefit from the naturalists' emphasis on the role of geographical factors and the need to take into account the ways in which migration and isolation can break up a once-uniform population.

The new perspective had a major impact on how naturalists approached the question of speciation (the splitting up of one species into a number of distinct but related species). Because of disagreements on this issue among the early Darwinians, saltationist theories had been able to gain ground at the expense of theories based on geographical isolation. The saltationists argued that a new species was produced whenever genetic mutation created individuals who would not (or could not) breed with the parent type. The field naturalists moved in exactly the opposite direction. Their work increasingly suggested that a species divides only when the original population is fragmented by physical barriers. Once isolated, the separate populations evolve in different directions and eventually become so distinct that interbreeding can no longer occur even if the barriers are removed. This interpretation was developed by the German naturalist Karl Jordan, who sidestepped the debate between Darwinism and Lamarckism in order to insist on the role of isolation and local adaptation. By 1905 the American biologist David Starr Jordan (1851–1931) was able to claim that most field workers accepted the role of geographical isolation in speciation. He lamented the laboratory workers' refusal take the evidence from the field seriously.

At the same time there was increased emphasis on the variation produced by climatic conditions even within a single species. It had long been known that wide-ranging species were subject to variation, and some generalizations had become apparent. Members of a species living in a cold climate, for instance, tend to have smaller extremities, which minimizes heat loss. The geneticists at first dismissed this kind of variation as purely somatic, i.e. it was produced by the effect of the environment on the individual animals in each generation and had no genetic component. The field workers came increasingly to doubt this; their evidence suggested that the differences between the individuals from different regions were inherited. The population, not just the individuals, was responding to the different environment. In America, Francis B. Sumner (1874–1945) worked on deer mice during the 1920s and confirmed that different races were maintained by a genetic mechanism. The German naturalist Bernhard Rensch (*b*. 1900) made extensive studies of geographical variation and came to the same conclusion. He believed that the division of a species into races was the prelude to full speciation. His work in turn influenced Ernst Mayr (*b*. 1904), who emigrated to the United States in 1930 after working on the birds of New Guinea and the Solomon Islands.

Sumner, Rensch and Mayr all began their work under the influence of the Lamarckian ideas still common in the early decades of the century. They assumed that the geographical races were formed by the cumulative and inherited effect of individual adaptations. By the early 1930s, though, they were all becoming convinced that natural selection of genetic variation was the true cause. This movement intensified once the theories of population genetics became available in a less mathematical format.

Here the leading role was played by Theodosius Dobzhansky, who came from Russia to work with T. H. Morgan's school and went on to recognize the crucial importance of synthesizing the selectionist model with the evidence derived from fieldwork. He was particularly attracted to Sewall Wright's theory, which included a much greater role for local isolation. The mathematics still had to be taken on trust, as this letter from Dobzhansky to Wright indicates[14]:

> Just read (or tried to read?) your paper ... I am delighted to see it although my mathematical understanding is far too insufficient to read and understand it completely. But I have done the same thing that I have with other papers: read the part of the text preceding and following the mathematics, skipped

the latter in assurance that to it the expression 'papa knows how' is applicable.

Despite this limitation, Dobzhansky's *Genetics and the Origin of Species* of 1937 was the most influential vehicle by which the mathematicians and the field workers were brought together. By opening up the prospect for fruitful collaboration, it allowed the two once-distinct branches of biology to integrate their research programmes around the new Darwinian theory.

The early 1940s saw the appearance of a number of influential books exploring the implications of the new synthesis. Mayr's *Systematics and the Origin of Species* of 1942 emphasized the role of geographical factors in speciation. In the same year Julian Huxley's *Evolution; the Modern Synthesis* also stressed the applicability of population genetics to fieldwork. Huxley (1887–1975) was a British field naturalist, the grandson of T. H. Huxley, who had done so much to promote Darwin's original theory. He had already done important work on animal behaviour (see chapter 11) and had collaborated with the popular writer H. G. Wells to produce a monumental work on *The Science of Life*. Paleontology was brought into the synthesis by George Gaylord Simpson's *Tempo and Mode in Evolution* of 1944. Here Simpson (1902–84) evaluated the evidence that paleontologists had traditionally offered for non-Darwinian evolution and concluded that it was largely illusory. The more that became known about the fossil record for any particular group, the more the allegedly linear trends broke up into a complex evolutionary tree, just as the Darwinians would expect. The new approach had an immediate impact on theories of human origins. As the old theories crumbled, so it became possible to see that the human family had been created by an adaptive breakthrough leading to the adoption of an upright posture, not by a rigid trend towards brain expansion. The fossils of Australopithecus (discussed above) fitted neatly into this new interpretation.

By the 1950s the 'modern synthesis' as Huxley called it came to dominate biology in the English-speaking world. The Darwinian emphasis of the new approach hardened as it gained influence. In the 1940s both Sewall Wright and G. G. Simpson had been prepared to admit small elements of random, non-adaptive change in isolated populations, creating a range of different characters that could be evaluated by natural selection when they spread into the species as a whole. In the course of the 1950s this residual element of non-adaptive evolution was eliminated from the theory. All levels of evolution, however trivial and localized, were now seen to be driven by natural selection.

Outside Britain and America the story was more complicated. The French had never had much time for Darwinism; its image of evolution as a haphazard chapter of local adaptations did not fit very well with their rationalist approach to science. Even classical genetics had failed to develop in France, where the highly centralized academic framework made it difficult for new disciplines to gain a foothold. The geographical dimension of the new synthesis did not have much impact, although the advent of molecular genetics allowed French biologists to play an important role at this level. The Darwinism of Jacques Monod's *Chance and Necessity* (translated 1971) is a rarefied, abstract model of trial-and-error selection that pays little attention to the real world of local adaptations.

In Germany too, classical genetics had never achieved the dominance it enjoyed in America. Paleontologists were able to work with biologists who adopted a less rigid position on the mechanism of heredity. Otto Schindewolf (1896–1971) continued to promote a theory of saltations worked out in the 1930s through into the post-war years. The geneticist Richard Goldschmidt (1878–1958) retained an interest in the process by which the genes are expressed in the development of the organism, and was sceptical of the claim that natural selection acting on minute character differences could produce new species. He too favoured a saltationist view, postulating evolution through the appearance of 'hopeful monsters', which were occasionally able to survive and reproduce. Goldschmidt moved to America in 1935 to escape the Nazis and remained a persistent critic of the emerging Darwinism, especially through the publication of his book *The Material Basis of Evolution* of 1940. Other German refugees were not so lucky and found it impossible to function within the very different scientific climate of America. The uniformity of opinion taken for granted by British and American biologists was thus something of an illusion concealing important national differences.

By eliminating the possibility that evolution is driven in a predictable direction by purely biological forces, modern Darwinism has turned biology into a historical science. To understand how living things came to be in the state we now observe, a host of contingent factors – geological changes, chance migrations, etc. – must be taken into account. Natural selection certainly does not (as its critics allege) reduce everything to chance, because local adaptation governs which genetic characters will be reproduced most frequently. But the historical/geographical dimension certainly implies that the course of evolution will be a complex, haphazard process that makes it impossible to see a single species (our own, for instance) as the goal towards which life on earth has been working over millions of years.

The Implications of Darwinism

Opinions have been divided on the full implications of the theory. Julian Huxley, for instance, tried to retain a sense of cosmic purpose by arguing that life must, in the long run, advance towards higher levels of complexity. Species tend to become less dependent on their environment, a trend continued by humankind's artificial technology. G. G. Simpson, on the other hand, emphasized the unpredictability of evolution and argued strongly against any attempt to salvage the idea that there was a purpose lying behind it all. He also challenged the claims of writers such as Arthur Koestler who wanted to endow animals with a creative power that allows them to direct the course of evolution. Huxley and Simpson disagreed over the evolutionary mysticism expressed in Pierre Teilhard de Chardin's *Phenomenon of Man* (translated 1959). Huxley wrote a preface for the English translation expressing cautious support for the idea of a cosmic purpose leading to the development of mind. Simpson wrote a review that dismissed this position as outdated nonsense and urged the need for modern society to face up to the implications of materialism. We shall see in the next section how the advent of sociobiology has allowed Darwinian ideas to be used once again to support an ideology of competitive individualism.

By emphasizing the lack of apparent purpose, Simpson highlighted just those implications of Darwinism which had long terrified religious thinkers and which had fuelled efforts to ban the teaching of evolution. With other biologists, Simpson campaigned to get evolution introduced into the biology taught in American schools. Despite the ridicule heaped upon the religious fundamentalists during the Scopes trial of 1925, the educators had, in fact, kept evolution out of the textbooks. With the confidence inspired by the new synthesis, the biologists of the post-war years demanded that evolution take its rightful place. The result has been a backlash by the creationists demanding equal time in the schools for what they call 'creation science' – usually a pastiche of anti-Darwinian arguments offering little or no positive research programme by which the idea of miraculous creation might be explored to make sense of the world.

Other groups also oppose modern Darwinism, usually on the grounds that it had been imposed as a kind of dogma that resists any challenge to its authority. The catastrophist theory of Immanuel Velikovsky was noted above (chapter 9), while Erich von Däniken's *Chariots of the Gods?* (translated 1970) argues for genetic engineering by extraterrestrials. The anti-Darwinists all use the same arguments (many of them borrowed from the original opponents of Darwin, and scarcely updated), but they seldom recognize the incompatibility of their many alternatives. Their objections are, however, symptomatic of the gulf that has arisen between the scientists and those sections of the community determined to resist materialism. The dinosaurs and other great denizens of our museums remain endlessly fascinating, but most non-scientists probably continue to think of evolution as the ascent of a ladder towards the high point represented by humankind. It is the opponents of Darwinism who most clearly recognize that this image has been undermined by the latest theories, and it is the opponents who continue to argue against the whole idea of evolution so that they can present their alternatives offering a more purposeful, or at least more romantic, story of our origins.

Even within scientific biology there have been controversies over particular aspects of evolutionism. In the 1970s great excitement was generated by the claim that the dinosaurs were not cold-blooded like modern reptiles, but may have been active, warm-blooded creatures like their descendants, the birds. Paleontologists also came into conflict with molecular biologists over efforts to date particular episodes in the evolution of life on earth. A 'molecular clock' was set up by relating the degree of genetic difference between modern species to the date of their evolutionary divergence. In the case of the separation of humans from apes, the molecular clock gave a much more recent date than paleontologists would allow, although interpretations of the fossil record were eventually modified to take account of the new information. Here was a classic example of two rival disciplines falling into disagreement over a topic of common interest.

There have also been efforts to modify or even challenge the Darwinian orthodoxy. The theory of mass extinctions by asteroid impacts leaves open the possibility that major episodes in evolution may be governed by the unusual conditions that must prevail while the earth is being restocked by the survivors. Much attention has also focused on the suggestion by two paleontologists, Stephen Jay Gould and Niles Eldredge, that evolution goes by fits and starts rather than by gradual transformations. Their 1972 theory of 'punctuated equilibrium' postulated rare episodes of relatively fast transformation among small, isolated populations, while most large populations remain unchanged over long periods of time. This idea is compatible with Darwinism, although it has been resisted by many biologists who fear that it represents the thin end of the wedge leading to a cornucopia of anti-Darwinian theories. There have been charges that punctuated equilibrium offers a Marxist alternative to Darwinism, stressing revolution rather than gradual progress – a charge denied by those who insist that it makes sense out of fossil evidence that otherwise has to be swept under the carpet in order to maintain the facade of gradualist orthodoxy.

There have also been heated debates in the fields of classification and biogeography. The evolutionary synthesis focused attention onto Darwin's claim that a good system for classifying species must correspond to genealogy, but taxonomists cannot agree on how the relationships should be expressed. Ernst Mayr and other orthodox Darwinians insist that the categories must take into account the degrees of change that have taken place within the branches of the evolutionary tree. If one line of development within the reptiles led to an entirely new level of organization – the birds – then the species originating from that line should be grouped into a distinct class. But a group of taxonomists known as the 'cladists', inspired by the work of Willi Hennig (1913–76), argue that classification should be based solely on the branching points of the tree (the name is derived from the Greek word for branch). When pushed to its extreme, cladism implies that – since the birds and the reptiles share a common ancestor – then birds *are* reptiles and should not be designated as a separate class. A small group of extremists known as the 'transformed cladists' have gone so far as to claim that taxonomy can throw no light on ancestral relationships, so that evolution theory cannot be tested scientifically. There was uproar in 1981 when cladistic ideas were incorporated into a display at the Natural History Museum in London.

Much ill-feeling was also generated by an attempt to popularize the anti-Darwinian 'panbiogeography' of the French-born Leon Croizat (1894–1982). Mayr, Simpson and the other Darwinians stressed the dispersal of species by migration from centres of evolution. But Croizat argued that the distribution of many species does not correspond to this model – some groups for which oceans ought to be a major barrier are found on separate continents. The advent of continental drift has, in fact, meant that some aspects of biogeography need to be rethought, but Croizat developed his opposition to Darwinism in the form of strongly (and often strangely) worded polemics. He was dismissed as a crank by the Darwinians, and the younger biologists who tried to rehabilitate his views were at first treated as dangerous lunatics by the establishment. D. E. Rosen recalled the events in 1975 when the 'young Turks' tried to get the American Museum of Natural History's gold medal awarded to Croizat[15]:

> That year, however, battle lines were instantly re-formed when Croizat's panbiogeography (and its consequences for Darwinian dispersalism) were brought into the debate. Following so closely on the heels of the stern challenge by cladistics, the advocacy of panbiogeography and its explicit criticism of traditional dispersalism proved to be intolerable to the museum's ardent Darwinians. Accusations were plentiful, but mostly they were about the morality and the intellectual competence of the advocates of vicariance [i.e

Croizat's] theory. It was to be many months before there were useful discussions about the conflicting ideas. In fact, one letter from an irate curator to the chairman of the museum's scientific council of curators expressed incredulity at the nomination of Croizat as a Corresponding Member of the museum and asked that the letter be disposed of lest it be used in a libel suit. Following a later recommendation by eleven staff members to award Croizat the museum's Gold Medal, another curator threatened resignation. A third curator stopped me one day in the museum's halls to ask why I (and others) continued to raise issues that were so clearly disrupting staff harmony. I in turn asked if he would keep to himself, or share, new ideas that might be viewed as fundamental to our science.

This long quotation gives some idea of the passions that can still be inflamed by scientific debate, and of the power that existing paradigms acquire to block criticism. Yet the new ideas did, in the end, make their way into the debate. Biology is not a monolithic Darwinian dogma, as its critics often claim. Radical alternatives are constantly being tried out, and although the evaluation process is conditioned by social pressures within the scientific community, there are always some who are willing to take on the challenge in the hope of founding a new research programme.

ANIMAL BEHAVIOUR

The study of animal behaviour (ethology) is essentially a twentieth-century science. Although natural theologians had thought of instincts as created by God to serve the animals' needs, they had made few systematic studies. Darwinism might have focused attention onto the topic, because evolution theory offered a new way of explaining how instincts were formed. But even Darwin's own work in this area was marred by a tendency to anthropomorphize animal behaviour, i.e. to interpret it in human terms wherever possible. By linking humankind with the animals, evolutionism turned the question of the origin of the human mind into a burning issue for late-nineteenth-century thinkers. Animals were studied for the light they could throw on the steps by which the various grades of mental function were created. Interest was thus deflected away from those aspects of animal behaviour which were controlled by instincts characteristic of particular species.

In the early twentieth century, psychology began to establish itself as a discipline in its own right. Those who wanted to ensure that psychology was seen as a science, rather than as a branch of philosophy, stressed the role of experimentation. But somewhat paradoxically, one consequence of this was to perpetuate a lack of interest in the instinctive behaviour of different species. Animals were studied in artificial situations in the hope that this would throw light on human learning processes. Behaviourists such as John B. Watson (1878–1958) wanted to banish the concept of mind from science, thereby making it possible to treat the rat finding its way through a maze as a model for all the processes by which animals and humans learn to deal with their environment. To apply this model generally was to assume that all behaviour is learned – there are habits, but no instincts imposed on the organism by its evolutionary past.

The science of ethology emerged from the demand by naturalists for more rigorous information about how species function in the wild. Instincts were studied in a wide range of species, and efforts made to explain them as adaptations to particular modes of life established (just like physical adaptations) by evolution. Even here, though, the link with human behaviour could not be ignored. The prospect that some instincts might be shared between humans and animals ensured that ethology would remain a controversial science. The claim that some aspects of human behaviour are conditioned by instincts offers a powerful tool for ideologues trying to argue that certain forms of society are more 'natural' than others. Psychologists and sociologists continue to regard all efforts to explain human behaviour in such terms as efforts to re-establish the case for social Darwinism. The question of whether the human personality is shaped by nature (inherited instinct) or by nurture (upbringing) is still debated today.

Mind and Brain

Darwin knew that a theory of evolution must explain instincts as well as physical adaptations. He was at first inclined to the view that instincts began as learned habits, which were gradually imprinted upon the species by the Lamarckian process of the inheritance of acquired characters. The habit is a character acquired by the adult organism, which, for a materialist, must correspond to a modification of the brain cells. If the Lamarckian effect works, then the modified brain structure can become inherited just like an acquired physical modification. Darwin soon realized that this would not explain the instincts of neuter insects such as the worker castes of wasps, bees and ants (if they do not breed, they cannot transmit any characters they have acquired). So he began to argue that instincts are subject to natural selection, again invoking a direct parallel between mental and physical evolution. Variations in instincts can be subject to selection just like anything else, and the species' behaviour can thus be adapted to any changes in its environment or to mating habits (by sexual selection).

In his efforts to understand the origin of human mental processes, Darwin gathered a mass of information on animal behaviour. His purpose was to show that no human mental function is absolutely unique; our minds are merely extensions – however dramatic in some areas – of mental processes common in the higher animals. Darwin was led to stress those accounts of animals which depict their behaviour as 'almost human'. He made no experiments of his own, and relied upon anecdotal information supplied by hunters, zoo-keepers and the like. It is small wonder that he was able to find evidence of all the human attributes, even moral behaviour and bravery[16]:

> Several years ago a keeper at the Zoological Gardens shewed me some deep and scarcely healed wounds on the nape of his own neck, inflicted on him, whilst kneeling on the floor, by a fierce baboon. The little American monkey, who was a warm friend of this keeper, lived in the same compartment, and was dreadfully afraid of the great baboon. Nevertheless, as soon as he saw his friend in peril, he rushed to the rescue, and by screams and bites so distracted the baboon that the man was able to escape, after, the surgeons thought, running great risk of his life.

By modern standards this was a sloppy technique, since it assumed that the animals were prompted by mental processes similar to our own. Darwin was rather more successful in explaining the origin of 'lower' human characteristics. The snarl of rage is obviously a continuation of the display of teeth used as a warning by many animals, providing clear evidence of our ape ancestry.

Darwin's disciple in this area was George John Romanes (1848–94), who published books on *Animal Intelligence* in 1881 and *Mental Evolution in Animals* in 1883. Romanes had done careful work on the nervous system of jellyfish to show how their pulsating motion is controlled. But his books on mental evolution are a classic expression of nineteenth-century progressionism. Following Darwin's technique of gathering anecdotal information, he produced an evolutionary scale showing at what point in the advance of life each mental function became apparent. Progress up the mental ladder was all that mattered, and Romanes paid little attention to the diversity of instincts brought about by adaptation to different conditions. To the extent that he did discuss such instincts, he was inclined to adopt a Lamarckian explanation. In his later years, Romanes became increasingly concerned over the relationship between mind and body, and returned to a theistic view in which evolution could be seen as the expression of divine purpose.

At the turn of the century, ideas on animal behaviour were still plagued by the tendency to invoke the higher mental functions. Where once religious thinkers had insisted that such functions were the guarantee of human uniqueness, in an evolutionary world view it seemed impossible to deny that at least some traces of intelligence and reasoning power must be possessed by the animals. A classic example of this is the case of the counting horse known as 'Clever Hans'. This horse appeared to be able to do simple arithmetic, tapping its hoof to indicate the number that was the solution to a problem put to its by its trainer. In 1904 a commission chaired by Professor Carl Stumpf of the Berlin Psychological Institute studied Clever Hans' performance and could detect no sign of fraud. Eventually Stumpf's student, Oskar Pfungst, designed a series of tests that showed that the horse was responding to minute visual cues given, quite unconsciously, by the trainer.

The case of Clever Hans showed that even trained observers could be fooled into seeing intelligent behaviour in animals. But the general rule that a higher mental function should never be invoked to explain a piece of behaviour that can be explained more simply had already been proposed by the psychologist Conway Lloyd Morgan (1852–1936). Now known as 'Lloyd Morgan's canon', this rule should guard against the kind of anthropomorphism that plagued the work of Darwin and Romanes. In his *Introduction to Comparative Psychology* of 1895, Morgan described experiments that showed that a dog, for instance, is incapable of solving problems by reasoning power. When asked to carry a stick through a narrow gap in a fence, the dog has to discover by trial and error that the only way to get through is by twisting the stick so that it is parallel to the gap. Morgan rejected Lamarckism and insisted that natural selection was the only force that could modify instinctive behaviour. To allow learned habits to play a role in directing evolution, he postulated the mechanism of organic selection, later known as the Baldwin effect (described above). Later on in his career, he developed the theory of 'emergent evolution', which supposed that entirely new phenomena – including new levels of mental function – were generated when evolution reached certain critical levels of complexity.

At the bottom end of the hierarchy of functions there were some aspects of animal behaviour so ingrained that they could hardly be called instincts – they seemed to be almost mechanical responses to environmental stimuli. The German–American biologist Jacques Loeb (1859–1924) studied these forced movements or 'tropisms' and used them as the foundation for a mechanistic philosophy of life. Loeb was trained by the botanist Julius von Sachs (1832–1897), who had studied the process by which plants orient themselves towards the light. Loeb found that the behaviour of many lower animals is similarly constrained by physical stimuli. Light, gravity, electric fields, chemicals, all had their power to impose an absolutely slavish response upon certain organisms. Although such activities seemed a long way from the more complex functions expressed in the behaviour of higher animals, Loeb nevertheless insisted that all creatures, including human beings, are nothing but elaborate mechanisms.

Loeb's materialistic philosophy was immensely controversial, and his techniques were appropriate more to the laboratory than to fieldwork. Like many other biologists of the time, he sought to put his subject on a more rigorously scientific footing, and saw a materialistic philosophy as the only suitable foundation for the experimental approach. But his was not the only way of studying behaviour, nor the only way of eliminating the concept of mind from Nature. Morgan's insistence that the higher animals solve problems by trial and error, rather than by reason, was to become a central feature of the school of psychology known as 'behaviourism'. The founder of this school was John B. Watson, who began as a psychologist in the evolutionary tradition but who soon went on to question the links that this tradition had preserved with philosophical speculations about the nature of mind. Watson became determined to establish psychology as a science totally independent of philosophy, and to do this he decided to turn it into an experimental discipline. The concept of mind was to be eliminated from scientific discussions, since all that could actually be observed was the behaviour of an animal or a human being.

The centrepiece of Watson's experimental programme was the investigation of learning, for which he developed the technique of studying how animals, especially rats, learn to run through mazes in order to gain a reward of food. But his conceptual programme went far beyond the mere study of learning – it elevated learning to the status of the only significant determinant of behaviour. Having become dissatisfied with the great variety of different instincts attributed to animals, Watson decided that there was no such thing as an instinct built into a species by evolution. All behaviour patterns are learned rather than inherited, and can be modified at will by placing the animal in a new environment where it must learn new habits. The learning process itself occurs by trial and error; when presented with a problem (e.g. a maze) the animal simply tries out new activities at random until eventually it is successful. The reward of food 'reinforces' the successful behaviour and, if repeated often enough, will confirm it as a habit. Here Watson drew upon the well known experiments by the Russian physiologist Ivan Pavlov (1849–1936), who had developed the concept of the 'conditioned reflex' by modifying the behaviour of dogs. If a bell was rung whenever food was presented, eventually the dog would salivate at the sound of the bell alone.

Watson's philosophy had immense implications because

he was quite prepared to apply it to human beings. In his book *Behaviorism* of 1924 he boasted that he could adapt any human being, whatever his or her origin, to any profession or way of life. This struck at the heart of the hereditarian philosophy underlying race theory and the eugenics movement: people were shaped entirely by nurture (upbringing), not by their biological or inherited nature. Psychology thus joined anthropology and sociology in severing its ties with evolution theory. Biology no longer had anything to say on the question of human behaviour. This was welcome news to the opponents of racism, but Watson's vision of human nature as totally plastic had its own darker side. It opened up the brave new world of social conditioning in which appropriate behavioural modifications would create people with any desired characteristic. In the original *Brave New World* of Aldous Huxley's futuristic novel (published in 1932), it is the state that makes use of this power to ensure that its citizens fit the social order. Watson himself left academic life to become an advertising executive on Madison Avenue, thus putting psychology at the service of private enterprise.

Evolution and Ethology

By turning psychology into an experimental science that minimized the role of instinct, Watson had severed any possible link with the study of animals in the wild. The emergence of ethology as a branch of natural history represents the opposite side of the fragmentation within the sciences of the mind. Naturalists were only interested in behaviour taking place in the natural environment, and they were necessarily more concerned with the instincts that characterize each species than with the learning power that all species possess to a greater or lesser degree. As field studies became more self-consciously scientific in the early twentieth century, it was inevitable that some naturalists would begin systematic observations of animal behaviour and would seek to explain the instincts they found in evolutionary terms. Yet even here there was an ideological dimension. This kind of work could be exploited by those who wanted to challenge Watson's environmentalist view of human behaviour and reinstate the case for evolutionary constraints. There would always be the possibility of arguing that human beings are also governed by instincts, thus placing limits on any proposals for social reform.

Typical of the early moves to create a science of ethology was Julian Huxley's study of the courtship rituals of the great crested grebe. Bird-watching was a well established branch of natural history, and efforts were already under way to turn it into a more scientific activity (chapter 8). In the spring of 1912 Huxley spent his holiday at Tring reservoir (northwest of London) observing the grebe's mating behaviour. His autobiography shows how he had to improvise his observation techniques[17]:

> We slept at Stocks Cottage, which had been lent to us by Aunt Mary, and bicycled off every morning to the reservoirs at Tring to watch and makes notes on the grebes' behaviour. This was facilitated by a contraption I invented – an ×30 telescope mounted between two strips of wood on a tripod with a ball-and-socket joint, which permitted observation and note-taking at the same time.

The paper that Huxley published on the grebes in 1914 was a pioneering move in the effort to incorporate the study of instincts into evolution theory. He discovered that many of the strange 'dances' performed by the birds seemed to be modified versions of their normal behaviour, and suggested that they had evolved in order to cement the pair-bonding necessary for successful mating. The crests surrounding the birds' heads had developed because this feature enhanced the effect of the dance. Huxley thus linked the evolution of both instinctive behaviour and physical structure. His work helped to revive interest in Darwin's theory of sexual selection, which had been ignored by the anti-Darwinian naturalists still active in the early years of the new century. Sexual selection focused attention onto successful breeding as the key to evolutionary change, thus foreshadowing the insights of sociobiology (see below). As we have already seen, Huxley went on to become one of the founders of the evolutionary synthesis, adopting a Darwinian theory to explain all aspects of adaptation and reproductive behaviour.

For many non-specialists it was the work of the Austrian biologist Konrad Lorenz (1903–86) that symbolized the emergence of ethology as a distinct area of study. His popular books, especially *King Solomon's Ring* (translation 1952), seemed to open up a new dimension of natural history, offering fascinating insights into the behaviour of many animals. Within science itself, Lorenz played a major role in establishing ethology as an independent discipline. Yet his work has not been free from controversy, because he made no secret of his desire to draw parallels between animal and human behaviour. His studies of aggression were certainly seized upon by some extremists to form a basis for the claim that human beings are innately aggressive.

Although trained as a doctor, Lorenz was inspired by Oscar Heinroth of the Berlin Zoological Gardens, who published a massive survey of the behavioural characteristics of European birds (1926–33). At his home outside Vienna, Lorenz established a centre for the detailed study of bird behaviour, rearing geese from chicks so they would be accustomed to his presence and would thus allow him to observe their natural habits. He studied the process by which a newly hatched goose chick will 'imprint' itself on the first moving object it sees, treating that object as its mother and following it about. He also studied the mating rituals of many species, confirming Huxley's view that the 'dances' are intended to show off the plumage during courtship. Following Heinroth, he assumed that such activities are based on instincts created by evolution. Each species has built-in motor patterns governing its behaviour in certain circumstances, and in some cases these patterns are quite beyond the learning capacity of the individual. Each instinct was a chain of reflexes forged by natural selection and thus transmitted uniformly to every member of the species. In a situation in which an instinctive action was thwarted, the individual might engage in 'displacement activity' resembling the behaviour appropriate for another situation entirely.

Lorenz scorned Lamarckism and claimed to be an ardent Darwinian. He certainly emphasized the adaptive nature of instincts and saw them as a guide to the species' evolution, but there are some aspects of his thought that do not mesh with the viewpoint of modern Darwinism. He had no interest in the possibility of individual variation within the instincts displayed by different members of the same species. This may have been a by-product of his desire to distance his researches from the studies of learning being performed by laboratory psychologists. Although he admitted that animals can learn new behaviour, he saw that learning would be a source of individual behavioural differences. Insisting upon the absolute uniformity of instinct was thus a move in the attempt to establish ethology as something quite distinct from normal animal psychology. Lorenz also believed that instincts could be used along with physical characteristics to reconstruct evolutionary relationships. Two species with very similar instincts must share a recent common ancestor. His programme for an evolutionary ethology was thus not really Darwinian – it was an extension of the old morphologists' effort to reconstruct the tree of life.

Lorenz's most controversial book was his *On Aggression*, which studied the aggressive behaviour of many different species. He showed that animals capable of severely injuring one another have evolved instincts that minimize the danger of confrontation between individuals. If two wolves get into a fight, the loser only has to make a 'surrender' gesture by exposing his throat and the winner will leave him alone.

If two pigeons are confined together, one may peck the other to death because the species has had no need to evolve such a surrender signal. Lorenz was prepared to extend this argument to the human race. He saw human aggression as particularly dangerous because we have not evolved surrender mechanisms, although modern technology has given us weapons far more dangerous than the wolves' teeth. Knowledge of how our instincts functioned would thus become crucial for the preservation of society[18]:

> I am really being very far from presumptuous when I profess my conviction that in the very near future not only scientists but the majority of tolerably intelligent people will consider as obvious and banal all that has been said in this book about instincts in general and intra-specific aggression in particular; about phylogenetic and cultural ritualization, and about the factors that build up the ever-increasing danger of human society's becoming completely disintegrated by the misfunctioning of social behaviour patterns.

Lorenz's work went beyond the claim that human beings fight because their ancestors were 'killer apes' and located human aggression within the fundamental characteristics of the animal kingdom. It was certainly not his purpose to glorify war; indeed he warned against the ease with which leaders could whip up antagonisms between different societies. His arguments could easily be misused, however, by writers who wanted to claim that aggression was innate to the human character. If such instincts really are built into our very nature, then any society that seeks to eliminate individual conflict altogether is doomed to failure. It is small wonder that those who picked up this notion of the 'anthropology of aggression' were accused of attempting to revive the ethos of social Darwinism.

Lorenz's interest in aggression has led some historians to investigate his past and to charge that he was both a Nazi and a follower of Haeckel's religion of Nature[19]. There can be no denying that Lorenz joined the Nazi party after the *Anschluss* (the annexation of Austria by Germany) and that he was a supporter of eugenics who feared that urban humanity was degenerating now that it was out of touch with Nature. Although ostensibly a Darwinist, his views tended to reflect an older version of evolutionism more in tune with Haeckel's approach. But such ideas were common in the early decades of the century, and Lorenz found no difficulty in promoting his field in the very different climate of opinion that flourished after the war. Instead of trying to oversimplify his position by making him out to be a Nazi, a Darwinian or a Haeckelian, we should perhaps use him as an example of the pervasive influence of ideology on science, and of the changing relationships that can be forged even within the thinking of a single individual as the social environment evolves.

If Lorenz began the process by which ethology was established as a recognized branch of biology, it was his Dutch student Nikolaas Tinbergen (1907–88) who completed the job. Tinbergen joined Lorenz in 1937 and pioneered an experimental approach to the study of behaviour in the wild. He tested the geese chicks' reaction to objects moving overhead to see what triggered the flight response appropriate when a predator was approaching. Even a model shaped roughly like a hawk triggered the response (we now know that it is triggered by any unusual shape overhead). Male sticklebacks display aggression when confronted with any object with a red underside, whether it is a rival male or a simple model. Tinbergen became much more concerned than Lorenz with the adaptive character of instincts and sought to explain them in terms of the environment in which each species lives. With his student, the Swiss ethologist Esther Cullen, he studied kittiwake nesting behaviour on the Inner Farne Islands, showing how this was adapted to the problems of rearing chicks on a crowded cliff-face.

Tinbergen moved to Oxford in 1949 to head the Animal Behaviour Research Group there. Still anxious to establish a role for ethology, he adapted his work to make use of the existing expertise at Oxford in genetics, evolution and ecology. Even at this late stage, it was still necessary for the youthful discipline to create its own niche by adapting to the environment in which it must function – a nice parallel with the explanatory model that Tinbergen was using. In the post-war decades ethology has grown to become a firmly established branch of biology, playing its own unique role in conjunction with related disciplines. As its studies have extended to include the great apes, it has attracted wide public interest that has challenged the image of 'social Darwinism' and focused attention on some of the most threatened animal species.

Primate Studies

If Lorenz sought to explain human behaviour in terms of instincts paralleling those of the lower animals, many ethologists assumed that the primates (the apes and monkeys) should throw most light on the question of human social behaviour. Now that evolutionism was generally accepted, it made sense to assume that our closest living relatives would be most likely to share the distinctive characters that determine our own family and social life. The suggestion that some aspects of human behaviour are shaped by instincts implanted by evolution offered primatologists (biologists specializing in the study of primates) an obvious lever by which they might be able to obtain funds for their research. But in making this bid to offer advice on human social interactions, the primatologists came into direct conflict with the newly independent disciplines of anthropology and sociology. These were committed to the view that cultural and social evolution was totally unconstrained by biology, allowing a wide range of different family and social structures to emerge in different parts of the world. From the start, then, the study of primate behaviour was controversial, and it became even more explosive when it seemed that the 'natural' instincts of apes seemed to parallel the stereotype of male dominance within a family group that was the norm for early-twentieth-century Europeans and Americans.

Critics were able to argue that the primatologists were simply reading their own preconceptions about what was 'natural' for humankind into the behaviour of the animals they observed. Since some of the early studies were performed on animals kept under artificial conditions, it was easy enough to argue that the observed behaviour was not natural at all. The problem of interpretation became more acute in the 1950s and 1960s when ethologists moved out to study chimpanzees and gorillas in the wild. The advent of women primatologists such as Jane Goodall injected a new factor into the situation. How would women respond to the stereotype of male dominance accepted by earlier workers? The projection of the mountain gorillas as essentially gentle creatures represents a direct challenge to the old image of the 'killer ape' – yet studies of chimpanzees have shown that they engage in hunting and even group violence. In such a sensitive area, all interpretations are open to the challenge that they reflect one preconception or another[20]. For many anthropologists the whole enterprise seems flawed, since they continue to argue that human social behaviour is undetermined by whatever instincts are found within our primate relatives.

In the 1920s and 1930s there were extensive studies of primates, performed mostly under artificial conditions. The South African born Solly Zuckerman (*b.* 1904), later Lord Zuckerman, architect of British science policy, made preliminary studies of baboons in the wild, but he moved to Britain in 1926 and continued to work with them at London Zoo. Unfortunately, conditions here were far from natural, because the 'Monkey Hill' had been stocked with a high proportion of the brightly coloured male baboons. In these circumstances, it was hardly surprising that there was considerable conflict among the males for access to the few available females. Zuckerman's book *The Social Life of Monkeys and Apes* (1932) offered detailed observations suggesting that sexual activity and grooming were important mechanisms holding baboon groups together. There was a strong emphasis on male domination of the groups. Zuckerman directly confronted the arguments that had raged among anthropologists seeking to chart the origin of family structures in human society, but he warned against the use of animal behaviour as a model for human societies[21]:

> The polygynous gorilla or baboon can guard his females from the attentions of other males while they all forage together for fruits and young shoots. Primitive man, who, as his Palaeolithic arts display, was an animal largely dependent on a diet of meat, would not have gone hunting if in his absence his females were abducted by his fellows. Reason may have forced the compromise of monogamy.

Zuckerman failed to realize that his warning nevertheless took certain aspects of male behaviour for granted. Nor could he have anticipated that the model of 'man the hunter' would eventually be challenged by feminist anthropologists determined to show that females had played a substantial role in early human societies.

At the same time the psychologist Robert Yerkes (1876–1956) was studying chimpanzee groups at the Yale Laboratory of Primate Biology in America. Yerkes was interested in the overall scale of mental development produced by evolution, and saw the primates as the next highest rung of the ladder beneath humankind. Their ability to participate in complex social groups was thus a model that could be extrapolated up to human beings. Here the model of the 'family' group was again taken to be natural for primates, thus allowing the studies to support traditional values of male dominance. Yerkes was closely associated with the movement to exploit psychology as a tool for social management. His apes would provide an experimental basis for studying techniques of behaviour modification.

Other primatologists followed the ethologist mode of investigation by trying to study behaviour in the wild. In 1931 the American psychologist Clarence Ray Carpenter (1905–75) began a survey of howler monkeys in Panama, finding little evidence of the rampant sexuality shown by Zuckerman's captive baboons. Later, as a member of the Asiatic Primate Expedition of 1937, he studied gibbons in Thailand. Here the natural group was a monogamous family of male, female and a few offspring. There was no sign of male dominance, and much social grooming within the family group. The groups were, however, extremely aggressive in defending territory. Carpenter borrowed techniques from sociology to record the behaviour of monkeys and apes in the wild, providing yet another avenue by which animal behaviour could be absorbed into the growing interest in social management.

After the war another American, George Schaller, was successful in gaining close access to mountain gorillas in central Africa. His work was followed up by Dian Fossey, who became world famous for her sympathetic rapport with these gentle creatures, thus dispelling the myth of the ferocious gorilla first propounded in Darwin's time. At the same time Jane Goodall began an extensive programme of research into the chimpanzees of the Gombe Stream Reserve on the shores of Lake Tanganyika. Her work showed that chimpanzees actually use primitive tools, such as a stick used to extract termites from a mound. The controversial anthropologist and fossil hunter Louis Leakey, who had promoted the work of the two women primatologists, insisted that this discovery altered our most basic conception of the distinction between humans and animals. Although noting the close family life of the chimps, Goodall was eventually forced to accept that there was a harsher side to their behaviour in the wild, including the organized hunting of small mammals and even conflict between rival groups.

Primate studies have offered a variety of models upon which to build an argument for what is 'natural' in human beings. Where once exponents of the nuclear family could argue that male dominance was built into the psychology of all primates, including human beings, their opponents could now offer evidence that, in the wild, the apes exhibit a less structured form of family life. For Goodall's chimpanzees, the mother and dependent offspring form the basic social group – the dominant male is nowhere to be seen. The presence of female primatologists has thus focused attention on the gender values that can so easily slip in to colour observations of creatures that seem to resemble us in so many respects.

The work of Goodall and Fossey also highlights the complex questions raised by the ability of the media to focus attention onto selected scientists. National Geographic made three films of Goodall's work between 1965 and 1984. All three project the study of animals in the wild as a route by which we can appreciate Nature directly; where the laboratory biologist eliminates Nature by reducing everything to a controlled environment, the ethologist sees Nature as a reality to be experienced and cherished. There is an emotional intensity generated by the image of Goodall as the white woman who makes the first real contact with a non-human intelligence in the depths of the African jungle, an emotion clear in her own words describing the first occasion on which she actually touched a wild chimpanzee[22]:

> . . . when I moved my hand closer he looked at it, and then at me, and then he took the fruit, and at the same time held my hand firmly and gently with his own . . . At that moment there was no need of any scientific knowledge to understand his communication of reassurance. The soft pressure of his fingers spoke to me not through my intellect but through a more primitive emotional channel: the barrier of untold centuries which has grown up during the separate

evolution of man and chimpanzee was, for those few
seconds, broken down.

The last film shows Goodall as an older and wiser woman
confronting the sombre facts of war, infanticide and canni-
balism among the animals she had grown to love. We are
still expected to believe that the studies reveal Nature as
it actually is, with no 'interpretation' placed on it by the
observer – yet the honeymoon of physical contact has been
replaced by the harsh 'facts' of life imposed by the logic of
reproductive rivalry.

Owing to similar publicity, Fossey's gorillas came to symbolize the plight of all those species threatened by hunting
and the destruction of their habitat. The film *Gorillas in the
Mist* (a title borrowed from Fossey's own book of 1983)
focused the world's attention on her murder in 1985 at the
hands of poachers. The image of inoffensive giants being
hounded to death by human greed for money and new
farming land was aimed directly at the new environmental
consciousness in the west. But such attitudes are not too
easy to square with the reality of life in Africa. Apart from
the pressure of population, there is a long-standing tradition
of local distrust of 'national parks', left over from the colonial days when such parks were used as a device to restrict
native (but seldom white) hunters. Ethology's move onto
centre stage in the campaign to protect the environment has
only pinpointed the tensions that underlie the developed
world's efforts to impose its values upon regions where the
practical realities of life may make them inappropriate.

Sociobiology

Primate ethology could only be exploited as a model for
human behaviour if it were assumed that human beings
do, indeed, possess instincts built into their brains by evolution, resembling those of our closest relatives. But the
apes and monkeys differ among themselves in their social

behaviour, and it has become clear that these differences represent adaptations to different environments and different modes of rearing offspring. In the last few decades the evolutionists themselves have attempted to make a direct input into discussions of social values by claiming that the new Darwinism offers techniques of direct relevance to the study of social instincts. As in Darwin's theory of sexual selection, successful reproduction is now recognized as the key to understanding evolution and behaviour. Sociobiology is the name of the new discipline created to exploit these techniques, and one of the most controversial topics within the modern relationship between science and society centres on whether or not the human race can be analysed in the same terms.

Where ethology can only offer individual examples of instinctive behaviour, which may or may not be relevant to the human situation, sociobiology offers a general explanation of all social instincts. If human beings are governed by such instincts, then the sociobiologist can pinpoint their exact nature by relating them to our natural reproductive strategies. Opponents insist that human beings have transcended all social instincts imposed by evolution, giving us unlimited freedom to construct whatever kind of society we wish.

Ethologists were not the only biologists interested in social behaviour. Ecologists were concerned with the problem of how populations remained in balance with the physical environment, and realized that in social creatures there must be instincts that govern mating and hence population size. In America, Warder C. Allee studied the populations of freshwater crustaceans and other species whose members tend to gather in large aggregations. His book *Animal Aggregations: A Study in General Sociology* (1931) used Sewall Wright's mathematics to support the claim that natural selection can act on groups as well as individuals, creating instincts that govern mating behaviour for the good of the species.

Similar views were later put forward by the fisheries ecologist V. C. Wynne-Edwards (*b*. 1906) in his *Animal Dispersion in Relation to Social Behaviour* of 1962. After spending some time in Canada, Wynne-Edwards moved to Aberdeen, Scotland, and became concerned about fish stocks in the North Sea. He began to doubt the plausibility of the Darwinian view that each individual invariably seeks to breed as frequently as possible. According to Wynne-Edwards, animals control their own population density through altruistic instincts formed by 'group selection'. The instinct to mate as widely as possible could be checked in periods of high population density by an instinct that sacrificed the good of the individual for the good of the whole species. The only way such an instinct could be formed was by selection favouring those groups with altruistic instincts at the expense of those in which all the individuals behaved selfishly.

The concept of group selection was unremarkable in an era when many biologists still thought of Nature as a system that transcended individual self-interest. But by the 1960s the new, synthetic Darwinism was beginning to dominate biology, and it was soon realized that group selection violated the rules of selection as now understood. Selection acting on groups cannot be responsible for building up altruistic instincts, because in any such group an individual who 'cheats' by failing to comply will be at an advantage. The gene that produces the instinct to cheat will thus spread rapidly in the group and undermine any effects of group selection. This point was made very effectively by George Williams (*b*. 1926) in his *Adaptation and Natural Selection* of 1966. Williams argued that most apparently altruistic or self-sacrificing instincts can be explained on the basis of individual selection if we take into account the fact that the altruism may be reciprocated later on. Helping your neighbour may seem like wasted effort now, but it may be repaid later when he helps you in your own time of need.

The attack on group selection was intensified when the

concept of 'inclusive fitness' was developed to take account of the fact that an individual's behaviour can affect the fortunes of its genetic relatives living in the same population. This possibility had been hinted at by J. B. S. Haldane, and was now formulated mathematically by W. D. Hamilton. Looking at selection from the gene's point of view, it will be advantageous (in terms of maximizing transmission to future generations) if the individual sacrifices itself for a certain number of genetic relatives who carry many of the same genes. The actual number would depend on the closeness of the relationship (two brothers or eight cousins was Haldane's off-the-cuff calculation). The individual does not breed, but its sacrifice enables the gene to be transmitted through the breeding of the relatives who are saved. 'Kin selection' acting at a purely individual level can thus build up altruistic instincts. Hamilton showed that this theory could account for the case of the neuter castes of ants, wasps and bees that had puzzled many naturalists including Darwin. Because of their unusual reproductive system, it is actually to the genetic benefit of the neuter castes to aid the reproduction of their fertile sister, the queen.

The science of sociobiology was established to use similar techniques to explain the social instincts of all species as adaptations to their particular reproductive circumstances. The Oxford biologist Richard Dawkins (*b.* 1941) gained much publicity by introducing the phrase *The Selfish Gene* as the title of his 1976 book. Genetic rather than individual selfishness was now the order of the day. But the high priest of sociobiology is the American, Edward O. Wilson (*b.* 1929), whose book *Sociobiology: The New Synthesis* was published amidst fanfares of publicity in 1975. Wilson surveyed social behaviour in a wide range of organisms from insects to Goodall's chimpanzees, developing the claim that kin selection would be able to account for the different instincts that had been evolved. Wilson had wider ambitions, however, and closed his book with a chapter proclaiming the power of sociobiology to account for many aspects of human social behaviour. The fact that this constituted biology's takeover bid for the human sciences was obvious from the chapter's opening paragraph[23]:

> Let us now consider man in the free spirit of natural history, as though we were zoologists from another planet completing a catalog of social species on Earth. In this macroscopic view the humanities and social sciences shrink to specialized branches of biology; history, biography, and fiction are the research protocols of human ethology; and anthropology and sociology together constitute the sociobiology of a single primate species.

Ethics, he insisted, should now be biologized by recognizing that moral judgements are merely rationalizations of deeply implanted instincts from our evolutionary past. This theme has been developed in a series of later books devoted to the biological foundations of human nature.

Others have been even more determined than Wilson to expound the message of competitive individualism that lies at the heart of the Darwinian explanation of behaviour. One of the most frequently quoted expressions of this view comes from the biologist and historian Michael Ghiselin (*b.* 1939)[24]:

> The economy of nature is competitive from beginning to end . . . No hint of charity ameliorates our vision of society, once sentimentalism has been laid aside. What passes for cooperation turns out to be a mixture of opportunism and exploitation . . . Given a full chance to act for his own interest, nothing but expediency will restrain [an animal] from brutalizing, from maiming, from murdering – his brother, his mate, his parent, or his child. Scratch an 'altruist' and watch a 'hypocrite' bleed.

To writers such as Ghiselin it is important that we do not let sentimentality blind us to the instincts that a harsh Nature may have impressed upon us. Other biologists who do not share his faith in Darwinian individualism dismiss this approach as nothing less than a renewed social Darwinism, a new bid to use biology as the foundation for a 'natural' model of human behaviour that just happens to express the values of the free-enterprise society.

The anthropologists and sociologists have also vigorously resisted this attempt to put them out of business. Some of them challenged the validity of animal sociobiology, despite its evident success in explaining the otherwise anomalous behaviour of species such as the insects with neuter castes. But the real conflict centred on the question of whether or not humankind could be equated with the animals in the first place. Given that the behaviour of animals is quite rigidly constrained by instincts, is it not obvious that human intelligence and adaptability have lifted us onto a different plane, where we are free to behave as we choose?

The campaign against Wilson took on the atmosphere of a crusade as social scientists and the political 'left' united against this new incursion of social Darwinism. Human sociobiology, they claimed, reduced us all to the level of automata driven by our genes. The old bogey of a rigid hereditarianism – the belief that the individual can never transcend the characters imposed on him or her by genetic inheritance – had been given a new lease of life by the fusion with Darwinism. Conservative forces would lap up this new means of demonstrating that a competitive society was natural and hence inevitable. Biologists should stick to biology and leave social management to those who understood the flexibility of human behaviour. The sociobiologists might well query the underlying assumptions of their rivals, who are also engaged in the business of social management, but who seek to shape individual behaviour through the control of learning. There are certainly deep moral and political divisions between the two approaches, but at least some of the heat in the academic debate has been generated by professional interests. The question of whether nature or nurture is the chief determinant of human behaviour has been caught up in the rivalry between different groups of scientists, each offering its own technique for the control of society to the appropriate political interests.

Ecology and Environmentalism

The ideological debates sparked off by sociobiology show that evolutionism is still a controversial issue. In recent years, though, it has been overtaken as a topic of public concern by the growing fear about the state of the environment. To many ordinary people, the increasingly obvious threat posed by human activity to the well-being of the world in which we live has come to symbolize the social tensions created by the expansion of science and technology. Everyone in the developed world comes face to face with the problem of pollution on an almost daily basis. Meanwhile the plight of Dian Fossey's mountain gorillas highlights the destruction of natural habitats throughout the world and warns us all that many species are being driven to extinction. Since the science of ecology deals directly with the interaction between species and their environment, it has been catapulted onto centre stage in the public debate.

The relationship between ecology and evolution theory is a complex one, but there can be little doubt that the advent of modern Darwinism has created part of the framework within which the debate over the environment will be conducted. In the early decades of the twentieth century there were still many biologists who did not believe that adaptation was an important factor in evolution. Such a view is no longer plausible now: the assumption that each species has been shaped by pressures from the environment is a key feature of modern Darwinism. The rise of ecology to prominence was obviously due to changing values in society at large, but it was also made possible by changes within biology that focused attention onto the relationship between the population and its environment. We accept that species can be wiped out by the destruction of their habitat because we realize the intimate and fragile nature of the link between the organism and its environment. At the theoretical level, this realization is, at least in part, a product of Darwinism. In this chapter we shall survey the other factors that have also influenced the rise of the science of ecology.

CHANGING VALUES

Fear of environmental degradation began in the mid nineteenth century (chapter 8) but did not become a major concern for most people until the later decades of the twentieth. A distinct science of ecology emerged in the 1890s and survived on a small scale until thrust into prominence by the concerns that began to develop in the 1960s. Many people now see ecology as a science whose subject matter must necessarily lead its practitioners to side with the environmentalists. The very word 'ecological' has come to denote a concern for the environment. In science, however, 'ecology' is merely the discipline that studies the interactions between organisms and their environment. History shows that such studies can be undertaken within a variety of different value systems. Many early ecologists thought that their studies would encourage a more scientific mode of exploitation. They certainly did not side with the radical demands of those who wanted to reject the values of modern industrial society. Only in more recent decades has the growth of environmentalism created a situation in which a significant number of ecologists are willing to use their science to support the assault on exploitation.

In its early form, ecology was influenced by the traditional link between science and the assumption that Nature was a passive system intended for humankind to develop for its own benefit. The study of relationships was just one more way of refining our ability to dominate the material world. Although the growth of environmentalism has now provided a counterweight to this tradition, there have been powerful forces at work tending to preserve the older, exploitative attitude. In many areas of biology, the twentieth century has seen the triumph of materialistic values and the success of models based on competition and exploitation (chapter 10). If evolutionists saw species as populations of competing individuals struggling to survive and reproduce, they were unlikely to favour a holistic interpretation of ecological relationships. Ecologists were also attracted to economic models that depicted Nature as a system for distributing resources. This in turn inspired a distrust of the romantic 'back-to-Nature' philosophy favoured by many environmentalists. If scientific ecology is to change in response to modern concerns, it must turn its back on some of the most important conceptual pressures that have influenced its growth in the past.

Subduing the Wild

As the last wilderness was conquered, so it became necessary to decide how humankind would govern the world it controlled. The late nineteenth century saw a growing tension between those who wanted unrestrained exploitation of Nature, those who wanted controlled exploitation to protect future resources, and a vocal minority who wanted to protect the wild at all costs. Theodore Roosevelt championed the cause of scientific exploitation in the early twentieth century, but the demands of the private sector were at first impossible to withstand. In later decades the damage to the environment became so obvious that state-controlled management was able to expand. To its critics, however, the notion of controlled exploitation was almost a contradiction in terms. From the 1950s onwards the demand that we should *protect* Nature has grown from a minority position to a major factor in the debate. Scientists have been active on both sides, and their research has often tended to support the conclusions for which they are fighting. Here, again, the 'objective' study of Nature turns out to be an illusion; when so much is at stake, people see what they want to see, especially in a complex situation where there are no clear-cut tests that can be applied.

Theodore Roosevelt's director of the Bureau of Biological Survey was C. Hart Merriam, whose scientific work centred on the life zones of North America. But Merriam was convinced that humankind has the right and the duty to alter the natural distribution of any species for our own purposes. He advocated the killing of birds and mammals that destroyed crops, and recommended poisoning as the best way of destroying 'pests' such as the prairie dog. In 1906 the Bureau became the centre for a national bounty system designed to eradicate many pests, with even the National Parks being incorporated into the programme. A campaign against the wolf was begun in 1915, with trained exterminators being sent out to speed up the process. Many farmers supported the policy, fearful of the effect that wolves had, especially upon sheep. When opponents of the extermination policies began to argue that coyotes, for instance, might have a beneficial effect by killing rats and mice, the Bureau actively set out to show that this was not the case. The argument was inconclusive, providing a classic example of the difficulty experienced by scientists seeking to provide absolutely clear-cut evidence of how the natural balance does, in fact, work.

The policy of scientific management was also applied to fisheries. In response to the increasingly obvious threat of over-fishing, marine biologists developed the concept of the 'maximum sustainable yield' and governments sought to impose appropriate limits on commercial fishermen. The application of these limits often had political and social implications, however. In California, the rules were deliberately framed so that they would apply most stringently to the fishing techniques used by Chinese immigrants. Later on, the sustainable yield available from the Pacific sardine harvest was set at a high level to suit the fishermen, but a series of poor breeding seasons in the late 1940s wiped the industry out. Without human interference, the population could survive such natural fluctuations in its environment, but it could not survive if fished at levels determined by its breeding capacity in good years[1].

For the farmers on land, the 'Dust-Bowl' of the American mid-west in the 1930s seemed to expose the weakness of the claim that unrestrained exploitation could go on unchecked. But even here it was possible to argue against the claim that Nature provided the only stable environment, any interference with which would spell disaster. European ecologists were well aware of the fact that many apparently natural environments were actually the stable products of human management over millennia. A. G. Tansley in Britain argued against the view that ecology must necessarily lead us to query the value of exploitation. Aware that there were few purely 'natural' environments left in Europe, he supported scientific management against the emotional supporters of a 'back-to-Nature' policy[2]:

> It is obvious that modern civilized man upsets the 'natural ecosystems' or 'biotic communities' on a very large scale. But it would be difficult, not to say impossible, to draw a natural line between the activities of the human tribes which presumably fitted into and formed parts of the 'biotic communities' and the destructive human activities of the modern world. Is man part of 'nature' or not? ... Regarded as an exceptionally powerful biotic factor which increasingly upsets the equilibrium of preexisting ecosystems and eventually destroys them, at the same time forming new ones of very different nature, human activity finds its proper place in ecology.

Other opponents of extreme environmentalism took up the same theme of looking for parallels between human and natural disturbances. By the 1950s some Americans were arguing that dust storms were a natural feature of the great plains, not necessarily the product of human interference. Perhaps the grassland itself was stable only because the native Indians had encouraged fires from time to time.

In the article quoted above, Tansley was arguing against the school of ecology founded by Frederic Clements. Despite its origins in the study of prairies that were being drastically modified by farming, Clements' theory stressed that the 'natural' climax vegetation of a region was somehow superior to anything else. By the 1930s this view was being explicitly linked to the holistic philosophy supported by many opponents of Darwinian evolutionism, including the South African statesman Jan Christiaan Smuts (see below). Tansley was afraid that by following this route ecology would ally itself with a view of Nature that was increasingly being identified with the critics who rejected the direction taken by modern science. To assume that there is a spiritual quality about certain natural states that makes them preferable to anything that humankind could produce is to sink into mysticism. Although such holistic ideas were to become the stock-in-trade of the environmentalist movement, many scientific ecologists continued to prefer the model of controlled management. An interest in natural relationships can be built on a desire to control Nature just as easily as upon a romantic view of the universe as a harmonious system that should be left untouched.

In this case, differences over the emotional question of whether human interference was a violation of Nature had a direct bearing on the underlying philosophy of ecological theory. The same interaction between theory and practice can be seen in animal ecology, where the question of what determines population density became crucial in the evaluation of pest-control techniques. Biological controls using artificially introduced predators to reduce insect pests had been tried out successfully as early as the late nineteenth century, and remained popular until the advent of the insecticide DDT swung the balance in favour of chemical control. By the 1950s the harmful side-effects of DDT were already becoming apparent, and a group of agricultural scientists in California appealed to theoretical models of the 'balance of Nature' to argue that biological methods should be reintroduced. They were opposed by Donald Chant of the Canadian Department of Agriculture, who insisted that theories of natural population balance were irrelevant in the artificial environment created by agriculture. In a monoculture (an area containing only a single plant species, the crop), explosions of insect pest populations were inevitable and would have to be controlled by chemicals[3].

Modern environmentalists often express a preference for biological pest control as being more 'natural', and this fits in with the claim that it exploits normal ecological relationships. Chant seems to have gone beyond Tansley's position to argue that some levels of human interference make natural models inapplicable. Since we cannot abandon modern farming techniques without causing mass starvation, the only solution was to adopt even more intrusive methods to defend the crops. But the debate went beyond the question of whether or not the 'balance of Nature' could be restored by further human interference. Chant accused his opponents of distorting the theory they claimed to be applying under the pressure of demands for practical results. Canada's Department of Agriculture enjoyed considerable independence from government interference and its scientists could thus claim to be more objective. This in turn led to the counter-charge that they were merely armchair researchers. Historians are increasingly aware of the fact that institutional differences play a role in shaping theoretical perspectives – although in this case the two sides may have been using the rhetoric of institutional rivalry as a weapon in what was actually a far more fundamental difference of opinion on basic issues. Opposition to the use of chemical pesticides was, in any case, moving outside the arena of scientific debate to become a central issue

in the growing demand for less interference with the
environment.

The Rise of Environmentalism

The twentieth century has seen a massive increase in the
general public's awareness of the damage being done to
the environment by human activities, and an increasingly
militant demand that something be done to halt the destruction. Environmentalism is, however, a complex movement
that has enjoyed the support of a wide range of intellectuals
whose positions on other issues are far from uniform. In
its most limited form, environmentalism demanded the
protection of selected areas of wilderness, especially those
of outstanding beauty, while acknowledging the need for
development elsewhere. The more militant supporters of
the Green movement have, in contrast, opposed the whole
framework of modern industrial society. They would like
to see the world returned to a more natural state by a
wholesale reduction in the level of industrialization. Such
extremism has guaranteed that a concern for Nature has
been linked to programmes that are equally radical with
respect to other issues.

The moderate form of environmentalism could be taken
up even by those committed to the idea that Nature should
be managed along rational lines. The great national parks
of America were established in the late nineteenth century
in order to provide recreational facilities for an increasingly
urbanized population. In Europe the large-scale preservation of wilderness was less practicable, but a growing
movement for the establishment of nature reserves had
emerged by the early years of the twentieth century. In
Britain, enough reserves existed by the 1940s for the
government to appoint a Nature Reserves Investigation
Committee to co-ordinate future developments. Immediately after the Second World War a Wild Life Conservation
Special Committee was set up by the government to identify

additional sites for conservation. The committee was chaired by Julian Huxley, but when Huxley moved to UNESCO in 1946 he was replaced by A. G. Tansley. Tansley's *Our Heritage of Wild Nature* of 1945 had already argued the need for more and better managed reserves, and in 1949 the government established the Nature Conservancy, chaired by Tansley, to establish and maintain reserves, and to promote ecological research into wild habitats. Scientific ecology thus threw its weight behind a programme based on the protection of limited areas, which could be used as a ground-base for the study of changes taking place in the rest of the environment.

The nature reserve programme illustrated how limited gains could be made by working within the existing system. But the real enthusiasts for a return to Nature wished to destroy the existing system and promote a new social order. In the course of the nineteenth century a few writers had begun to assert that unspoilt landscapes had an emotional significance for humankind. There was a stable natural order out there in the world, an order that could only be disrupted by human interference. The consequences of unrestrained exploitation were disastrous both in material and in spiritual terms, since human-kind would become alienated from a world that had been poisoned by the consequences of our own activity. This radical alternative now began to gain widespread attention. Many people became suspicious of unrestrained exploita-tion and campaigned for a limit to be placed on industrial development. A few extremists argued that the whole foundation of modern civilization had been corrupted by the materialist ethic of domination over Nature. As more people became involved, environmentalism took on the character of a political movement, paving the way for the emergence of the Green parties that now flourish in many countries. But the politics of environmentalism stood outside the conventional left–right split on the question of state control, since from the Green perspective exploitation is bad whether encouraged by the state or by private enterprise.

One historian has suggested that hard-line environmentalism flourished most actively in those countries with a middle class influenced by Protestant values, especially in Britain, Germany and America[4]. Since this is precisely the social framework that has normally been associated with individualism and exploitation, we must see environmentalism as a backlash within this tradition based upon a deliberate reversal of its original values. There is a strong link to the kind of anti-materialism that sees the whole as something more than the sum of its parts. Nature is not a machine whose components can be studied or modified individually; it is a comprehensive system governed by laws that interrelate all the various parts. The human race is a part of this system, and can flourish only if it respects the laws governing the whole. This does not mean exploiting knowledge of ecological relationships to further our control of Nature – it means backing away from the whole philosophy of dominance that has sustained the rise of modern science and industry.

In Germany, this attitude was encouraged by the followers of Ernst Haeckel, whose evolutionism was based on an almost mystical philosophy in which the physical universe had a spiritual dimension built into it. There was also a vitalist strand in physiology promoted by one of Haeckel's students, Hans Driesch (1867–1941), who insisted that living things were driven by a non-physical force or entelechy. Although Driesch was an active scientist in the early years of the twentieth century, his vitalist position was soon marginalized as the majority of biologists moved towards a materialistic position. But outside science these ideas formed an intellectual current that fuelled both opposition to Darwinian materialism and a growing tendency to argue that Nature was being violated by human activity. Although a product of Nature, the human race had gained an unnatural power that was threatening to upset the balance of the whole. Nature was unforgiving, and would punish this transgression unless human culture returned to a non-exploitative mode that respected the needs of the whole.

Within this tradition, the mystic philosopher Rudolf Steiner (1861–1935) promoted organic or 'bio-dynamic' farming, and his ideas were taken up by some members of the Nazi party. Once again, we encounter the alleged link between the Haeckelian 'religion of Nature' and right-wing policies. Environmentalism fitted into the Nazi's ideology because they encouraged a suspicion of urban values and saw a renewed peasantry as the foundation of their social order. They established nature reserves – on land cleared of Jews and Poles sent to the death camps. Modern environmentalists may dislike being reminded of the link that once existed between their movement and a right-wing ideology that has deservedly become a focus of hatred in the post-war years. The implication of this episode is not that environmentalism is inherently tainted with racism or any other unpleasant philosophy. On the contrary, we need to recognize the immense flexibility that may allow those whose chief concern is the environment to establish links with a wide range of other positions depending upon the political and cultural dimension within which they operate.

This point can be illustrated by developments in a country that chose the opposite course of political development, Russia. There had been some moves towards conservationism in Tsarist times, and supporters of such policies were able to gain the ear of Lenin after the Bolshevik revolution in 1917. The mineralogist Vladimir Ivanovich Vernadsky (1863–1945) coined the term 'biosphere' as part of his programme for a holistic view of Nature. Through the 1920s conservation enjoyed government support, with benefit both to techniques of scientific management and appeals to the value of wild Nature. The rise of Stalin threw the emphasis onto demands for mass industrialization and the collectivization of agriculture. Respect for wild Nature vanished in the face of demands that science be used to transform the world for human needs. One beneficiary of the new atmosphere of exploitation was T. D. Lysenko, who destroyed Soviet genetics in the name of a Lamarckian programme designed to improve agriculture (chapter 10). Only with the death of Stalin did Soviet environmentalism begin to stir again, and even then the demands of industrialization continued to limit its effectiveness. It seems clear that governments of the extreme right and the extreme left can both support environmentalism – and withdraw that support when it suits them. The Lysenko affair also reveals that Darwinism has no monopoly within the exploitative view of science's role in society.

In America, demands for the establishment of national parks suitable for recreation had flourished since the late nineteenth century. But what about the lands that were not protected from human interference? The philosophy of controlled management was increasingly challenged by a holistic, anti-interventionist view following the tragedy of the Dust-Bowl in the 1930s. In 1936 the Great Plains Committee presented a report on *The Future of the Great Plains* to Franklin D. Roosevelt, arguing that the dust storms were a disaster caused by humans and challenging the philosophy that assumed that humankind could conquer Nature. The policy of exterminating coyotes and other 'pests' was opposed by some biologists who began to see that these species played a role in any balanced ecology.

By the mid century there were defections from among the ranks of the scientists themselves, the most notable being that of Aldo Leopold (1886–1948), originally responsible for game management in Wisconsin, who in 1935 helped to found the Wilderness Society and who sought his own place in the wilderness where he could study Nature unpolluted by human activity. Leopold's *Sand County Almanac*, published posthumously in 1949, has become a founding document of modern American environmentalism. The message of scientific ecology was that Nature is a complex system of relationships, but Leopold saw that conservationism required the scientific knowledge to be supplemented with an ethical judgement based on recognizing the right of all components in the ecological chain to exist[5]:

> Conservation is getting nowhere because it is incompatible with our Abrahamic concept of land. We abuse land because we regard it as a commodity belonging to us. When we see land as a community to which we belong, we may begin to use it with love and respect. There is no other way for land to survive the impact of mechanized man, nor for us to reap from it the esthetic harvest it is capable, under science, of contributing to culture.
>
> That land is a community is the basic concept of ecology, but that land is to be loved and respected is an extension of ethics. That land yields a cultural harvest is a fact long known, but latterly often forgotten.

Leopold's essay 'The Land Ethic' expressed at length the need for the human race to treat Nature as something to which which we belong, not as a commodity to be exploited.

Leopold was no Darwinian, but he recognized that struggle was an inescapable part of Nature's activity. The important message to extract from ecology was the complexity of the system, not a lesson about how to behave. In fact, however, theories about how Nature operates can seldom be separated so clearly from our value-judgements about how we should behave towards the system. Ecological theory had certainly made use of holistic ideas about the coherence of the natural community, and such ideas made it easier to see Nature as something that should be respected. In the post-war years the holistic perspective was maintained within scientific ecology by Warder Allee's group at Chicago (see below). But after Allee's retirement in 1950, ecology was increasingly dominated by the view that natural systems must be treated as differing modes of exploiting the energy provided by the sun, and that natural relationships are determined almost exclusively by competition.

The 1960s was a decade of widespread protest against established values. Young people everywhere sensed that a militaristic and exploitative society seemed to be driving the human race towards destruction, and took to the streets to protest. Opposition to science as a branch of the military–industrial complex was widespread, and in these circumstances the holistic viewpoint characteristic of environmentalism could flourish. Rachel Carson's *Silent Spring* of 1962 became a bestseller by highlighting the blight of the environment by pesticides. If the scientific ecologists were becoming ever more interested in economic models, the environmentalists in society as a whole were turning their backs on economics and demanding that Nature be treated with respect. Eventually the concern became widespread enough to attract the attention of less radical thinkers. The United Nations sponsored a conference on the problems of the environment in 1972, the year in which the Club of Rome brought together a group of influential figures concerned about the growing threat to the world order caused by overexploitation. By the 1980s the Green party had become a major force in German politics and was beginning to attract support elsewhere. On the scientific front, the International Biological Programme (1964–74) gathered much valuable information on the state of the living environment.

The sciences have necessarily been caught up in this explosion of interest, and there is now far more scope for biologists to develop an active concern for the environment. Television brings the study of natural history into every home, often with an environmentalist message included. The personalities who present this message – Carl Sagan, David Attenborough and the like – have become familiar to everyone.

Yet the holistic philosophy that inspires many opponents of materialism has been slow to gain headway in science. Great public interest was aroused by James Lovelock's 'Gaia' hypothesis, which supposes that the whole earth is a self-regulating system that will preserve an environment suitable for life against all threats – including any too obtrusive challenge from the human race[6]. The whole biosphere participates in a complex network of interactions that has resisted physical changes, including a thirty per cent increase in the sun's radiation since life first appeared. Yet although Lovelock's credentials as a scientist are impeccable (he has developed remote sensing techniques used by NASA), his thesis was greeted with considerable scepticism by the majority of scientists. Whatever the public interest in an alternative to materialism, many scientists remain wedded to the mechanistic paradigm that served them so well in the age of expansion. The claim that the whole earth is, in effect, alive or aware still sounds too much like mysticism for the majority of scientists to take it seriously.

Lovelock's approach also runs counter to the professional fragmentation that has become typical of modern science. Like many environmentalists, he insists that the earth and its biosphere can only be understood as a network of interacting processes. The scientist who studies one of these processes in isolation will inevitably fail to appreciate the overall picture. Gaia represents the kind of hypothesis that can only be taken seriously by scientists who have transcended the modern tendency to break everything down into discrete units. It remains to be seen whether the threat of global environmental catastrophe can reintroduce the synthetic approach once favoured on philosophical grounds by Humboldt and his generation.

Any account of modern developments within the environmental sciences must thus end on a note of paradox and uncertainty. There is no need to reiterate here the predictions of global catastrophe that have been made by Lovelock and many other environmentalists. We are all aware that our planet is threatened by massive problems of pollution and the exhaustion of natural resources. The Greens insist that to deal with these problems we must turn our backs on the drive towards increased industrialization. To do this, they argue, we must reject the whole materialistic ethos that has pervaded western culture and return to a world view that treats the earth as a living system upon which we all depend. Yet during the very decades in which this change of attitude has begun to make some headway among the public at large, the science of ecology has committed itself ever more firmly to materialistic models based on the concept of exploitation and Darwinian individualism. Does this mean that science has, indeed, sold its soul to the exploiters? Must the Greens urge a rejection of science along with all other manifestations of the old attitude of indifference to Nature, or can science itself change course and begin to move in the direction sketched out by Lovelock? To answer these questions we must move on to look more carefully at the development of the environmental sciences in the mid twentieth century.

ECOLOGY COMES OF AGE

Ecology had emerged as a coherent discipline at the turn of the century. Plant ecology flourished most actively at first, although by the 1930s animal ecology was becoming the more active field. Neither attracted a great deal of public attention, however, until the middle decades of the century. By then, expanding public awareness of the threat to the environment focused attention onto those branches of biology that were most likely to provide information on what was happening and to advise on what should be done. In today's world of pollution and environmental exhaustion, ecology has entered the world of big science, with all the attendant opportunities and outside pressures.

Ecology had created a niche for itself at the beginning of the twentieth century (chapter 8). The British Ecological

Society was founded in 1913, followed two years later by the Ecological Society of America (whose journal *Ecology* was established in 1920). But growth in the inter-war years was slow. In 1923 A. G. Tansley blamed the First World War for stifling what had appeared a promising start for the new discipline. Many young enthusiasts had been killed, and international links had still not been re-established[7]. British ecology remained on the margins of science until the social upheavals associated with the Second World War allowed it to gain a more secure foothold in the scientific and academic communities.

In America the initial development had been more secure, but membership of the Ecological Society of America remained largely static through the inter-war years. In 1936 the society's president, Walter Taylor, admitted that ecology was hardly the most attractive field for an ambitious young biologist to specialize in[8]. The expansion of ecology into one of the most active areas of biology is very much a function of the last half of the twentieth century. The number of ecologists tripled between 1945 and 1960, and doubled again in the next decade.

The shift of emphasis from plant to animal ecology in the 1930s alerts us to the fact that 'ecology' is not a unified discipline. It is a collection of distinct research programmes that present a public facade of unity based on their common interest in the interactions between organisms and their environment. Plant and animal ecology formed two partially distinct disciplines, while the ecology of the seas was studied independently by oceanographers. Even within fields such as plant ecology, different schools of thought have emerged, each with its own conceptual foundations, and each with its own institutional home. In the early decades F. E. Clements' openly holistic philosophy played a dominant role. The idea that the ecological relationships between the species occupying a particular territory form a coherent system that can be studied as a unit has certainly remained popular, but doubts were soon raised about the wisdom of describing such a unit as a super-organism. Tansley coined the term 'ecosystem' in 1935, but became a persistent critic of Clements' philosophy.

Similar divisions can be seen within animal ecology. The founders of this discipline were field researchers interested in population densities. They were soon challenged by newcomers who constructed mathematical models in an effort to understand natural fluctuations – although some field workers (as in evolution theory) complained about the incomprehensibility of the mathematics. Serious links between population ecology and evolution theory only developed after the emergence of the Darwinian synthesis in the 1940s. A rival school of 'systems ecology' developed in the post-war years.

Historians have begun to ask questions about the observational, intellectual and social factors that maintain the separate programmes. There is no global uniformity of plant communities, yet the biologist will often take his or her home territory as the model upon which theoretical concepts are based. The vast prairies and woodlands of North America create an impression quite unlike that of the small local variations typical of Europe. But there are other factors arising from the social environment that can also shape scientists' perceptions. National and institutional differences affect what a biologist sees as the most important features of the world he seeks to describe, and thus affect both theoretical perspective and the kind of observations that seem relevant. In some cases – as with evolution theory and ethology – models of Nature can have a direct bearing upon political and social issues. Scientists may be inclined to favour a particular view of Nature because they can see a parallel with their preferred image of how human society functions.

The various schools of thought certainly flourished in their own social environments. Clements' school of grassland ecology was a product of the agricultural research and education network in midwestern America. Rival schools emerged in the academic atmosphere of universities dedicated to pure research, such as Chicago or Oxford. The oceanographers were able to draw on different sources of funding. But apart from these professional factors, the rival schools of thought were often divided over the political implications of their theories. Ecologists could not debate whether Nature was to be pictured as a living organism, a machine, or an economic system without conveying messages about the significance of humankind's relationship to the rest of the world. These ideological debates centred on the relative significance of co-operation and competition in Nature and in society. They may seem remote from the concerns of environmentalists who see ecology's chief social function as the foundation for a new respect for Nature. But our current concern with pollution and other environmental problems is a very modern phenomenon. Until recently, most ecologists saw their science as the basis for improved management of Nature. The Green movement has merely added a new social dimension to those which had already been developed in earlier decades.

Plant Ecology

These factors can be seen at work in the different schools of plant ecology that established themselves in the early years of the twentieth century (chapter 8). These schools emerged from plant physiology or from plant geography, the differing institutional backgrounds being reflected in alternative philosophies of Nature. Some ecologists adopted a materialistic approach which saw natural communities as based on the struggle for space among rival species capable of living in the same habitat, while others adopted a more holistic view in which the plant community was to be treated as a super-organism with a life of its own.

In America, one of the most active schools was the grassland ecology of Frederic Clements. Founded within an institutional framework dedicated to practical research on great-plains agriculture, Clements' approach nevertheless represented a direct application of holistic, almost vitalistic, ideas to ecology. The natural vegetation of a region, its 'climax', had the status of a mature living organism. Variations on this state in the same region were to be explained as immature stages in the development towards the climax, maintained by local factors that disturbed the natural succession. Human interference was one such disturbing influence, and the farmed prairie thus had to be seen as an unnatural state that would revert to grassland as soon as cultivation ceased.

An important spin-off from Clements' position was his involvement in what became known as 'experimental taxonomy'. The traditional science of plant classification drew its origins from the work of Linnaeus and identified the relationships between species purely on the basis of structural resemblances (chapters 5 and 7). Clements dismissed this as an old-fashioned approach to botany that encouraged hair-splitting over minute structural differences. He insisted that a true determination of what constituted a good species would have to take into account the ability of a single species to adapt itself to a number of different environments. He carried out transplantation experiments designed to show that the same species took on different forms when grown in different conditions, thus blurring what the orthodox taxonomists took to be rigid distinctions. These experiments were linked to an explicitly Lamarckian interpretation of adaptation that was quite in tune with Clements' holistic philosophy.

The Swedish botanist Göte Turesson performed similar experiments by growing varieties from different regions in the same garden and showing that they maintained their differences. He argued that a species has immense potential for genetic variation in different environments and coined the term 'coenospecies' to denote this theoretical potential. His ideas were taken up by some geneticists (effectively undermining Clements' Lamarckian interpretation) and a wholesale campaign was mounted to persuade taxonomists that they should adopt the experimental method to determine what they meant by a species. The conventional taxonomists were not impressed, and the resulting argument dragged on for decades. The drive to import experiment was part of a deliberate effort to move biology beyond the old tradition of natural history – but many trained in the old tradition felt that it was still performing effectively. It was difficult to resolve the issues separating the two approaches because each defined the kind of evidence that was significant in its own terms. Even if they agreed on what was actually observed, they disagreed over which were the crucial observations upon which the framework of analysis should be based.

In ecology, Clements' research school continued to flourish into the 1930s. The organismic interpretation of the climax was taken up by a number of his students, including John Phillips. In articles published in 1934 and 1935, Phillips both declared his loyalty to the idea of the climax as a super-organism and linked it to the holistic philosophy developed by writers such as J. C. Smuts. The holists were vehemently opposed to the Darwinian selection theory, and tended to favour Lamarckism. Clements was able to exert considerable influence on land-use policies during the Dust-Bowl episode of the 1930s, being asked to serve with the Soil Erosion Service and other government agencies. He urged that the drier areas of the plains be returned to grazing in order to minimize the damage done by farming. Ecology could present itself as a science that would help humankind to make the best use of the land by pointing out those modes of exploitation that were most in tune with what was natural for the region. This advice could only be implemented by governments who would override individual greed for the benefit of society as a whole. The image of a vegetational super-organism thus had a direct parallel with the reformers' model of society as a unified whole with the government as its brain.

In the long run, however, the effect of the Dust Bowl was to cast doubts on Clements' philosophy. Farmers were angry that they were being blamed for the disaster and took refuge in arguments designed to undermine the claim that there was a 'natural' climax for the region, which they had destroyed. At the opposite end of the spectrum of environmental politics, the very fact that the usable soil had blown away was taken as evidence that Nature could *not* recover from human interference. The transition to what was known as 'range management' was associated with a rejection of the organismic philosophy.

This diversity of opinion was reflected within ecology itself. Clements' research programme had never dominated the whole of American ecology. Most other workers in the field shared his opinion that the vegetation occupying a region constituted a system or formation that could be treated as a unified whole, but many disagreed with the claim that it represented a true organism. From the beginning of the century H. C. Cowles had studied the succession of vegetation in the zones surrounding Lake Michigan and had argued that a geographical feature such as the lake allowed a permanent violation of the natural climax for the region. Although admitting that (in the absence of such disturbing factors) there was a natural climax for each region, Cowles focused attention on the physical conditions that determined where a species could find a foothold. If those conditions were subject to local variation, then so was the natural vegetation.

Other ecologists were even more actively opposed to Clements' super-organism concept. The most extreme alternative was the individualistic ecology of Henry Allan Gleason (1882–1973). In 1917 Gleason challenged Clements by arguing that the whole concept of a unified system of vegetation was meaningless. Each plant lived where it could, given the local physical conditions and existing vegetation, and any similarities between the vegetation of different areas was due merely to the fact that the same conditions would allow the same kinds of plants to flourish. Far from there being vast uniform climaxes dominating whole regions, local changes in the environment produced a continuous variation in the precise composition of the vegetation. Changes could be unpredictable, as when new species were accidentally transported to a region where they could gain a foothold.

Although few ecologists took up this extreme position at the time, Gleason's ideas were later adopted by one of the most persistent critics of the social applications of Clements' philosophy, the agricultural historian James C. Malin, who defended the farmers against the government's attempt to impose a monolithic solution to the problem of the great plains. Clements began to back away from the more overtly holistic version of the super-organism concept in the late 1930s. The analogy between the individual and the plant association was expressed much less explicitly in a paper of 1936, and was omitted altogether from the book, *Bio-Ecology*, which Clements co-authored with Victor E. Shelford in 1939. When the Darwinian synthesis began to dominate biology in the 1950s, the individualistic approach at last came into its own.

In Britain, A. G. Tansley imported many of Clements' techniques for monitoring vegetation but turned actively against the super-organism concept. In 1935 he attacked Phillips' uncritical use of the idea[9]:

> Phillips' articles remind one irresistibly of the exposition of a creed – of a closed system of religious or philosophical dogma. Clements appears as the major prophet and Phillips as the chief apostle, with the true apostolic fervour in abundant measure.

From a European perspective it seemed ridiculous to speak of natural climaxes that were inherently superior to any other form of vegetation in the region. Many of the most stable European environments were maintained by constant human activity. Tansley's introduction of the term 'ecosystem' was meant to denote an interpretation of ecological relationships based on a more materialistic view of how the parts of the system interacted.

Continental European botanists developed an entirely different approach to ecology. From the turn of the century, the Zurich–Montpellier school had promoted the analysis of vegetation samples from different locations based on the precise taxonomic identification of all the species present in each sample. This technique was extended by the Swiss botanist Josias Braun-Blanquet (1884–1980), who became director of a newly established Station Internationale de Géobotanique Mediterranéene et Alpine at Montpellier in 1930. The resulting school of thought, often called Sigmatism (from the station's acronym), has flourished ever since. Braun-Blanquet's followers were encouraged to pick out representative quadrats from each location, and to identify all the species as accurately as possible. Ecology was dependent upon taxonomy, and worked through the creation of tables of association between the species. Characteristic plant associations were named and classified almost like the species themselves. The concept of the association as a super-organism was simply meaningless within this view of the ecologist's goals.

Sigmatism was an attempt to preserve and extend an older tradition of botany based on classification against the inroads of the new botany in which physiology (the model used by many rival ecologists) was dominant. Where Clements dismissed orthodox taxonomy as useless, Braun-Blanquet elevated it to the central tool of ecology and was equally dismissive of botanists who did not know how to classify the plants they collected. The professional gulf between the Sigmatists and Clements' school was sustained by their very different roles in society. Clements' position required him to promote ecology as a practical discipline capable of solving the problems created by agriculture, but the Montpellier station shared the autonomy enjoyed by most French academic institutions. The Sigmatists did not have to engage in research of practical value and could present their approach as a 'pure' science free from the distortion of outside pressures. There was no actual rivalry between the Sigmatists and the American ecologists – they simply lived in different worlds and ignored one another completely. The supposedly international character of modern science was certainly not apparent in the field of plant ecology.

Animal Ecology

Although there were some early efforts to apply the models of plant physiology to animal communities, the real development of animal ecology did not take off until the 1920s. At the University of Chicago, Victor E. Shelford (1877–1968) began to apply the concepts of community and succession to animals, basing his ideas on those pioneered by Clements and Cowles. His book *Animal Communities in Temperate America* (1913) adopted the analogy of the community as a super-organism that tended to develop towards a mature state typical for the region. Clements and Shelford's *Bio-Ecology* of 1939 symbolized the unity between their disciplines, although Clements still insisted that the animal community was determined by the vegetation upon which it depended. By its very nature, however, the study of animal ecology raised questions that did not occur in botany. Shelford took an interest in the sequence of predator–prey relationships making up the food chain, and in the causes of population fluctuations in animals.

Study of these topics within a generally organismic framework was continued into the 1930s and 1940s at Chicago under Warder Clyde Allee (1885–1955), who specialized in the forces determining the social groupings or aggregations exhibited by many species. Allee's *Animal Aggregations* of 1931 promoted group selection as the explanation of how animals came to co-operate with one another. The Chicago school led by Allee served as the focus for a view of both ecology and human society based on the claim that co-operation was a natural function of animal behaviour. The book *Principles of Animal Ecology* of 1949 co-authored by Allee and his disciples, Alfred Emerson, Orlando Park, Thomas Park and Karl Schmidt, summed up the group's philosophy and made explicit use of ecology as the source of a holistic value system (it was popularly known as the 'Great AEPPS' book, after the initial letters of the authors' names).

For the Chicago school, ecological interdependence symbolized Nature's overall tendency to evolve in the direction of greater integration, a tendency that was continued by the development of more co-operation in human society. Following the philosophy of the biologist William Morton Wheeler (1865–1937), the closely knit community of the ant or termite colony was proclaimed as the ideal product of this trend towards greater integration. But Allee criticized Wheeler and others who saw sexual relationships as the source of all social activity. For him, the tendency to aggregate was already present in the single-celled Protozoa, and had steadily increased as animals became more complex.

Allee also challenged studies of animal behaviour that had emphasized the existence of a 'pecking order' or dominance hierachy within the group. He insisted that such pecking orders were by no means universal, and were a dangerous model upon which to base human affairs. Dominance hierarchies were often seen as the product of natural selection, since the individual at the top of the pecking order has a better chance to breed. Allee's interest in group selection and social co-operation represented a direct challenge to the model of competitive individualism being projected by many biologists. The Chicago group worked actively to promote co-operation as a model for international relations and social reform. After Allee's retirement in 1950, however, the group seemed to lose cohesion. The Second World War had inevitably thrown doubts on its optimistic message, while the spectre of totalitarianism revealed that the concept of human integration into a well ordered society could be taken too far.

Meanwhile the British biologist Charles Elton (*b.* 1900) had made important moves towards establishing animal ecology as an independent discipline. Trained by Julian Huxley, Elton spent the years 1925–31 working with the records of the Hudson's Bay Company to uncover the fluctuations in the population of fur-bearing animals in the Canadian Arctic over long time periods. In 1932 he moved to the newly founded Bureau of Animal Population at Oxford, which was to play a vital role in the development of animal ecology by encouraging visits from foreign scientists. Elton's *Animal Ecology* of 1927 provided a textbook for the field, defining its central problems and techniques. This book popularized the concept of the 'niche' (introduced originally by Joseph Grinnell in 1913) to denote the species' role in the local ecology[10]:

> It is ... convenient to have some term to describe the status of an animal in its community, to indicate what it is *doing* and not merely what it looks like, and the term used is 'niche'. Animals have all manner of external factors acting upon them – chemical, physical and biotic – and the 'niche' of an animal means its place in the biotic environment, *its relations to food and enemies*.

The niche, in Elton's formulation, was equivalent to the job or profession of each of the human beings making up a society.

Studying the Arctic led Elton to realize that here ecological relationships were simpler and hence easier to identify. He was particularly concerned with fluctuations in animal populations, which were often dramatic in the arctic environment. His later book *Voles, Mice and Lemmings* (1942) charted such explosive fluctuations and sought to explain them in terms of the forces governing reproduction. Faced with this kind of evidence, he rejected the concept of the community as a nicely balanced super-organism and looked instead at factors such as the competition for food. He wrote[11]:

> The 'balance of nature' does not exist, and perhaps never has existed. The numbers of wild animals are constantly varying to a greater or less extent, and the variations are usually irregular in period and always irregular in amplitude. Each variation in the numbers of one species causes direct and indirect repercussions on the numbers of the others, and since many of the latter are themselves independently varying in numbers, the resultant confusion is remarkable.

The model in use here was essentially an economic one: animals were consumers dependent upon their food supply and hence upon anything that affected that supply.

As a student of Huxley's, Elton retained an interest in evolution. He included a chapter on this in *Animal Ecology* and then wrote a whole book on the subject. The plant ecologists talked vaguely about development and progress, but they were not inclined to think about how stable communities became established. Elton's interest in population size and the factors limiting it led him towards a Darwinian model, just as this was becoming more acceptable elsewhere in biology. But his special interests allowed him to suggest modifications to the Darwinian mechanism, with its relentless pressure for adaptation. Many naturalists still retained the belief that at least some of the characters identifying species are non-adaptive. Elton argued that such characters might be established in the population during a time when it was booming; the pressures of the struggle for existence would then be relaxed owing to the overabundance of food.

Elton approached the question of population size from the angle of a field naturalist who had little interest in mathematical theories. But others were becoming aware of the possibility that ecology might join other branches of biology in creating mathematical models that would predict how populations would vary in certain circumstances. In Australia, the British-trained entomologist Alexander John Nicholson (1895–1969) explored the different kinds of factors that might affect population size. He emphasized the distinction between density-dependent and density-independent factors. A density-independent factor, such as temperature or moisture, exerts its effect on the organisms whether the population is large or small. A density-dependent factor varies with the population size and thus acts as a regulator – if the population gets too large, it will be brought back down to the normal level.

Competition is a density-dependent factor, since it increases with overcrowding, and Nicholson argued that such factors were by far the most important in regulating population size. Unlike Elton, he thought that there *was* a balance of Nature, although it involved fluctuations about a mean value. He compared density-dependent controls to the action of the governor on a steam engine, which regulates the flow of steam so as to cancel out accidental variations in speed. The effect of parasites is also density-dependent, and Nicholson joined with a physicist colleague to produce calculations showing that in such a relationship the populations of parasite and host should fluctuate wildly.

The American physical chemist Alfred J. Lotka (1880–1949) became interested in the mathematical theory of population size and published a book on the topic in 1925. Although promoted by the population geneticist Raymond Pearl, Lotka's formulae had little impact on ecology until they became associated with the work of the eminent Italian mathematical physicist Vito Volterra (1860–1940). Having become interested in the problem of fish stocks, Volterra calculated a formula for the relationship between the populations of a predator and its prey, using as his model the mathematical theory governing the behaviour of molecules in a gas (contact between individual predator and prey was equivalent to collisions between gas molecules). The results were sufficiently impressive for Volterra to produce a series of papers on this theme in the late 1920s. In the 1930s the Russian biologist G. F. Gause (1910–86) performed experiments with Protozoa to test the 'laws' of population suggested by mathematics. The 'Lotka–Volterra equations', and Gause's efforts to test them, would play a vital role in the development of post-war ecology (see below).

As in the case of population genetics, many field workers, including Elton, were suspicious of the mathematical formulae. One early supporter, the Canadian-born entomologist William R. Thompson (1887–1972), turned against the new approach in the 1930s, arguing that natural relationships were essentially unpredictable and hence not susceptible to mathematical treatment. Thompson wanted to stress that the effects of external factors make it impossible for the biologist to treat the population as a system subject to predictable laws. He thought that density-independent factors such as changes in the physical environment were more important in controlling numbers – again denying the validity of the model in which the population is a self-regulating system. Field studies had shown that with the corn-borer (a notorious pest insect), a warm winter resulted in an explosion of numbers the following year – yet the natural predators upon the corn-borer did not follow the trend. Nicholson's theory that the interaction of predator and prey would tend to restore the balance was dismissed as a product of abstract mathematics that bore no relationship to real life.

These debates had a series of hidden agenda based on both the social and the practical implications of the various theories. But many different pressures are involved, and it is impossible for the historian to identify simple one-to-one relationships between theories and external factors. A scientist's choice of theoretical model may be influenced both by political views and by interests arising from the practical applications of the theory – but these two pressures may not act in the same direction. Different aspects of the same theory can be emphasized to give quite different implications. Individual scientists may have idiosyncratic backgrounds that affect their thinking in unusual ways.

Historians' efforts to uncover the network of interactions between scientific theories and the social environment are thus beset with problems. It seems easy enough to demonstrate that there *are* links between the way a particular scientist thinks and his or her social or institutional background. But as in the case of evolution theory it is difficult to find testable generalizations allowing us to 'predict' how an individual will react to a certain new idea. For this reason, attempts to show that scientific knowledge is determined by social values have been widely criticized – often by scientists who want to go to the opposite extreme by arguing that scientific knowledge is purely objective. 'Determinism' is the wrong word to use in this context: scientific thought is certainly *influenced* by external factors, but it is not *determined* by the scientists' social and cultural background because there are too many different and often conflicting sources of influence for us to predict which will be decisive in a particular situation.

The theories of Clements and Allee show a very fundamental ideological influence: a holistic value system in which co-operation is important tends to generate a theory in which the population is seen as a super-organism produced by a general trend towards integration in Nature. Conversely, theoretical models that present the population as a collection of individuals lend support to an individualistic and competitive value system. But there are notable exceptions to this rule. Thompson, for instance, saw the population as a collection of individuals – but this was certainly not because he supported competitive individualism. In fact he was a devout Catholic who studied Aristotelian philosophy. His opposition to Nicholson's mathematical biology was based on Aristotle's distinction between the modes of reasoning appropriate in mathematics and in natural science.

There were also more direct pressures. Thompson and

Nicholson were economic entomologists dependent on funds devoted to pest control. Thompson was afraid that mathematical ecology might undermine support for the kind of careful fieldwork upon which his own conclusions were based. The theoretical debate thus became mixed up with the controversy over the most practicable methods of pest control. Opinions on the best method of control were influenced by preferences for density-dependent or density-independent regulation of the population. Biological control of pests by the deliberate introduction of parasites was favoured by those ecologists who preferred density-dependent models, because parasites are more effective in dense populations of their hosts. The effectiveness of chemical insecticides does not increase with population size, and this method of control is likely to be favoured by ecologists committed to the model of the population as a collection of individuals. A whole range of factors, scientific, institutional and ideological, were thus interacting to shape the opinions of ecologists confronting these complex technical questions.

Marine Ecology

Important developments were also being made in the study of ecological relationships in the oceans. German scientists had studied the role of plankton in determining populations of larger marine creatures during the early years of the century, but their work was largely forgotten in the post-war years. The same area of study was revived in Britain during the 1920s and led to notable advances in our understanding of the factors that control the sea's productivity. Because government support for the study of fisheries science offered an independent professional environment for oceanographers, this work developed in isolation from that of both the plant and animal ecologists, and exploited different theoretical models.

The Marine Biological Laboratory at Plymouth had endured difficult times in the early decades of the century, but renewed government support for fisheries research opened up new opportunities during the 1920s. Laboratory facilities were improved and vessels were made available for oceanographical study. Led by E. J. Allen (1866–1942) – a student of the biometrician W. F. R. Weldon – a group of oceanographers began to develop improved techniques for studying the factors that controlled the populations of marine organisms, including fish. The Germans had shown that the plankton was crucial, since these minute plants and animals served as the basis of the food chain. In effect, the production of fish in the sea could be compared to land-based agriculture, where the productivity of the land determined the number of cattle, etc., that could be supported. Allen adopted a holistic, but non-vitalist philosophy that stressed the interrelatedness of all the sea's inhabitants. His group applied the methods of chemistry and physics to study the factors that controlled the minute plants which served as the most basic food resource. By using photometers they showed how variations in the sun's light affected the growth of these plants. The circulation of chemical nutrients was also studied, and a classic paper of 1935 by H. W. Harvey and his co-workers outlined how all of these factors interacted to shape the seasonal fluctuations of the plankton.

Although the Plymouth group continued its work in later decades, the theoretical initiative passed to America in the 1940s. Mathematical models would now be brought in to marine ecology, as they had into terrestrial animal ecology (see below). The impulse that led to this development arose from new conceptual tools that were presented as the basis for a major revolution in ecological thinking.

MODERN ECOLOGY

The external pressures upon ecologists were hardly likely to diminish as concern for the environment quickened in the post-war years. The field was flourishing as governments responded to public pressure by looking for scientific advice on how to deal with the avalanche of environmental problems. The new mathematical techniques were expanded to create a more 'scientific' image for ecology, while the emergence of the synthetic theory of evolution opened up the possibilities of a fruitful co-operation between the two areas. If ecology was the study of interactions between organisms and their environments in the modern world, evolution was the study of how those interactions had developed and had shaped the organisms themselves. Expansion did not necessarily generate unity, however, since the ever-growing community of ecologists divided itself into rival research schools and theoretical orientations.

These divisions still reflect differing social and institutional pressures. But the pressures combine to give a situation that is far more complex than most non-scientists appreciate. It would be easy to fall into the trap of thinking that the only serious external pressures acting upon ecologists derive from the competing demands of industry and the environmentalist lobby. The industrialists want to find evidence that their activities are not as damaging as their critics imply, while the critics see ecology as the foundation for a more caring approach to the environment. Many non-scientists see ecology as a subject devoted to the promotion of a holistic perspective that must encourage a return to a more natural way of life. But many ecologists derive their funds from industry or from government, and will thus tend to favour models that endorse controlled exploitation.

The tension between industry and the environmentalist lobby is only one of the pressures to which scientific ecologists are subject. The link with a revived Darwinian evolution theory has ensured that ecology has been drawn into the still-active debates about materialism and the search for a 'natural' basis for human behaviour. Many ecologists accept theoretical models that are quite explicitly opposed to the holistic perspective of the radical environmentalists. Population ecology favours the individualistic approach of

Darwinism, while systems ecology looks for models of natural relationships based on the economic networks existing within human society. The development of scientific ecology cannot be equated with the rise of environmentalism, nor have the tensions between these two areas diminished in the modern world.

Populations and Systems

Following his experimental tests of the Lotka–Volterra equations, G. F. Gause in 1939 began to formulate a different concept of the niche, focusing attention on the fact that in a world dominated by inter-species competition the best-adapted species would drive out any rivals from its chosen niche. The claim that two different species could not occupy the same niche attracted the attention of both ecologists and evolutionists. In 1960 Garrett Hardin introduced the term 'principle of competitive exclusion' to denote what became a central theme of post-war ecological thinking.

One of the first biologists to make use of the new idea was the British ornithologist David Lack (1910–73). In 1938 Lack visited the Galapagos islands using research funds generated by Julian Huxley. The islands' finches had been discussed by biologists ever since Darwin had used them to bolster his case for speciation produced by geographical isolation. The different species had evolved on different islands, but there had been a certain amount of subsequent migration. On some islands several almost identical (yet reproductively distinct) species could be observed feeding together. Opponents of Darwinism used this fact to discredit the selection theory, and it could be taken as a direct violation of the principle of competitive exclusion. Lack showed that the very similar species actually had significant differences in feeding habits. They might all feed together, but they were exploiting different sources of food in the same environment. Lack's classic study,

Darwin's Finches of 1947, helped to establish the case both for the synthetic theory of evolution and for the principle of competitive exclusion. It also focused historians' attention onto Darwin's encounter with the finches, thus creating the legend of his conversion to evolutionism in the Galapagos.

At the same time the British-trained ecologist G. Evelyn Hutchinson (*b*. 1903), who had moved to America in 1928, began to make out a new case for the role of mathematical models in ecology. In 1946 Hutchinson gave a paper in which he promoted the view that ecological relationships should be seen as systems governed by causal interactions. Although drawing upon the long tradition of viewing communities as super-organisms, Hutchinson transformed this approach by replacing the organic metaphor with an economic one. The transfer of chemicals and energy through the system is governed by feedback loops that create stability in the face of environmental fluctuations. Hutchinson picked up the 'biogeochemical' approach pioneered by the Russian, V. Vernadsky, who had coined the term 'biosphere' earlier in the century. One could construct budgets for the circulation of chemicals through the ecological networks of the oceans and the land surface, showing how changes in the availability of nutrients affected the populations of the organisms that exploit them. These circulation systems were themselves governed by feedback loops that tended to restore the equilibrium – any system that was not thus stabilized would sooner or later be stressed beyond its limits and would disintegrate.

The concept of the feedback loop as a means by which a physical system could maintain itself in balance gained wide popularity in the post-war years. Vitalism was now unnecessary because machines could be made that would also exhibit purposeful behaviour. The new science of cybernetics was founded by Norbert Weiner to exploit the idea in the construction of self-regulating machines. Hutchinson saw the same principles at work on a global scale, maintaining the systems that allow vast numbers of living things to obtain their raw materials. He was also aware of the possibility that the same concept could be extended to the social sciences, and participated in conferences on this theme organized by the Macy Foundation between 1946 and 1953. Where Clements had once justified government control of the environment by appealing to the image of society as a super-organism, the new systems theory offered the prospect of social control through the setting up of stable feedback loops of human interactions. In an atmosphere of post-war optimism, science seemed to offer the prospect of creating a new and more secure world.

These insights were subsequently developed by the brothers Eugene P. and Howard T. Odum (*b.* 1913 and 1924) to become the basis for what is now sometimes known as 'systems ecology'. H. T. Odum gained his Ph.D. in 1950 for a study that showed that the chemical balance in the oceans had remained constant over millions of years. He went on to develop a technique for representing ecological systems in terms of energy flow, converting all other natural resources into equivalents in this, the most basic resource of all. A pioneering step in this direction had been taken by another of Hutchinson's students, Raymond Lindemann, in a classic 1942 paper analysing the flow of energy within a specific environment, Cedar Bog Lake in Minnesota. He showed how the transfer of the sun's energy from one part of the lake's ecology to another was always accompanied by a loss. But by the time the paper was published Lindemann was dead at the age of twenty-seven. His insights were built upon by the Odums, who created models of the energy flow within a variety of ecological systems, drawing charts modelled on the social scientists' representations of the flow of resources within the economy.

Research on the flow of energy and resources within ecological systems formed a major part of the International Biological Program of the late 1960s. The Odums urged the study of particularly stable systems as models that could help humankind to understand how to manage the stresses imposed by our own activities. A major study of the rain-forest in Puerto Rico was funded by the Rockefeller Foundation – and by the US Atomic Energy Commission, interested in the susceptibility of natural systems to major catastrophes. The Odums also made a study of the coral-reef ecosystem at Eniwetok atoll, the site of the American H-bomb tests. H. T. Odum's *Environment, Power and Society* of 1971 expounded the technocrats' dream of a society constructed to a carefully manufactured pattern, and argued that science offered the answers to the future problem of how to live with diminished energy supplies. As far as systems ecology was concerned, the human economy was simply one aspect of the global network of resource utilization that science hoped to understand and control.

In his 1946 paper Hutchinson also seized upon the possibility of constructing mathematical models of population regulation to argue that here, too, was a phenomenon governed by feedback loops. He had already become dissatisfied with the lack of rigorous analysis present in Elton's studies of animal populations, entitling his review of Elton's *Voles, Mice and Lemmings* 'Nati sunt mures, et facta est confusio' [Mice are born, and the result is confusion].[12] Hutchinson urged the creation of a new ecology based on a fruitful interaction between mathematics and experiment. Where the critics wanted to reject Gause's work because it did not fit the facts, Hutchinson wanted to tinker with the mathematics in the hope of getting a better fit, and of learning something in the process. He also interpreted the niche concept in mathematical terms, in the hope of throwing light on his own work with lake plankton. The fact that different species of plankton could co-exist in the uniform environment of a lake was explained by extending Gause's arguments on competition between the species to include other factors. It was possible to show that changes in the physical environment were the starting point for a whole sequence of complex interactions between the different species. In a far-reaching address given at a

Cold Spring Harbor conference on animal ecology in 1957, Hutchinson set out a programme for integrating ecology and evolution theory by means of the new techniques, a theme expressed in the title of one of his later books, *The Ecological Theatre and the Evolutionary Play* (1965).

Hutchinson's student, Robert MacArthur (1930–72), expanded this approach to create a powerful new science of community ecology. In a 1958 study of the warblers in the forests of New England, MacArthur confirmed the principle of competitive exclusion by showing that similar species actually had different feeding patterns. He went on to develop mathematical models for predicting the numbers of the various species inhabiting a given area, refining traditional insights such as the 'pyramid of numbers', which relates the size of the predator population to that of its prey. Hutchinson and MacArthur asked whether the population size was governed by the physical conditions or by competition from other species trying to exploit the same resources. How closely could species be 'packed together' in the ecology and still each have its own niche? Did the niches themselves evolve along with the species that occupy them? These and other questions were brought under the wing of a new mathematical ecology that MacArthur presented as the driving force for the science's development in the 1960s. His theory was based on the assumption that the relationships between species occupying the same area are determined by competition.

One specific application of the new approach was to the study of island biogeography. Here MacArthur linked up with E. O. Wilson to develop a theory based on a different set of questions to those traditionally asked in the field. As in the case of the Galapagos finches, most studies had concentrated on the evolutionary processes that had created the particular species inhabiting groups of islands. But MacArthur and Wilson were not interested in the history of a particular group: they wanted to know what factors governed the number of species that inhabited any oceanic island. It turned out that the diversity of species was proportional to the island's area, and that the precise number of species on an island represented a balance between immigration and extinction (always a threat to small, newly established populations). The two biologists also considered the question of what reproductive strategy gave a newly arrived species the best chance of establishing itself. From this developed Wilson's more general interest in reproductive strategies and his subsequent creation of the science of sociobiology (discussed above).

Hutchinson's preference for mathematical modelling was applied to marine ecology by his student Gordon A. Riley (1911–85). From the mid 1930s onwards Riley attempted to introduce mathematical rigour into the study of the ways in which light and nutrient supply governed the populations of plankton species, as pioneered by German and British oceanographers. During the 1940s he applied the techniques of population ecology to this problem, quantifying the interactions of the different species that made up the food chain. He believed that his equations would allow the oceanographer to predict fluctuations in the populations, given a knowledge of changes in the physical factors that controlled the system. Owing to the massive government investment in oceanography during and following the Second World War (see chapter 9), marine ecology was given the tools that would enable it to move into the new world of mathematics and quantification. Riley's work nevertheless remained largely ignored in other areas of ecology, thus perpetuating the professional isolation that had for so long maintained a barrier between oceanography and other disciplines.

The 1960s was an exciting time in the development of ecology. Money was pouring into the field and the numbers of scientists specializing in the area was increasing dramatically. New techniques such as those offered by MacArthur offered the prospect that the whole field might be revolutionized and given a more scientific foundation. A unified science of community ecology and evolutionary theory might be created as both areas began to exploit the mathematical models based on the assumption that relationships were governed by individual competition. But in fact the expansion generated fragmentation and debate rather than unity of purpose. Ecology split up into numerous research schools with different theoretical perspectives, each violently critical of its rivals. Field naturalists were still suspicious of mathematics, while the mathematicians disagreed among themselves over the best models to use. MacArthur himself complained about the consequence of these divisions[13]:

> Unfortunately there are propaganda efforts by insecure members of the various schools aimed at others, and it would not be the first time in history if one of these efforts succeeded in temporarily putting one school out of favor . . . In the interests of freedom and diversity, even these destructive attacks must be tolerated, but it is well to recognize that they tell us more about the attacker than the attacked. However, it is a pity that several promising ecologists have been wasting their lives in philosophical nonsense about there being only one way – their own way, of course – to do science.

In fact, it was MacArthur's own far-reaching proposals that were the source of much of the debate. Although he was soon removed from the scene by his death from cancer in 1972, the controversy over how and whether to exploit his suggestions continued, and ecology is still very far from being a unified science today.

Apart from the divisions within population ecology, there remains a more fundamental gulf between this area and systems ecology. Population studies have linked up with evolution theory to exploit the view that competition determines both the structure of a region's ecology and the evolution of the species themselves. The emphasis is very much on individualism and the lack of any 'purposeful' direction in Nature – the classic posture of true Darwinism. Systems ecology, in contrast, stresses the role of feedback loops and the ability of natural systems to regulate themselves in an apparently purposeful manner. The tension between these two approaches emerges most clearly when they are pushed to extremes, thus reviving the long-standing conflict between the teleological (purposeful) and the 'trial-and-error' philosophies of Nature.

This tension can be seen in the scorn that was heaped upon James Lovelock's 'Gaia' hypothesis, in which the whole earth is treated as a self-regulating system designed to maintain conditions suitable for life. Lovelock certainly talks about cybernetics and the properties of feedback loops in mechanical systems, but his critics feel that he has extended the analogy so far that he has reintroduced the old organismic philosophy that modern science has only just succeeded in eliminating after centuries of wishful thinking. If the whole earth behaves like a system that protects itself against disturbances, have we not, in the end, re-created the image of the earth mother or the earth goddess? Lovelock's choice of the name 'Gaia' – the Greek earth goddess – surely symbolizes this return to the old view that Nature, or at least the earth itself, is a living, purposeful entity. Lovelock professes not to understand why his approach has been so thoroughly rejected by the scientific establishment[14]:

I had a faint hope that *Gaia* might be denounced from the pulpit; instead I was asked to deliver a sermon on *Gaia* at the Cathedral of St John the Divine in New York. By contrast *Gaia* was condemned as teleological by my peers and the journals, *Nature* and *Science*, would not publish papers on the subject. No satisfactory reasons for rejection were given; it was as if the establishment, like the theological establishment of Galileo's time, would no longer tolerate radical or eccentric notions.

The claim that science no longer permits anything that smacks of purpose into its thinking is also a stock-in-trade of those writers who attack the dogmatism of modern Darwinism. Lovelock is certainly no Lamarckian, but his efforts to reintroduce the concept of purposeful activity in Nature run directly counter to one of the most powerful trends in modern biology.

Darwinians such as Richard Dawkins made the most determined effort to show that a self-regulating earth could not be produced by natural processes. Lovelock claims to have countered this objection, but the language with which his critics assailed him makes it clear just how far he has gone beyond the bounds of what is considered acceptable. In a review entitled 'Is Nature Really Motherly?', W. Ford Doolittle dismissed Gaia as 'group selection writ large' and linked the theory directly to old-fashioned wishful thinking[15]:

> The good thing about the engaging little book by Jim Lovelock is that reading it gives one a warm, comforting feeling about Nature and man's place in it. The bad thing is that this feeling is based on a view of natural selection ... which is unquestionably false.

Modern Darwinism (and modern population ecology) have created a model of Nature based on individual struggle producing change only by trial and error. For its supporters, this approach offers the only way of freeing science (and ultimately our whole way of thinking) from outdated concepts based on the hope that we are the preordained goal of an essentially purposeful Nature. Only by facing up to that harsh reality can we hope to come to terms with the real problems that confront us. To its opponents, however, the Darwinian way of thinking is merely the end-product of a scientific materialism created to serve the ideology of the market economy.

Ecology and the Environmental Crisis

From the environmentalist perspective, Lovelock's is far from a comforting view of humankind's position on the earth. His main purpose is to warn of the extent to which human activity is threatening to destroy the stability of a system that has regulated itself for millions of years. Lovelock uses his hypothesis to warn that certain acts of destruction, especially the burning of the rain-forests, may have consequences far beyond anything we can predict. He thus joins the ranks of the prophets of doom who predict an environmental catastrophe through exhaustion of the natural resources needed to sustain industry and the pollution of the air and water that make life itself possible. It will make little difference to us if the events that wipe us out are merely Gaia's belated effort to preserve the earth for other forms of life.

To an outsider it does indeed seem strange that, at the time when there is an increasing call for humankind to respect the global environment or face extinction, the science of ecology has moved firmly away from the organic model of Nature that seems most likely to encourage the right kind of behaviour. Radical environmentalists actively encourage the view that the earth should be treated as an organic, feminine system that nurtures all life including our own. As concern for the environment grows among the general public, this world view is likely to expand its influence. Yet paradoxically science itself has committed itself ever more strongly to a model of Nature based on individual struggle, a devil-take-the-hindmost philosophy that offers little support for harmonious inter-species relationships and openly endorses competition as the mechanism of progress within species. That, at least, is the opinion of its critics.

There was a time in the 1960s and 1970s when many

scientists genuinely believed that 'improved' education would eliminate the old ways of thought to leave technocratic materialism in charge of our society. The creationist revolt against Darwinism was the first sign that, even in the world's most technically advanced country, traditional values were making a stand. The rise of environmentalism has established a quite different line of philosophical opposition to materialism, one that – unlike creationism – may draw support from everyone who sees the harmful effects of over-exploitation on the world. As we contemplate the destruction of the rain-forests and the ozone layer, the extinction of species, the climatic disturbances that seem to herald the onset of global warming and the overcrowding now accepted as normal in our cities, we may be moving towards a clash between science and the broader values of society that will make the debates over Darwinism seem tame by comparison. Will ordinary people adopt Green values and turn against professional scientists, blaming them for providing both the tools and the moral authority used by the military–industrial complex, which is destroying the earth? It is not the business of the historian to predict, but history may nevertheless provide insights that prevent us from oversimplifying the issues involved.

One point that seems self-evident from our survey of the environmental sciences is that we can no longer see science as the source of value-free information that can be used in whatever way society prefers. The use of scientific ideas to uphold social values has been so obvious in this area that more perceptive scientists have given up pretending that they have a method for gathering purely objective knowledge. Scientific theories are models of reality constructed by human beings, and those models inevitably reflect, and are used to endorse, the values and interests of those who create them. Those interests may range all the way from the protection of research funding to the broadest moral and social principles, but their role seems inescapable. The debate between the organic and the mechanistic models of the biosphere is merely the most general example of the conceptual differences that result from building different principles into our models of Nature.

One consequence of this is that we must recognize the existence of underlying values in *all* scientific theories, not just in those we do not like. It is very easy for the opponent of materialism to lambast the Darwinians for their worship of brute force, but all too easy to forget that anti-Darwinian theories are themselves projections of alternative value systems. When the Green party complains that scientists employed by industry are distorting their findings in order to cover up the extent of pollution, it may be slow to admit that its own scientists have also constructed a model that will highlight the environmentalist interpretation of the situation. Disputes over the 'facts' about pollution merely confirm that there are no truly objective facts to be had. Measurements are made within the context of a theoretical approach to the question; change the theory and you change what counts as a relevant fact. If we accept that all science is based on theories that can be related to human values, then we can no longer use science to decide on the values we wish to adopt. No one can prove that a value-system is right by appealing to the facts of Nature, because the facts are always filtered through the value-systems of the underlying theories. Whether you support the free-enterprise system, or see industry as a curse that must be removed, you should do so because that is how you feel about the situation in which you live, not because you think science offers unequivocal support for your position.

There is thus a real sense in which science is value-free. Not the old sense in which it was supposed to offer objective knowledge of the real world, but a more sophisticated sense in which we must accept that all sorts of different values can be used as the basis for scientific theories. Conflict between theories can sometimes be resolved when one shows itself to be superior to another in the generation of fruitful ideas for experimental study. But in all too many of the cases we have seen, the conflict could not be resolved by the discoveries themselves: alternative theories continued as rivals because both could make some sense out of the bewildering array of information that Nature presents to our senses. Theories fell into the background because scientists lost interest in them, often for reasons that lay outside the scope of science itself. In many cases a once-rejected line of thought re-emerged later on in a new guise because its conceptual scheme could be reformulated to meet new needs (Lovelock's Gaia hypothesis would fit into this category).

This view of science requires us to be very suspicious of claims that science itself is necessarily a component of the problems that plague the modern world. It would be easy to construct an interpretation of history in which science was presented as nothing more than the handmaiden of rampant commercialism. From Francis Bacon's call for us to 'put Nature to the question' to the rise of modern Darwinism, science has moved steadily further towards a materialistic model of Nature and the endorsement of struggle as the only means of changing the world. At the same time it has destroyed our ability to gain an overview of the world and its inhabitants by encouraging a narrow specialization that focuses attention only onto neatly compartmentalized phenomena.

It would be silly to pretend that this line of development does not represent one of the main thrusts of modern science. But we have also seen enough in our survey to realize that science is not a monolithic network of interconnected theories that all support the same set of values. Science has advanced by exploiting different models of Nature in different circumstances, and the links that have been drawn between those theories and the values of the time are many and various. To pretend that science can *only* function by exploring materialistic values is to ignore the many episodes in which holistic and organismic models have played an active role in scientific research. The predominance of materialistic ideas in modern theories does not place a straitjacket on future scientific thinking. On the contrary there are many working scientists who are quite willing to try out alternative views. If *Nature* and *Science* would not print Lovelock's articles in the 1970s, the latest edition of his book appears with glowing endorsements on the cover reprinted from the pages of *New Scientist* and *Scientific American*.

At the same time, environmentalist concerns have re-emphasized the need for the kind of synthetic studies represented by the Gaia hypothesis. Lovelock demonstrates the complex web of interactions that link all aspects of the earth's physical structure and the living things that inhabit it. The kind of narrow specialization favoured by early-twentieth-century science may have achieved short-term gains in specific areas, but it is no longer suitable for studying the broad range of interactions that must be taken into account when dealing with the global environmental crisis. Humboldt and an earlier generation of environmental scientists actively sought to take account of these interactions because they had a philosophical commitment to a holistic view of Nature. If materialism has encouraged fragmentation, its days may be numbered in a world that has very practical reasons for reviving the Humboldtian approach.

Even where materialistic theories predominate in a certain area of science, we must beware of the assumption that they must inevitably tend to promote an inhumane or exploitative view of Nature and society. No doubt Darwinism, in both its original and its modern forms, has been used to endorse the free-enterprise system and the value of competition. But the image of Nature as a self-regulating machine was also offered as a version of materialism linked to the idea that science could reorganize society from the top downwards. A Darwinian is also in a strong position to emphasize the fragility of the environment and the ease with which species may be driven to extinction, exactly the position that would be favoured by an environmentalist. From the Darwinian perspective, it is precisely because Nature is based on an unplanned, trial-and-error mode of evolution that each one of the resulting products is so precious. The unpredictability of a truly historical mode of explanation need not lead to the assumption that all the results are worthless. Rather, it leads us to value all the results because we know that we ourselves have no right to assume that we are 'higher' than any of the others.

We must also be extremely careful of claims that alternative, holistic or organismic theories will necessarily reinforce more humane values and a greater respect for the environment. When such theories have gained a dominant role in science they have not always been used to support the moral or social applications that their modern supporters assume to be inevitable. Clements' use of the organismic model of the plant community was used as the basis for his claim that governments can use ecological knowledge to regulate humankind's exploitation of Nature – not to endorse the radical environmentalists' rejection of human domination. Anti-Darwinian evolutionary theories were linked to racism and other social policies based on biological determinism, exactly the kind of unpleasantness that anti-determinists frequently imagine to be the hallmark of Darwinism itself. Creationists extol the virtues of free enterprise more enthusiastically than any social Darwinist, and some of them reject environmentalism because they think God will create new resources when we need them. The possibility of a link between Nazism and certain aspects of radical environmentalism suggests that even the Greens need to watch out for the anti-democratic tendencies that can easily be associated with any movement claiming to know what is right for society.

Let us hope that we can weather the coming environmental crisis without sacrificing the humane values that most of us want to preserve. But in any case, let us not fall into the trap of thinking that science can only be used to endorse a single set of values. If we are to survive without allowing vast numbers of people to starve, science will have to be used rather than rejected. For all the input from subjective factors outlined above, science remains the only way of demonstrating that certain ideas do *not* seem to have their expected consequence when used as models for Nature. Although it seems impossible for scientists to devise crucial tests that will distinguish between fundamentally different conceptual schemes (at least where very complex phenomena are concerned), it is still possible *within a single scheme* to eliminate potential applications that do not seem to square with how Nature itself operates. This is the basis of science's growing ability to help us control Nature – and to demonstrate the consequences of exploitation.

In a world where everyone accepts that excessive exploitation is wrong, it may thus be possible for science to help us avoid the errors of the past. The evidence of history suggests that scientists are able to generate fruitful reseach programmes based on a wide variety of different conceptual schemes. They are also remarkably adept at sensing the possibility of new research funds offered by different interests. In the middle decades of the twentieth century there were strong pressures encouraging the development of theories based on the model of economic exploitation. But many biologists now work for environmentalist groups and will create theoretical models that highlight the problems of pollution. These models may be no closer to an absolute truth about Nature than those proposed by rival scientists working for industry. But the fact that both are present in the debate will allow the effects of human activity to be monitored more critically; neither side will be able to get away with pretending that its own interpretation is the only conceivable one. If the environmentalists are right (as most of us suspect they are), professional and empirical pressures will combine to force scientists in the direction of theories embodying a more responsible attitude towards Nature. Science's very adaptability to social influence, rather than its imagined objectivity, will allow it to be used constructively in a world that has seen the Green light.

NOTES

CHAPTER 1

1 Yi-Fu Tuan, *Space and Place: The Perspective of Experience* (London: Edward Arnold, 1977), p. 13.
2 Michel Foucault, *The Order of Things: The Archaeology of the Human Sciences* (New York: Pantheon, 1970), p. xv.
3 Mary Douglas, *Purity and Danger: An Analysis of Concepts of Pollution and Taboo* (London: Routledge and Kegan Paul, 1966), chap. 3.
4 Karl Popper, *The Logic of Scientific Discovery* (London: Hutchinson, 1959).
5 Popper's original attack appears in his article 'Darwinism as a Metaphysical Research Programme', in *The Philosophy of Karl Popper*, ed. Paul A. Schilpp (La Salle, Ill.: Open Court, 1974, 2 vols), vol. 1, pp. 133–43. For a response see Michael Ruse, *Darwinism Defended: A Guide to the Evolution Controversies* (Reading, Mass.: Addison-Wesley, 1982).
6 Robert C. Olby, *A History of Biology* (London: Fontana Press; and New York: W.W. Norton, 1993)
7 Thomas S. Kuhn, *The Structure of Scientific Revolutions* (Chicago: University of Chicago Press, 1962).
8 Anna Bramwell, *Ecology in the Twentieth Century: A History* (New Haven, Conn.: Yale University Press, 1989).

CHAPTER 2

1 Ernst Mayr, *The Growth of Biological Thought: Diversity, Evolution, and Inheritance* (Cambridge, Mass.: Harvard University Press, 1982), pp. 45–7 and 304–5.
2 Lucretius, *On the Nature of the Universe*, transl. R. E. Latham (Harmondsworth: Penguin Classics, 1951), pp. 195 and 197.
3 Plato, *Timaeus*, introductory conversation, from *Timaeus and Critias*, transl. Desmond Lee (Harmondsworth: Penguin Classics, 1971), p. 38. This edition has an appendix on the Atlantis myth.
4 Aristotle, *Meteorologica*, book 1, 14, 352a. From

the translation in Jonathan Barnes (ed.), *The Complete Works of Aristotle: The Revised Oxford Translation* (Princeton, NJ: Princeton University Press, 1984, 2 vols), vol. 1, p. 574.

5 Aristotle, *Historia Animalium*, book 5, 6, 541b, transl. *ibid.*, vol. 1, p. 855.

6 T. H. White, *The Bestiary: A Book of Beasts* (New York: Putnam, 1954), pp. 42–3.

7 Albertus Magnus, *Book of Minerals*, transl. Dorothy Wyckoff (Oxford: Clarendon Press, 1967), p. 52.

8 *Ibid.*, p. 22.

CHAPTER 3

1 Translated by Sir Charles Sherrington, *The Endeavour of Jean Fernel* (Cambridge: Cambridge University Press, 1946), p. 136.

2 Adam Zaluzniansky, *Methodi Herbariae Libri Tres*, translated in Agnes Arber, *Herbals: Their Origin and Evolution: A Chapter in the History of Botany, 1470–1670* (Cambridge: Cambridge University Press, 1912), p. 117.

3 John Ray, *Synopsis Methodica Stirpum Britannicarum* (1690), preface, translated in C. E. Raven, *John Ray, Naturalist: His Life and Works* (Cambridge: Cambridge University Press, 1942), p. 241.

4 Francis Bacon, *The New*

Atlantis, from Bacon *The Advancement of Learning and The New Atlantis* (London: Oxford University Press, 1951), p. 288.

5 Robert Boyle, *A Free Enquiry into the Vulgarly Receiv'd Notion of Nature* (London, 1686), pp. 18–19.

6 William Derham, *Physico-Theology: or, a Demonstration of the Being and Attributes of God, from His Works of Creation* (2nd edn, London, 1714), p. 55.

CHAPTER 4

1 See Michel Foucault, *The Order of Things: the Archaeology of the Human Sciences* (New York: Pantheon, 1970). For further discussion of Foucault's views see the material on biological taxonomy in chapter 6.

2 Translated from Pierre Perrault, *De l'origine des fontaines* (Paris, 1674) in K. F. Mather and S. L. Mason (eds), *A Source Book in Geology* (New York: McGraw-Hill, 1939), p. 22.

3 See Margarita Bowen, *Empiricism and Geographical Thought: from Francis Bacon to Alexander von Humboldt* (Cambridge: Cambridge University Press, 1981).

4 John Dennis, *Miscellanies in Verse and Prose* (London, 1693), p. 139, quoted in

G. R. Davies, *The Earth in Decay: a History of British Geomorphology, 1578–1878* (New York: Science History Publications, 1969), p. 36.

5 Ray to Lhwyd, 1695, from *Further Correspondence of John Ray*, ed. R. W. T. Gunther (London: Ray Society, 1928), letter 154.

6 Robert Hooke, 'A Discourse of Earthquakes', in *The Posthumous Works of Robert Hooke* (London, 1705; reprinted New York: Johnson Reprint Corporation, 1969), pp. 279–450, see p. 449.

7 Translated from Buffon, *Les époques de la nature* (*Histoire naturelle*, suppl. vol. 5, Paris, 1778), p. 179. See Jacques Roger (ed.), *Mémoires du Muséum National d'Histoire Naturelle* (new series C, vol. 10, 1962), pp. 151–2.

8 Robert Jameson, *Elements of Geognosy* (Edinburgh, 1808; reprinted New York: Hafner, 1976), p. 78.

9 James Hutton, *Theory of the Earth with Proofs and Illustrations* (Edinburgh, 1795, 2 vols), vol. 1, p. 200.

10 John Playfair, *Illustrations of the Huttonian Theory of the Earth* (Edinburgh, 1802; reprinted New York: Dover, 1964), pp. 14–15.

CHAPTER 5

1 A good example of this approach is Aram Vartanian, *Diderot and Descartes: A Study of Scientific Naturalism in the Enlightenment* (Princeton, NJ: Princeton University Press, 1953).

2 William Bligh, *A Voyage to the South Seas* (London, 1792), p. 5.

3 John Ray, *The Wisdom of God Manifested in the Works of Creation* (11th edn, Glasgow, 1744), preface.

4 Alexander Pope, *An Essay on Man*, epistle 1, in *The Works of Alexander Pope* (London, 1752, 7 vols), vol. 3, p. 18.

5 Gilbert White, *The Natural History and Antiquities of Selborne* (London, 1789), p. 233.

6 William Derham, *Physico-Theology* (3rd edn, London, 1714), p. 171.

7 Isaac Bilberg, 'The Oeconomy of Nature', transl. in Benjamin Stillingfleet, *Miscellaneous Tracts relating to Natural History, Husbandry, and Physick* (3rd edn, London, 1775), pp. 39–129, see p. 114.

8 White, *The Natural History and Antiquities of Selborne* (note 5), pp. 139–40. Note, however, that White still found it difficult to throw off the traditional belief that swallows hibernate in Britain over the winter, see pp. 138–9.

9 Erasmus Darwin, *The Temple of Nature* (London,

1803), canto I, lines 295–302.

CHAPTER 6

1 See for instance Susan F. Cannon, *Science in Culture: The Early Victorian Period* (New York: Science History Publications, 1978), chap. 3.

2 Alexander von Humboldt, *Cosmos: A Sketch of a Physical Description of the Universe*, transl. E. C. Otté (London: Henry G. Bohn, 1864, 5 vols), vol. 1, p. 1.

3 The term 'gentlemanly specialist' is used by Martin Rudwick; see his *The Great Devonian Controversy: The Shaping of Scientific Knowledge among Gentlemanly Specialists* (Chicago: University of Chicago Press, 1985).

4 Brongniart's paper was translated the following year as 'General Considerations on the Nature of the Vegetation which Covered the Earth at the Different Epochs of the Formation of its Crust', *Edinburgh New Philosophical Journal*, **6** (1829): 349–71.

5 See William Thomson, 'On the Secular Cooling of the Earth', *Philosophical Magazine*, 4th ser., **25** (1863): 1–14, p. 8. Other physicists had already made this point, as stressed by William Hopkins in his 'President's Address', *Quarterly Journal of the Geological Society of London*, **8** (1852): xxix–lxx, see p. lviii.

6 Charles Darwin to W. H. Fitton, June 1842, in Frederick Burckhardt and Sydney Smith (eds), *The Correspondence of Charles Darwin*, (Cambridge: Cambridge University Press, 1986), vol. II, pp. 321–2. A copy of this letter was sent to Agassiz by Buckland, who quoted it against Murchison at the 1842 meeting of the British Association.

7 From Lyell's notebook of 1829, as quoted in Leonard G. Wilson, *Charles Lyell: The Years to 1841* (New Haven, Conn.: Yale University Press, 1972), p. 253.

8 Charles Lyell, *Principles of Geology: Being an Attempt to Explain the Former Changes of the Earth's Surface by Reference to Causes Now in Operation* (London: John Murray, 1830–3, 3 vols), vol. 1, p. 80.

CHAPTER 7

1 The term 'biologie' was coined independently by Lamarck and G. G. Treviranus in 1802. On the problem of deciding when a distinct science of 'biology' actually emerged see Joseph A. Caron, '"Biology" in the Life Sciences: A Historiographical Contribution', *History of Science*, **26** (1988): 223–68.

2 William Paley, *Natural*

Theology, in *The Works of William Paley, D. D.* (London: F. C. and J. Rivington, 1819, 5 vols), vol. 4, p. 371.

3 Richard Owen, *On the Nature of Limbs* (London: Van Voorst, 1848), p. 3.

4 Hewett C. Watson, *Cybele Britannica: Or British Plants and their Geographical Relations* (London: Longman, 1845–59, 4 vols), vol. 4, p. 449.

5 Charles Lyell, *Principles of Geology* (London: John Murray, 1830–3, 3 vols), vol. 2, p. 131; translating Augustin de Candolle's 'Géographie botanique', in F. G. Levrault (ed.), *Dictionnaire des sciences naturelles* (Strasbourg and Paris: Le Normant, 1820), vol. 28; pp. 359–418, see p. 384.

6 Charles Darwin, 'Essay of 1844', in Charles Darwin and Alfred Russel Wallace, *Evolution by Natural Selection* (Cambridge: Cambridge University Press, 1958), p. 180.

7 Louis Agassiz, 'On the Succession and Development of Organized Beings at the Surface of the Terrestrial Globe', *Edinburgh New Philosophical Journal*, **33** (1842): 388–99, p. 399.

8 Richard Owen, 'Report on British Fossil Reptiles, Part 2', *Report of the British Association for the Advancement of Science*, 1841 meeting; pp. 60–204, see p. 202.

9 Hugh Miller, *Footprints of the Creator; or the Asterolepis of Stromness* (3rd edn, London: Johnston and Hunter, 1850), p. 179.

10 Owen, *On the Nature of Limbs* (note 3), p. 89.

11 See Frank Sulloway, 'Darwin and his Finches: the Evolution of a Legend', *Journal of the History of Biology*, **15** (1982): 1–54.

12 Darwin, 'Essay of 1844', in Darwin and Wallace, *Evolution by Natural Selection* (note 6), pp. 119–20.

13 See M. J. S. Hodge, 'Darwin as a Lifelong Generation Theorist', in David Kohn (ed.), *The Darwinian Heritage* (Princeton, NJ: Princeton University Press, 1985), pp. 207–43.

CHAPTER 8

1 George Perkins Marsh, *Man and Nature*, ed. David Lowenthal (Cambridge, Mass.: Harvard University Press, 1965), pp. 42 and 43.

2 See for instance the elephant hunt in chapter four of Rider Haggard's *King Solomon's Mines*.

3 W. S. Jevons, *The Coal Question: an Inquiry Concerning the Progress of the Nation, and the Probable Exhaustion of our Coal Mines*, 3rd edn, ed. A. W. Flux (London: Macmillan, 1906), p. 9; Jevon's italics.

4 Huxley later turned against

Spencer's social evolutionism in his *Evolution and Ethics* of 1893, but in the 1860s and 1870s they worked together to promote the evolutionary philosophy.

5 Samuel Butler, 'The Deadlock in Darwinism' [1890], reprinted in Butler, *Essays on Life, Art and Science*, ed. R. A. Streatfield (London: Fifield, 1908), p. 308.

6 Sir J. F. W. Herschel, *Physical Geography* (Edinburgh: A. & C. Black, 1861), p. 12. On the 'higgledy-piggledy' comment see Darwin's letter to Lyell, 12 December 1859, in *The Life and Letters of Charles Darwin*, ed. Francis Darwin (London: John Murray, 1887, 3 vols), vol. 2, p. 241.

7 Richard Owen, *On the Anatomy of Vertebrates* (London: Longmans, Green, 1866–8, 3 vols), vol. 3, p. 808.

8 E. Ray Lankester, *Degeneration: a Chapter in Darwinism* (London: Macmillan, 1880), p. 33.

9 E. W. MacBride, *A Textbook of Embryology: Invertebrates* (London: Macmillan, 1914), p. 663.

10 W. J. Sollas, *Ancient Hunters and Their Modern Representatives* (London: Macmillan, 1911), p. 383.

11 A. R. Wallace, *Darwinism: an Exposition of the Theory of Natural Selection with Some of its Applications* (London: Macmillan, 1889), pp. 352–3. On the site of human origins see *ibid.*, pp. 459–60.

12 William Bateson, *Materials for the Study of Variation, Treated with Especial Regard to Discontinuity in the Origin of Species* (London: Macmillan, 1894), p. v.

13 J. S. Burdon–Sanderson, 'President's Address', *Report of the British Association for the Advancement of Science*, 1893 meeting, pp. 3–31, see pp. 8–9.

14 Charles Darwin, *On the Origin of Species by Means of Natural Selection; or the Preservation of Favoured Races in the Struggle for Life* (London: John Murray, 1859), p. 74.

15 Stephen A. Forbes, 'The Lake as a Microcosm', [1887], reprinted in *Illinois Natural History Survey Bulletin*, **15** (1925): 537–50.

16 Andreas Schimper, *Plant-Geography upon a Physiological Basis*, transl. W. R. Fisher (Oxford: Clarendon Press, 1903), p. vi.

17 Eugenius Warming, *Oecology of Plants: an Introduction to the Study of Plant Communities*, transl. P. Groom (Oxford: Clarendon Press, 1909), pp. 91–2.

18 *Ibid.*, p. 348.

19 Frederic E. Clements, *Plant Succession: an Analysis of the Development of Vegetation*

(Washington: Carnegie Institution, 1916), p. 6.

CHAPTER 9

1 Vilhelm Bjerknes, 'The Structure of the Atmosphere when Rain is Falling', *Quarterly Journal of the Royal Meteorological Society*, **46** (1920): 119–30, see p. 127.

2 The clearest expression of this view is Robert Muir Wood, *The Dark Side of the Earth* (London: Allen and Unwin, 1985).

3 Quoted in Arthur S. Eve, *Rutherford* (New York: Macmillan, 1939), p. 107.

4 Alfred Wegener, *The Origin of Continents and Oceans* (New York: Dover, 1966), p. 77.

5 Bailey Willis, 'Continental Drift: Ein Märchen', *American Journal of Science*, **242** (1944): 509–13.

6 Philip Lake, 'Wegener's Displacement Theory', *Geological Magazine*, **59** (1922): 338–46, see p. 338.

7 Alexander Du Toit, *Our Wandering Continents: An Hypothesis of Continental Drifting* (Edinburgh: Oliver and Boyd, 1937), p. 3.

8 Interview with Richard Doell transcribed in William Glen, *The Road to Jaramillo: Critical Years of the Revolution in Earth Science* (Stanford, Calif.: Stanford University Press, 1982), p. 265.

9 Lawrence Morley's recollections, as recorded in an interview quoted *ibid.*, p. 299.

10 Interview with Fred Vine quoted *ibid.*, p. 310.

CHAPTER 10

1 For a discussion of the symbolism expressed by Akeley's dioramas and the Roosevelt Memorial Atrium see Donna Haraway, *Primate Visions: Gender, Race and Nature in the World of Modern Science* (New York: Routledge, 1989), chap. 3.

2 The possible link between Haeckel's philosophy of Nature and Nazism is explored by Daniel Gasman, *The Scientific Origins of National Socialism: Ernst Haeckel and the Monist League* (New York: American Elsevier, 1971); for an alternative view see Alfred Kelly, *The Descent of Darwinism: the Popularization of Darwinism in Germany, 1860–1914* (Chapel Hill, NC: University of North Carolina Press, 1981). The link between Haeckel and radical environmentalism is one theme of Anna Bramwell, *Ecology in the Twentieth Century: A History* (New Haven, Conn.: Yale University Press, 1989).

3 Vernon L. Kellogg, *Headquarters Nights: A Record of Conversations and Experiences at the Headquarters*

of the German Army in France and Belgium (Boston: Atlantic Monthly Press, 1917). Kellogg's *Darwinism Today* (New York: Henry Holt, 1907) is one of the best contemporary accounts of the debates then raging over Darwinism and the alternative evolutionary mechanisms.

4 Henry Fairfield Osborn, 'Why Central Asia?', *Natural History*, **26** (1926): 263–9, see pp. 266–7.

5 Raymond Dart, 'The Status of *Australopithecus*', *American Journal of Physical Anthropology*, **26** (1940): 167–86, see p. 178.

6 Stephen Jay Gould, *Wonderful Life: The Burgess Shale and the Nature of History* (London: Hutchinson Radius, 1989).

7 H. F. Osborn, *The Titanotheres of Ancient-Wyoming, Dakota and Nebraska*, Monograph No. 55 (Washington: US Geological Survey, 1929, 2 vols), vol. 2, p. 845.

8 George Bernard Shaw, *Back to Methuselah: A Metabiological Pentateuch* (London: Constable, 1921), p. liv.

9 Arthur Koestler, *The Case of the Midwife Toad* (London: Hutchinson, 1971). See Paul Kammerer, *The Inheritance of Acquired Characteristics* (New York: Boni and Liveright, 1924).

10 See Peter J. Bowler, 'E. W. MacBride's Lamarckian Eugenics and its Implications for the Social Construction of Scientific Knowledge', *Annals of Science*, **41** (1984): 245–60.

11 T. H. Morgan, *A Critique of the Theory of Evolution* (Princeton, NJ: Princeton University Press, 1916), pp. 87–8.

12 J. B. S. Haldane, 'Natural Selection', *Nature*, **124** (1929): 444.

13 For a more balanced approach arising out of the debate between Mayr and the historians of genetics see Ernst Mayr and William B. Provine (eds), *The Evolutionary Synthesis: Perspectives on the Unification of Biology* (Cambridge, Mass.: Harvard University Press, 1980).

14 Theodosius Dobzhansky to Sewall Wright, 25 July 1937; quoted in William B. Provine, *Sewall Wright and Evolutionary Biology* (Chicago: University of Chicago Press, 1986), p. 346.

15 D. E. Rosen, 'Introduction', in G. Nelson and D. E. Rosen (eds), *Vicariance Biogeography: A Critique* (New York: Columbia University Press, 1981), pp. 1–5, see p. 3.

16 Charles Darwin, *The Descent of Man and Selection in Relation to Sex* (2nd edn, revised, London: John Murray, 1885), p. 103. The *Descent of Man* first appeared in two volumes in 1871;

Darwin's second book on
the relationship between
humans and animals was his
*Expression of the Emotions in
Man and the Animals* of 1872.

17 Julian Huxley, *Memories*
(London: Allen
and Unwin, 1970),
pp. 85–9.

18 Konrad Lorenz, *On
Aggression*, transl.
Marjorie Latzke (London:
Methuen, 1966),
pp. 237–8.

19 See Theodora J. Kalikow,
'Konrad Lorenz's Ethological
Theory: Explanation and
Ideology', *Journal of the
History of Biology*, **16** (1983):
39–73; for a critique of the
alleged links between Lorenz
and Nazism see Robert J.
Richards, *Darwin and the
Emergence of Evolutionary
Theories of Mind and Behavior*
(Chicago: University of
Chicago Press, 1987), pp.
529–36. On Lorenz and
Haeckel's philosophy of
nature see Bramwell, *Ecology
in the Twentieth Century* (note
2), pp. 56–60 (Richards
objects to this too).

20 See Haraway, *Primate Visions*
(note 1).

21 Solly Zuckerman, *The Social
Life of Monkeys and Apes:
Reissue of the 1932 Edition
together with a Postscript*
(London: Routledge and
Kegan Paul, 1981), p. 316.

22 Jane van Lawick-Goodall
[her married name], *In the
Shadow of Man* (London:

Collins, 1971), pp. 240–1.
For an account of the
National Geographic films
see Haraway, *Primate Visions*
(note 1), pp. 179–85.

23 E. O. Wilson, *Sociobiology:
The New Synthesis*
(Cambridge, Mass.:
Harvard University Press,
1975), p. 547.

24 Michael T. Ghiselin,
*The Economy of Nature
and the Evolution of Sex*
(Berkeley, Calif:. University
of California Press,
1974), p. 24.

CHAPTER 11

1 These examples are drawn
from Arthur F. MacEvoy,
*The Fisherman's Problem:
Ecology and the Law in the
California Fisheries, 1850–1980*
(Cambridge: Cambridge
University Press, 1986).

2 A. G. Tansley, 'The Use
and Abuse of Vegetational
Concepts and Terms',
Ecology, **16** (1935): 284–307,
see p. 303.

3 See Paolo Palladino,
'Ecological Theory and
Pest-Control Practice: A
Study of the Institutional
and Conceptual Dimensions
of a Scientific Debate', *Social
Studies of Science*, **20** (1990):
255–81.

4 Anna Bramwell, *Ecology
in the Twentieth Century*
(New Haven, Conn: Yale
University Press, 1989),
e.g. p. 5.

5 Aldo Leopold, *A Sand County Almanac: with other Essays on Conservation from Round River* (New York: Oxford University Press, 1966), foreword, p. x. In this edition, 'The Land Ethic' is pp. 217–41.

6 See James Lovelock, *Gaia: A New Look at Life on Earth* (new edn, Oxford: Oxford University Press, 1987).

7 A. G. Tansley, *Practical Plant Ecology* (London: Allen and Unwin, 1923), preface.

8 Walter P. Taylor, 'What is Ecology, and What Good is It?', *Ecology*, **17** (1936): 333–46.

9 A. G. Tansley, 'The Use and Abuse of Vegetational Concepts and Terms' (note 2), p. 285.

10 Charles Elton, *Animal Ecology* (Oxford: Oxford University Press, 1927), pp. 63–4.

11 Elton, *Animal Ecology and Evolution* (Oxford: Oxford University Press, 1930), p. 17.

12 G. Evelyn Hutchinson, 'Nati Sunt Mures, et Facta est Confusi', *Quarterly Review of Biology*, **17** (1942): 354–7.

13 Robert MacArthur, 'Coexistence of Species', in J. Behnke (ed), *Challenging Biological Problems* (New York: Oxford University Press, 1972), p. 259.

14 Lovelock, *Gaia* (note 6), preface. On cybernetics see *ibid.*, chap. 4.

15 W. Ford Doolittle, 'Is Nature Really Motherly?', *CoEvolution Quarterly*, no. 29 (1981): 58.

BIBLIOGRAPHICAL ESSAY

The history of science is an active discipline, with a number of journals publishing scholarly articles. These include *Annals of Science*, the *British Journal for the History of Science* (published by the British Society for the History of Science) and *Isis* (the journal of the American-based History of Science Society). Of particular interest to historians of the environmental sciences are *The Journal of the History of Biology*, *The Archives of Natural History* (published by the Society for the History of Natural History), *Earth Sciences History* and *History and Philosophy of the Life Sciences*. The *Critical Bibliography* published each year by *Isis* lists virtually every article and book published in the field. See also the *Isis Cumulative Bibliography, 1913–1965* (London: Mansell, 1971–6, 3 vols) and the *Isis Cumulative Bibliography, 1966–1975* (London: Mansell, 1980–5, 2 vols).

David Knight's *Sources for the History of Science, 1660–1914* (London: The Sources of History, 1975) describes the sources of information used by historians of science. A more recent survey of topics and issues is R. C. Olby *et al.* (eds), *Companion to the History of Modern Science* (London: Routledge, 1990). Biographical information on many scientists may be found in C. C. Gillispie (ed.), *Dictionary of Scientific Biography* (New York: Charles Scribner's Sons, 1979–80, 16 vols). E. Marcorini (ed.), *The History of Science: A Narrative Chronology* (New York: Facts on File, 1988, 2 vols) offers many factual details and has sections on areas of the environmental sciences that are not covered in standard histories of science.

The following account of the specialist literature begins with general surveys of the various environmental sciences. More detailed works will then be listed in sections linked to each chapter in the text; where necessary, references will be repeated in later sections in shortened form (author and short title only). References to primary sources will concentrate on modern reprints of classic texts, which should be widely available in good libraries.

Histories of geography include J. N. L. Baker, *A History of Geographical Discovery and Exploration* (London: Harrap, 1931) and Preston E. James, *All Possible Worlds: A History of Geographical Ideas* (Indianapolis: Bobbs-Merrill, 1972). See also G. R. Crone (ed.), *The Explorers: An Anthology of Discovery* (London: Harrap, 1931). On map-making, see Leo Bagrow, *History of Cartography*, rev. R. A. Skelton (London: Watts, 1964) and Norman J. W. Thrower, *Maps and Man: An Examination of Cartography in Relation to Culture and Civilization* (Englewood Cliffs, NJ: Prentice-Hall, 1972). Bibliographies include Garry S. Dunbar, *The History of Modern Geography: An Annotated Bibliography* (New York: Garland, 1985) and Stephen G. Brush and Helmut E. Landsberg, *The History of Geophysics and Meteorology: An Annotated Bibliography* (New York: Garland, 1984).

General histories of oceanography include Clarence P. Idyll (ed.), *Exploring the Ocean World: A History of Oceanography* (New York: Crowell, 1969) and Margaret B. Deacon, *Scientists and the Sea, 1650–1900: A History of Marine Science* (New York: Academic Press, 1971). Deacon has also edited a useful sourcebook, *Oceanography: Concepts and History* (Stroudsberg, Pa.: Dowden, Hutchinson and Ross, 1978).

On geology, there are several classic surveys written many decades ago that are comprehensive but need to be treated with caution because they express views that are no longer accepted by modern historians: see F. D. Adams, *The Birth and Development of the Geological Sciences* (reprinted New York: Dover, 1954); Archibald Geikie, *The Founders of Geology* (reprinted New York: Dover, 1962); and Karl von Zittel, *History of Geology and Palaeontology* (reprinted Weinheim: J. Cramer, 1962). For surveys of the modern literature, see William Sarjeant, *Geologists and the History of Geology: An International Bibliography from the Origins to 1978* (London: Macmillan, 1980) and Roy Porter, *The Earth Sciences: An Annotated Bibliography* (New York: Garland, 1983). A good modern survey is Anthony Hallam, *Great Geological Controversies* (2nd ed, Oxford: Oxford University Press, 1989). Modern histories of paleontology include M. J. S. Rudwick, *The Meaning of Fossils: Episodes in the History of Palaeontology* (2nd edn, New York: Science History Publications, 1976) and Eric Buffetaut, *A Short History of Vertebrate Palaeontology* (London: Croom Helm, 1987).

A classic survey of the history of the biological sciences is Erik Nordenskiöld, *The History of Biology* (reprinted New York: Tudor, 1946), although this was written early in the century and adopts very strange views on post-Darwinian issues; see also Charles Singer, *A History of Biology to About the Year 1900* (London: Abelard–Schuman, 1959). Another survey written by an eminent modern evolutionist is Ernst Mayr, *The Growth of Biological Thought: Diversity, Evolution, and Inheritance* (Cambridge, Mass.: Harvard University Press, 1982). On botany, see A. G. Morton, *History of Botanical Science* (London: Academic Press, 1981). The pictorial representation of Nature is covered by Wilfrid Blunt, *The Art of Botanical Illustration* (London: Collins, 1950) and David Knight, *Zoological Illustration: An Essay Towards a History of Printed Zoological Pictures* (Folkestone: Dawson, 1977). See also Judith A. Overmier, *The History of Biology: A Selected, Annotated Bibliography* (New York: Garland, 1989).

CHAPTER 1

On the general question of the relationship between culture and perception of the environment, see the works of Yi-Fu Tuan, especially his *Space and Place: The Perception of Experience* (London: Edward Arnold, 1977) and *Topophilia: A Study of Environmental Perception, Attitudes, and Values* (Englewood Cliffs, NJ: Prentice-Hall, 1974). Michel Foucault's views on the relationship between language and the world can be found in his *The Order of Things: The Archaeology of the Human Sciences* (New York: Pantheon, 1970). The relationship between mythology and perception of the animal and plant kingdoms is explored in Claude Lévi-Strauss, *The Savage Mind* (London: Weidenfeld and Nicolson, 1976) and *The Raw and the Cooked* (London: Jonathan Cape, 1970) and in Mary Douglas, *Purity and Danger: An Analysis of Concepts of Pollution and Taboo* (London: Routledge and Kegan Paul, 1966). On the relationship between traditional and scientific concepts of Nature, see Scott Atran, *Cognitive*

Foundations of Natural History: Towards an Anthropology of Science (Cambridge: Cambridge University Press, 1990).

The claim that Christianity's view of Nature has encouraged modern attitudes of exploitation is expressed by Lynn White Jr, 'The Historical Roots of Our Ecological Crisis', *Science*, **155** (1967): 1203–7. This is reprinted along with a number of related articles in David and Eileen Spring (eds), *Ecology and Religion in History* (New York: Harper, 1974). For a critique suggesting that the Greek philosophy of Stoicism played a role in creating Christianity's human-centred vision of Nature, see part 1 of John Passmore's *Man's Responsibility for Nature: Ecological Problems and Western Traditions* (London: Duckworth, 1974).

Other cultures' perceptions of time are explored in Mircea Eliade, *The Myth of the Eternal Return* (London: Routledge and Kegan Paul, 1951). On western ideas, see G. J. Whitrow, *Time in History: the Evolution of Our General Awareness of Time and Temporal Perspective* (Oxford: Oxford University Press, 1988). On the development of the 'historical' sciences from antiquity to Darwinism, see Stephen Toulmin and June Goodfield, *The Discovery of Time* (New York: Harper and Row, 1965). Paolo Rossi's *The Dark Abyss of Time: The History of the Earth and the History of Nations from Hooke to Vico* (Chicago: University of Chicago Press, 1984) argues that the expansion of the geological timescale was only made possible through a breakdown of faith in the biblical story of human origins.

For a conventional account of the hypothetico-deductive method, see Carl Hempel, *Philosophy of Natural Science* (Englewood Cliffs, NJ: Prentice-Hall, 1966). Sir Karl Popper's philosophy of science is developed in his *The Logic of Scientific Discovery* (London: Hutchinson, 1959). On the need for a new philosophy of science to include evolutionary biology, see Ernst Mayr, *Toward a New Philosophy of Biology* (Cambridge, Mass.: Harvard University Press, 1988), and for a brief survey, Michael Ruse, *Philosophy of Biology Today* (Albany: State University of New York Press, 1988). Thomas S. Kuhn's views of the development of science can be found in his *The Structure of Scientific Revolutions* (Chicago: University of Chicago Press, 1962). For the debate between the followers of Popper and Kuhn, see Imre Lakatos and Alan Musgrave, *Criticism and the Growth of Knowledge* (Cambridge: Cambridge University Press, 1970).

On the sociology of science, see Michael Mulkay, *Science and the Sociology of Knowledge* (London: Allen and Unwin, 1979); Barry Barnes and Stephen Shapin (eds), *Natural Order: Historical Studies of Scientific Culture* (Beverly Hills and London: Sage, 1979); and Stephen Yearley, *Science, Technology and Social Change* (London: Unwin Hyman, 1988). For a pioneering effort to uncover the ideological dimensions of the Darwinian revolution, see Robert M. Young's essays, collected in his *Darwin's Metaphor: Nature's Place in Victorian Culture* (Cambridge: Cambridge University Press, 1985). A good example of the latest technique of minute sociological analysis is Adrian Desmond, *The Politics of Evolution: Morphology, Medicine and Reform in Radical London* (Chicago: University of Chicago Press, 1989).

CHAPTER 2

The geographical knowledge of the ancient civilizations is described by E. H. Bunbury, *A History of Ancient Geography* (London: John Murray, 1879, 2 vols) and J. Oliver Thomson, *History of Ancient Geography* (Cambridge: Cambridge University Press, 1948). See also M. Cary and E. H. Warmington, *The Ancient Explorers* (London: Methuen, 1929). A detailed and broad-ranging survey of attitudes towards the environment from ancient times onwards is Clarence J. Glacken, *Traces on the Rhodian Shore: Nature and Culture in Western Thought from Ancient Times to the End of the Eighteenth Century* (Berkeley: University of California Press, 1967).

For surveys of ancient science, see George Sarton, *A History of Science: Ancient Science Through the Golden Age of Greece* (Cambridge, Mass.: Harvard University Press, 1952) and *A History of Science: Hellenistic Science and the Culture of the Last Three Centuries B.C.* (Cambridge, Mass.: Harvard University Press, 1959). Excellent modern surveys are G. E. R. Lloyd, *Early Greek Science: Thales to Aristotle* (London: Chatto and Windus, 1970) and *Greek Science after Aristotle* (New York: Norton, 1973). See also Charles Singer, *Greek Biology and Greek Medicine* (Oxford:

Clarendon, 1922). On the influence of Greek culture on early science, see G. E. R. Lloyd's *Science, Folklore and Ideology: Studies in the Life Sciences in Ancient Greece* (Cambridge: Cambridge University Press, 1983). Arthur O. Lovejoy, *The Great Chain of Being: A Study of the History of an Idea* (reprinted New York: Harper, 1960) outlines the history of this particular ancient view of Nature.

On the pre-Socratics, see G. S. Kirk, J. E. Raven and M. Schofield, *The Presocratic Philosophers: A Critical History with a Selection of Texts* (Cambridge: Cambridge University Press, 1983). Several relevant texts by classical writers are translated in the Penguin Classics series: *Early Greek Philosophy* (introd. Jonathan Barnes, 1987); Cicero, *The Nature of the Gods* (introd. J. M. Ross, 1972); Herodotus, *The Histories* (introd. A. R. Burn, 1972); *Hippocratic Writings* (introd. G. E. R. Lloyd, 1978); Lucretius, *On the Nature of the Universe* (introd. R. E. Latham, 1951); and Plato, *Timaeus* and *Critias* (introd. Desmond Lee, 1971).

On atomism, see Cyril Bailey, *The Greek Atomists and Epicurus* (Oxford: Clarendon, 1928); and on Empedocles, see D. O'Brien, *Empedocles' Cosmic Cycle: A Reconstruction from the Fragments and Secondary Sources* (Cambridge: Cambridge University Press, 1969). Plato's view of the universe is discussed in A. E. Taylor, *A Commentary on Plato's Timaeus* (Oxford: Clarendon, 1928) and F. M. Cornford, *Plato's Cosmology* (London: Routledge and Kegan Paul, 1937). A traditional account of Aristotle's philosophy is W. D. Ross, *Aristotle* (rev. edn, London: Methuen, 1949). Ross also edited the standard edition of *The Works of Aristotle* (Oxford: Clarendon, 1908–52, 12 vols); a compact version of this translation is edited by Jonathan Barnes, *The Complete Works of Aristotle: The Revised Oxford Translation* (Princeton, NJ: Princeton University Press, 1984, 2 vols). On the conflict between atomism and the Aristotelian viewpoint, see David Furley, *The Greek Cosmologists*, vol. 1 (Cambridge: Cambridge University Press, 1987) and *Cosmic Problems: Essays on Greek and Roman Philosophy of Nature* (Cambridge: Cambridge University Press, 1989).

For a discussion of the difficulties attending evaluation of Aristotle's biological observations, see Liliane Bodson, 'Aristotle's Statement on the Reproduction of Sharks', *Journal* of the History of Biology, **16** (1983): 391–407. For surveys of recent scholarship on Aristotle, see Alan Gotthelf (ed.), *Aristotle and Living Things: Philosophical and Historical Studies Presented to David M. Balme on His Seventieth Birthday* (Pittsburgh: Mathesis / Bristol: Bristol Classical Press, 1985), and Alan Gotthelf and James G. Lennox (eds), *Philosophical Issues in Aristotle's Biology* (Cambridge: Cambridge University Press, 1987). See also James G. Lennox, 'Aristotle on Genera, Species and "the More and the Less"', *Journal of the History of Biology*, **13** (1980): 321–46, and Pierre Pelegrini, *Aristotle's Classification of Animals* (Berkeley: University of California Press, 1986). On botany, see *The Greek Herbal of Dioscorides, Illustrated by a Byzantine*, A.D. *512*, ed. Robert T. Gunther (New York: Hafner, 1959) and John Scarborough, 'Theophrastus on Herbals and Herbal Remedies', *Journal of the History of Biology*, **11** (1978): 353–85.

Medieval knowledge of the earth is outlined in G. H. T. Kimble, *Geography in the Middle Ages* (London: Methuen, 1938). The claim that medieval explorers knew far more about the world outside Europe than is commonly supposed is supported by J. R. S. Phillips, *The Medieval Expansion of Europe* (Oxford: Oxford University Press, 1988). On medieval science, see A. C. Crombie, *Augustine to Galileo* (2nd edn, London: Heinemann, 1959, 2 vols). See also Charles E. Raven, *English Naturalists from Neckham to Ray* (Cambridge: Cambridge University Press, 1947).

On Albertus Magnus, see James A. Weisheipl (ed.), *Albertus Magnus and the Sciences: Commemorative Essays, 1980* (Toronto: Pontifical Institute of Medieval Studies, 1980). Albert's *Book of Minerals* has been translated by Dorothy Wyckoff (Oxford: Clarendon, 1967). A modern edition of an important part of Albertus Magnus' biological work is his *Man and Beasts: De Animalibus (Books 22–26)*, transl. J. J. Scanlon (Binghampton, NY: Center for Medieval and Early Renaissance Studies, 1987). See also *The Art of Falconry: Being the De Arte Venandi cum Avibus of Frederick II Hohenstaufen*, transl. Casey A. Wood and F. Marjorie Fyfe (Stanford, Calif: Stanford University Press, 1943) and *The Herbal of Rufinus*, ed. Lynn Thorndike (Chicago: Medieval Academy of America and the University of Chicago, 1946). A popular translation of a bestiary is T. H. White's *The Bestiary:*

A Book of Beasts (New York: Putnam, 1954). For reassessments
of bestiaries, see Wilma George, 'The Bestiary: A Handbook of
the Local Fauna', *Archives of Natural History*, **10** (1981): 187–203
and W. B. Yapp, 'Medieval Knowledge of Birds as shown in
Bestiaries', *ibid.*, **14** (1987): 175–210.

On herbals, see J. Stannard, 'The Herbal as a Medical
Document', *Bulletin of the History of Medicine*, **43** (1969): 212–26
and 'Medieval Herbals and their Development', *Clio Medica*, **9**
(1974): 23–33; see also John M. Riddle, 'Pseudo-Dioscorides
Ex Herbis Feminis and Early Medieval Medical Botany', *Journal
of the History of Biology*, **14** (1981): 43–81. Agnes Arber,
*Herbals, Their Origin and Evolution: A Chapter in the History
of Botany, 1470–1640* (Cambridge: Cambridge University Press,
1912; new edn, 1987) deals only with printed herbals. Other
general works include Wilfrid Blunt and Sandra Raphael, *The
Illustrated Herbal* (London: Hutchinson, 1979) and Frank J.
Anderson, *An Illustrated History of Herbals* (New York: Columbia
University Press, 1977).

CHAPTER 3

On the expansion of geographical knowledge in the Renaissance, see E. G. R. Taylor, *Tudor Geography, 1485–1583* (London:
Methuen, 1930) and *Late Tudor and Early Stuart Geography,
1583–1650* (London: Methuen, 1934); more generally on the
history of navigation, see the same author's *The Haven-Finding
Art: A History of Navigation from Odysseus to Captain Cook* (London:
Hollis and Carter, 1956). On the relationship between geography and other cultural developments, see David N. Livingstone,
'Science, Magic and Religion: A Contextual Reassessment of
Geography in the Sixteenth and Seventeenth Centuries', *History
of Science*, **26** (1988): 269–94. John Dee's life is described
by Peter J. French, *John Dee: the World of an Elizabethan
Magus* (London: Routledge and Kegan Paul, 1972). On the
consequences of the spread of European plants, animals and
people around the globe, see Alfred W. Crosby, *Ecological Imperialism: the Biological Expansion of Europe, 900–1900* (Cambridge:
Cambridge University Press, 1986).

On Renaissance science generally, see Crombie, *Augustine to*

Galileo, vol. 2; Marie Boas, *The Scientific Renaissance, 1450–1630* (London: Collins, 1962); and Alan Debus, *Man and Nature in the Renaissance* (Cambridge: Cambridge University Press, 1978). More generally on the Renaissance and seventeenth century, see A. Rupert Hall, *The Scientific Revolution, 1500–1800* (London: Longmans, 1954) and Hugh Kearney, *Science and Change, 1500–1700* (London: Weidenfeld and Nicolson, 1971). Useful articles may be found in John W. Shirley and F. David Hoeniger (eds), *Science and Arts in the Renaissance* (Washington, DC: Folger Shakespeare Library, 1985) and in Allan Ellenius (ed.), *The Natural Sciences and the Arts: Aspects of Interaction from the Renaissance to the Twentieth Century* (Stockholm: Almquist and Wiksell, 1985). Keith Thomas, *Man and the Natural World: Changing Attitudes in England, 1500–1800* (London: Allen Lane, 1983) surveys both scientific and general attitudes towards Nature in this period. R. T. Gunther's *Early Science in Oxford*, vol. 3 (Oxford: for the subscribers, 1925) deals mainly with the sixteenth and seventeenth centuries. Raven's *English Naturalists from Neckham to Ray* also gives valuable details for this period.

Edward Topsell's edition of Gesner's *Historie of Foure-Footed Beastes* referred to in the text is reprinted in facsimile (Amsterdam and New York: Da Capo, 1973). For a sixteenth-century botanical text, see the reprint of William Turner, *Libellus de Re Herbaria (1538) / The Names of Herbes (1548)* (London: Ray Society, 1965). Agnes Arber's *Herbals* is especially useful on this period; see also R. T. Gunther, *Early British Botanists and their Gardens* (Oxford, 1922; reprinted New York: Kraus, 1971) and Karen Meier Reeds, 'Renaissance Humanism and Botany', *Annals of Science*, **33** (1976): 519–42. The symbolic value of animals and plants is discussed by William B. Ashworth Jr, 'Natural History and the Emblematic World View', in David C. Lindberg and Robert S. Westmann (eds), *Reappraisals of the Scientific Revolution* (Cambridge: Cambridge University Press, 1990), pp. 303–32. See also Ashworth's forthcoming book *Emblematic Imagery of the Scientific Revolution*.

On the effect of new discoveries on botany, see Donald F. Lack, *Asia in the Making of Europe*, vol. 2, book 3 (Chicago: University of Chicago Press, 1977), chap. 9; see also chap. 10 on cartography and geography. There is an English translation of Oviedo y Valdes' *Natural History of the West Indies* by S. A. Stroudemire (Chapel Hill, NC: University of North Carolina Press, 1959); see Antonello Gerbi, *Nature in the New World: From Christopher Columbus to Gonzalo Fernandez de Oviedo* (Pittsburgh: Pittsburgh University Press, 1985). Mea Allen, *The Tradescants: Their Plants, Gardens and Museum, 1570–1662* (London: Michael Joseph, 1964) describes the work of John Tradescant, whose collection became the basis for the Ashmolean Museum at Oxford; see also Prudence Leith-Ross, *The John Tradescants: Gardeners to the Rose and Lily* (London: Owen, 1984). F. D. Adams' *The Birth and Development of the Geological Sciences* contains information on Renaissance ideas about minerals, and on fossils see M. J. S. Rudwick's *The Meaning of Fossils*, chap. 1.

A. Rupert Hall's *From Galileo to Newton, 1630–1720* (London: Collins, 1963) offers a general survey of seventeenth-century science. There is an enormous modern literature exploring the cultural and social context within which the Scientific Revolution occurred; see for instance Lindberg and Westman (eds), *Reappraisals of the Scientific Revolution*. On the role played by the Hermetic tradition, see Charles Webster, *From Paracelsus to Newton: Magic and the Making of Modern Science* (Cambridge: Cambridge University Press, 1982); R. S. Westman and J. E. McGuire, *Hermeticism and the Scientific Revolution* (Los Angeles: Clark Memorial Library, 1977); and Frances A. Yates, *The Rosicrucian Enlightenment* (London: Routledge and Kegan Paul, 1972). William Coleman, 'Providence, Capitalism, and Environmental Degradation: English Apologetics in an Era of Economic Revolution', *Journal of the History of Ideas*, **37** (1976): 27–44 argues that Christian attitudes were adapted to the new economic climate of the seventeenth century. More generally on religion and the new science, see Richard S. Westfall, *Science and Religion in Seventeenth-Century England* (New Haven, Conn.: Yale University Press, 1958).

The suggestion that Puritanism was linked to the origins of modern science came in Robert K. Merton's article 'Science, Technology, and Society in Seventeenth-Century England', *Osiris*, **4** (1938): 360–632. This provoked extensive debates. For more recent studies, see for instance Michael Hunter, *Science and*

Society in Restoration England (Cambridge: Cambridge University Press, 1981); Charles Webster, *The Great Instauration: Science, Medicine and Reform, 1626–1660* (London: Duckworth, 1975); and Margaret C. Jacob, *The Newtonians and the English Revolution* (Ithaca, NY: Cornell University Press, 1976). Jacob's *The Radical Enlightenment: Pantheists, Freemasons, and Republicans* (London: Allen and Unwin, 1981) carries the story through into the eighteenth century. On support for science in France, see Alice Stroup, *A Company of Scientists: Botany, Patronage, and Community at the Early Parisian Academy of Sciences* (Berkeley: University of California Press, 1990). For a feminist perspective, see Carolyn Merchant, *The Death of Nature: Women, Ecology, and the Scientific Revolution* (London: Wildwood House, 1980).

The claim that natural history was an integral part of the scientific revolution is developed by Joseph M. Levine, 'Natural History and the History of the Scientific Revolution', *Clio*, **13** (1983): 57–73. Margareta Bowen's *Empiricism and Geographical Thought: From Francis Bacon to Alexander von Humboldt* (Cambridge: Cambridge University Press, 1981) argues that the methodology of the new science worked against the production of a synthesis of geographical knowledge.

CHAPTER 4

On the problems encountered by historians trying to evaluate eighteenth-century science, see G. S. Rousseau and Roy Porter (eds), *The Ferment of Knowledge: Studies in the Historiography of Eighteenth-Century Science* (Cambridge: Cambridge University Press, 1980). A useful collection of articles is L. Jordanova and R. Porter (eds), *Images of the Earth: Essays in the History of the Environmental Sciences* (Chalfont St Giles, Bucks: British Society for the History of Science, 1979). Margareta Bowen's *Empiricism and Geographical Thought* is again important on the decline of eighteenth-century geography.

There is a considerable literature on the explorations of Cook and others; see for instance J. C. Beaglehole (ed.), *The Journals of Captain Cook on his Voyages of Discovery* (Cambridge: Cambridge University Press, 1955–74, 4 vols). The last volume of this set is a biography of Cook by Beaglehole, also published separately as *The Life of Captain James Cook* (London: Black, 1974). More generally, see Robin Fisher and Hugh Johnston (eds), *Captain James Cook and his Times* (Seattle: University of Washington Press, 1979); David MacKay, *In the Wake of Cook: Exploration, Science and Empire, 1780–1801* (London: Croom Helm, 1985); Roy MacLeod and Phillip F. Rehbock (eds), *Nature in its Greatest Extent: Western Science in the Pacific* (Honolulu: University of Hawaii Press, 1988); and Derek Howse (ed.), *Background to Discovery: Pacific Exploration from Dampier to Cook* (Berkeley: University of California Press, 1990). More generally, see P. J. Marshall and Glyndwr Williams, *The Great Map of Mankind: British Perceptions of the World in the Age of Enlightenment* (London: Dent, 1982), although this concentrates on how the human inhabitants of other regions were perceived. Oceanography in this period is covered by Deacon, *Scientists and the Sea*.

On early botanical gardens as attempts to recreate a geometrical vision of paradise, see John Prest, *The Garden of Eden: The Botanic Garden and the Re-Creation of Paradise* (New Haven, Conn.: Yale University Press, 1981). More generally on gardens as indications of changing attitudes towards Nature, see Christopher Thacker, *The History of Gardens* (London: Croom Helm, 1979). On changing attitudes towards mountains, see Marjorie Hope Nicolson, *Mountain Gloom and Mountain Glory: The Development of the Aesthetics of the Infinite* (Ithaca, NY: Cornell University Press, 1959).

Many of the classic theories of the earth have been reprinted. There is an edition of Thomas Burnet's *The Sacred Theory of the Earth* introduced by Basil Willey (London: Centaur, 1965). Facsimiles of William Whiston's *A New Theory of the Earth* (1696), John Woodward's *An Essay Toward a Natural History of the Earth* (1695), and John Ray's *Three Physico-Theological Discourses* (1713) have been issued (all New York: Arno, 1977). Ray's *Miscellaneous Discourses concerning the Changes of the World* of 1692 is reprinted (Hildesheim: Georg Olms, 1968). Hooke's work on earthquakes appears in *The Posthumous Works of Robert Hooke* (1705; reprinted New York: Johnson, 1969) and separately in his *Lectures and Discourses of Earthquakes and Subterranean Eruptions* (New York: Arno, 1977). There is a translation of *The Prodromus of Nicholas Steno's Dissertation* concerning a Solid Body enclosed by process of Nature within a Solid (reprinted New York: Hafner, 1968). De Maillet's *Telliamed* has been translated by Albert V. Carozzi (Urbana: University of Illinois Press, 1968). The various editions of William Smellie's translation of Buffon's *Natural History, General and Particular* contain his original account of earth history, but not the later *Epochs of Nature*. The original French *Les époques de la nature* is reprinted ed. Jacques Roger (Paris: Muséum d'Histoire Naturelle, 1962).

There is a translation of Werner's *Short Classification and Description of the Various Rocks* by Alexander Ospovat (New York: Hafner, 1971). The most detailed description of the Wernerian theory in English is Robert Jameson's *Wernerian Theory of the Neptunian Origin of Rocks* (reprinted New York: Hafner, 1976). Hutton's *Theory of the Earth* has been reprinted (Weinheim: H. R. Engelmann (J. Cramer)/ Codicote, Herts: Wheldon and Wesley, 1960) as has Playfair's *Illustrations of the Huttonian Theory of the Earth* (New York: Dover, 1964).

Older histories of geology that must be treated with some caution include F. D. Adams' *Birth and Development of the Geological Sciences* and A. Geikie's *Founders of Geology*. C. C. Gillispie's *Genesis and Geology* (reprinted New York: Harper, 1959) discusses the Neptunist–Vulcanist debate in the context of religious issues. A. Hallam's *Great Geological Controversies* also has a chapter on this debate. Other studies include John C. Greene, *The Death of Adam: Evolutionism and its Impact on Western Thought* (Ames: Iowa State University Press, 1959); Francis C. Haber, *The Age of the World: Moses to Darwin* (Baltimore: Johns Hopkins University Press, 1959); and G. R. Davies, *The Earth in Decay: a History of British Geomorphology, 1578–1878* (New York: Science History Publications, 1969).

The link between theories of the earth and changing ideas on human origins is explored by Paolo Rossi, *The Dark Abyss of Time: The History of the Earth and the History of Nations from Hooke to Vico* (Chicago: University of Chicago Press, 1984). John Woodward's work in archaeology and fossil collecting is described by J. E. Levine, *Dr Woodward's Shield: History, Science and Satire in Augustan England* (Cambridge, Mass.: Harvard University Press, 1977). Ray's geological work is described in Charles E.

Raven, *John Ray, Naturalist: His Life and Works* (Cambridge: Cambridge University Press, 1942). On the emergence of geology as a discipline, see Roy Porter, *The Making of Geology: Earth Sciences in Britain, 1660–1815* (Cambridge: Cambridge University Press, 1977). The mineralogical tradition is described by Rachel Laudan, *From Mineralogy to Geology: The Foundations of a Science, 1650–1830* (Chicago: University of Chicago Press, 1987). On paleontology, see M. J. S. Rudwick's *The Meaning of Fossils.* Cecil J. Schneer (ed.), *Towards a History of Geology* (Cambridge, Mass.: MIT Press, 1969) contains a number of useful articles on this period. Stephen Jay Gould's *Time's Arrow, Time's Cycle* (Cambridge, Mass.: Harvard University Press, 1987) emphasizes the cyclic component in the world views of Burnet and Hutton.

CHAPTER 5

There are modern editions of a number of important eighteenth-century texts, including the 1716 edition of Derham's *Physico-Theology* (New York: Arno, 1978). The 1717 edition of John Ray's *Wisdom of God* has been reprinted (New York: Arno, 1977), while the third edition (1724) of his *Synopsis Methodica Stirpum Britannicum* is reprinted along with Linnaeus' *Flora Anglica* (London: Ray Society, 1973). Other reprints of Linnaeus' works include the first edition of the *Systema Naturae* (Nieuwkoop: De Graaf, 1964) and the tenth edition (vol. 1, London: Ray Society, 1956; vol. 2, Weinheim: J. Cramer, 1960); the *Philosophia Botanica* (Codicote, Herts: Wheldon and Wesley / New York: Stechert Hafner Service Agency, 1966); the *Species Plantarum* (London: Ray Society, 1955–9, 2 vols); and the *Genera Plantarum* (Weinheim: J. Cramer, 1960). There are innumerable editions of Gilbert White's *Natural History of Selborne*; the first edition of 1789 has been reprinted (Menston: Scolar, 1970). There was a contemporary English translation of Buffon's *Natural History* by William Smellie (several editions); for a modern selection see Phillip Sloan and John Lyon (eds), *From Natural History to the History of Nature: Readings from Buffon and his Critics* (Notre Dame, Ind.: University of Notre Dame Press, 1981). Erasmus Darwin's *Zoonomia: or the Laws of*

Organic Life has been reprinted (New York: AMS Press, 1974) and there is a selection of his writings in Desmond King-Hele (ed.), *The Essential Erasmus Darwin* (London: McGibbon and Kee, 1968). King-Hele has also edited *The Letters of Erasmus Darwin* (Cambridge; Cambridge University Press, 1981). There is a translation of Lamarck's *Zoological Philosophy* by Hugh Elliot (reprinted New York: Hafner, 1963).

On the problem of interpreting the eighteenth-century life sciences, see Jacques Roger, 'The Living World', in Rousseau and Porter (eds), *The Ferment of Knowledge*, pp. 255–83. For those who read French, Roger's *Les Sciences de la vie dans la pensée francaise du XVIIIe siècle* (Paris: Armand Colin, 1963) provides a detailed survey. Phillip C. Ritterbush's *Overtures to Biology: The Speculations of the Eighteenth-Century Naturalists* (New Haven, Conn.: Yale University Press, 1964) provides useful information. Foucault's interpretation of this period is defended in Francois Jacob, *The Logic of Life* (London: Penguin, 1988) and in Lorin Anderson's *Charles Bonnet and the Order of the Known* (Dordrecht: Reidel, 1982).

David E. Allen's *The Naturalist in Britain: A Social History* (London: Allen Lane, 1976) is an invaluable source of information on the social framework within which natural history developed. On Sir Joseph Banks' role in promoting natural history, see H. C. Cameron, *Sir Joseph Banks: The Autocrat of the Philosophers* (London: Batchworth, 1952) and, more recently, Harold B. Carter, *Sir Joseph Banks, 1743–1820* (London: British Museum (Natural History), 1988) and Patrick O'Brian, *Joseph Banks: A Life* (London: Collins Harvill, 1988). On Banks' first explorations, see A. M. Lysaght, *Joseph Banks in Newfoundland and Labrador, 1766: His Diary, Manuscripts and Collections* (London: Faber, 1971). On Kew Gardens, see W. B. Turrill, *The Royal Botanic Gardens, Kew* (London: Jenkins, 1959). See also A. T. Gage and W. T. Stearn, *A Bicentenary History of the Linnean Society of London* (London: Academic Press, 1988). The practical implications of exploration are surveyed in Mackay's *In the Wake of Cook* and in MacLeod and Rehbock, *Nature in its Greatest Extent*. On the Europeans' discovery of Australia, see G. Pizzey, *A Separate Creation: Discovery of Wild Australia by Explorers and Naturalists* (London: Croom Helm, 1984) and

M. Steven, *First Impressions: The British Discovery of Australia* (London: British Museum (Natural History), 1988).

On classification, see David Knight, *Ordering the World: A History of Classifying Man* (London: Burnett, 1981). On John Ray's contributions, see C. E. Raven's *John Ray, Naturalist* and Phillip R. Sloan, 'John Locke, John Ray, and the Problem of the Natural System', *Journal of the History of Biology*, **5** (1972): 1–53. See also Sloan's 'The Buffon-Linnaeus Controversy', *Isis*, **67** (1976): 356–75. On Linnaeus himself, see Knut Hagberg, *Carl Linnaeus* (New York: Dutton, 1953); James L. Larson, *Reason and Experience: The Representation of Natural Order in the Work of Carl von Linné* (Berkeley: University of California Press, 1971); F. Stafleu, *Linnaeus and the Linnaeans: The Spreading of their Ideas in Systematic Botany* (Utrecht: Oosthoek, 1971); and Tore Frängsmyr (ed.), *Linnaeus: The Man and his Work* (Berkeley: University of California Press, 1984).

For a general history of ecology, see Donald Worster, *Nature's Economy: A History of Ecological Ideas* (Cambridge: Cambridge University Press, 1985). See also Frank N. Egerton, 'Richard Bradley's Understanding of Biological Productivity: A Study of Eighteenth-century Ecological Ideas', *Journal of the History of Biology*, **1** (1968): 1–22 and 'Changing Concepts of the Balance of Nature', *Quarterly Review of Biology*, **48** (1973): 322–50. On migration and biogeography, see Janet Browne, *The Secular Ark: Studies in the History of Biogeography* (New Haven, Conn.: Yale University Press, 1983) and James Larson, 'Not Without a Plan: Geography and Natural History in the Late Eighteenth Century', *Journal of the History of Biology*, **19** (1986): 447–88.

On materialism in the life sciences, see Aram Vartanian, 'Trembley's Polyp, La Mettrie, and Eighteenth-Century French Materialism', *Journal of the History of Ideas*, **11** (1950): 259–86 and *Diderot and Descartes: A Study of Scientific Naturalism in the Enlightenment* (Princeton, NJ: Princeton University Press, 1953). For general histories of evolutionism, see John C. Greene, *The Death of Adam* and Peter J. Bowler, *Evolution: The History of an Idea* (Berkeley: University of California Press, rev. edn, 1989). There are several useful articles on eighteenth-century evolutionism in Bentley Glass *et al.* (eds), *Forerunners of Darwin, 1745–1859* (Baltimore: Johns Hopkins

University Press, 1959). On Buffon, see Paul Farber, 'Buffon and the Problem of Species', *Journal of the History of Biology*, **5** (1972): 259–84 and Peter J. Bowler, 'Bonnet and Buffon: Theories of Generation and the Problem of Species', *ibid.*, **6** (1973): 259–81. On Erasmus Darwin, see M. McNeil, *Under the Banner of Science: Erasmus Darwin and his Age* (Manchester: Manchester University Press, 1987) and Roy Porter, 'Erasmus Darwin: Doctor of Evolution?', in J. R. Moore (ed.), *History, Humanity and Evolution: Essays for John C. Greene* (Cambridge: Cambridge University Press, 1989), pp. 39–69. On Lamarck, see Richard W. Burkhardt Jr, *The Spirit of System: Lamarck and Evolutionary Biology* (Cambridge, Mass.: Harvard University Press, 1977); L. Jordanova, *Lamarck* (Oxford: Oxford University Press, 1984); and Pietro Corsi, *The Age of Lamarck: Evolutionary Theories in France, 1790–1830* (Berkeley: University of California Press, 1988).

CHAPTER 6

A good introduction to the new framework of thought in the nineteenth century is Maurice Mandelbaum, *History, Man and Reason: A Study in Nineteenth-Century Thought* (Baltimore: Johns Hopkins University Press, 1971). The links between the various levels of historical thinking, including archaeology and geology, are explored in Peter J. Bowler, *The Invention of Progress: The Victorians and the Past* (Oxford: Blackwell, 1989). On the world of Victorian science, see David Knight, *The Age of Science: The Scientific World View in the Nineteenth Century* (Oxford: Blackwell, 1986). Susan F. Cannon, *Science in Culture: The Early Victorian Period* (New York: Science History Publications, 1978) gives much background on British science of the period and explores the notion of 'Humboldtian science'; on the latter topic, see also Malcolm Nicolson, 'Alexander von Humboldt, Humboldtian Science, and the Origins of the Study of Vegetation', *History of Science*, **25** (1987): 167–94.

Translations of Humboldt's *Cosmos* are freely available, and there is a modern reprint of his *Personal Narrative of Travels to the Equinoctal Regions of the New Continent during the Years 1799–1804* (7 vols, 1814–29; reprinted New York: AMS Press,

1966). A good example of early-nineteenth-century geology is W. D. Conybeare and W. Phillips, *Outlines of the Geology of England and Wales* (1822; reprinted New York: Arno, 1977). Buckland's *Reliquiae Diluvianae: Or Observations of the Organic Remains Contained in Caves, Fissures, and Diluvial Gravel, and Other Geological Phenomena Attesting the Action of a Universal Deluge* (1823) has been reprinted (New York: Arno, 1977). Cuvier's *Discours sur les révolutions de la surface du globe* was translated by Robert Jameson under the title *Essay on the Theory of the Earth*; two editions of this have been reprinted, the 1813 edition (Farnborough: Gregg, 1971) and the 1817 edition (New York: Arno, 1977). The second edition of Cuvier and Brongniart's *Description géologiques des environs de Paris* of 1825 is reprinted (Brussels: Culture et Civilization, 1969). Agassiz's *Studies on Glaciers* is translated by Albert V. Carozzi (New York: Hafner, 1967). The 1872 edition of Murchison's *Siluria: The History of the Oldest Known Rocks Containing Organic Remains* has been reprinted (Millwood, NY: Kraus). The first edition of Charles Lyell, *Principles of Geology: Being an Attempt to Explain the Former Changes of the Earth's Surface by Reference to Causes Now in Operation* (3 vols, 1830–3) has been reprinted with an introduction by M. J. S. Rudwick (Chicago: University of Chicago Press, 1990–1).

James and Martin, *All Possible Worlds* contains valuable material on nineteenth-century geography; see also Thomas W. Freeman, *A History of Modern British Geography* (London: Longman, 1980). Many of the works on exploration cited for chapter 4 above deal also with the nineteenth century. See also J. Mirsky, *To the Arctic: the Story of Northern Exploration from the Earliest Times to the Present* (Chicago: University of Chicago Press, 1970). There is a popular history of the Royal Geographical Society by Ian Cameron, *To the Farthest Ends of the Earth: 150 Years of World Exploration by the Royal Geographical Society* (London: Macdonald, 1980). On geography, geology and imperialism, see Robert A. Stafford, *Scientist of the Empire: Sir Roderick Murchison, Scientific Exploration and Victorian Imperialism* (Cambridge: Cambridge University Press, 1989) and James Secord, 'King of Siluria: Roderick Murchison and the Imperial Theme in Nineteenth-Century British Geology',

Victorian Studies, **25** (1982): 413–42. On the British Geological Survey, see Harold E. Wilson, *Down to Earth: One Hundred and Fifty Years of the British Geological Survey* (Edinburgh: Scottish Academic Press, 1985); and on Ireland, Gordon Herries-Davies, *Sheets of Many Colours: The Mapping of Ireland's Rocks, 1750–1890* (Dublin: Royal Dublin Society, 1983). Nicolaas A. Rupke (ed.), *Science, Politics and the Public Good* (London: Macmillan, 1988) contains papers on the British Geological Survey and on Owen's role in the founding of the Natural History Museum.

The American surveys are discussed in A. Hunter Dupree, *Science in the Federal Government: A History of Policies and Activities to 1940* (reprinted New York: Harper, 1967); see also Thomas G. Manning, *Government in Science: The United States Geological Survey, 1867–1894* (Lexington: University of Kentucky Press, 1967) and more recently Manning's *US Coast Survey Vs. Naval Hydrographic Office: A 19th-Century Rivalry in Science and Politics* (Tuscaloosa: University of Alabama Press, 1988). On the problems of the state geological surveys, see Walter B. Hendrickson, 'Nineteenth-Century State Geological Surveys: Early Government Support of Science', *Isis*, **52** (1961): 357–91. On the role of geology and the magnetic survey in the origins of Canadian science, see Suzanne Zeller, *Inventing Canada: Early Victorian Science and the Idea of a Transcontinental Nation* (Toronto: University of Toronto Press, 1987); see also Morris Zaslow, *Reading the Rocks: The Story of the Geological Survey of Canada* (Toronto: Macmillan, 1975). Susan Sheets-Pyenson explores the role of museums in developing colonial science in *Cathedrals of Science: The Development of Colonial Natural History Museums during the Late Nineteenth Century* (Montreal: McGill-Queen's University Press, 1989).

On oceanography, see Deacon, *Scientists and the Sea*, and on this later period, J. R. Dean, *Down to the Sea: A Century of Oceanography* (Glasgow: Brown, Son and Ferguson, 1966) and, more substantially, Susan Schlee, *Edge of an Unfamiliar World: A History of Oceanography* (New York: Dutton, 1973). Mary Sears and Daniel Merriman (eds), *Oceanography: The Past* (New York: Springer, 1980) includes papers on American oceanography. Eric Linklater's *The Voyage of the Challenger* (London: John Murray, 1972) offers a well illustrated account of the first oceanographic expedition. A partial reprint of the survey of the expedition's findings is available: John Murray, *Report of the Scientific Results of the Voyage of H.M.S. Challenger . . . A Summary of the Scientific Results* (1895; reprinted New York: Arno, 1977).

Biographies of Humboldt include L. Kellner, *Alexander von Humboldt* (London: Oxford University Press, 1963) and Douglas Botting, *Humboldt and the Cosmos* (London: Sphere, 1973), the latter well illustrated. See also Bowen, *Empiricism and Geographical Thought*. On the links between Humboldt and Romanticism, see Malcolm Nicolson, 'Alexander von Humboldt and the Geography of Vegetation', in Andrew Cunningham and Nicholas Jardine (eds), *Romanticism and the Sciences* (Cambridge: Cambridge University Press, 1990), pp. 169–85.

A number of books cited in chapter 4 above are again useful on nineteenth-century geology. See Geikie, *The Founders of Geology*, although this was written very much in the shadow of Lyell. Schneer (ed.), *Toward a History of Geology* contains several useful papers. Porter, *The Making of Geology* and Laudan, *From Mineralogy to Geology* are useful on the early decades of the century, the latter especially for the influence of Werner. On the debate with religion, see Gillispie's *Genesis and Geology*. Hallam's *Great Geological Controversies* has chapters on the uniformitarian–catastrophist debate, the ice age and (in the second edition only) on stratigraphy. The complex link with industry is explored in Roy Porter's 'The Industrial Revolution and the Rise of the Science of Geology', in M. Teich and R. M. Young (eds), *Changing Perspectives in the History of Science* (London: Heinemann, 1973), pp. 320–43. On the development of geological maps and cross-sections, see M. J. S. Rudwick, 'The Emergence of a Visual Language for Geological Science, 1760–1840', *History of Science*, **14** (1976): 149–95. See also Nicolaas A. Rupke, *The Great Chain of History: William Buckland and the English School of Geology (1815–1849)* (Oxford: Oxford University Press, 1983).

For a revision of the traditional interpretation of the uniformitarian–catastrophist debate, see R. Hooykaas, *Natural Law and Divine Miracle: The Principle of Uniformity in Geology, Biology and History* (Leiden: Brill, 1959) and the same author's *Catastrophism in Geology: Its Scientific Character in Relation to Actualism and Uniformitarianism* (Amsterdam: North-Holland, 1970). See also two articles by M. J. S. Rudwick, 'The Strategy of Lyell's *Principles of Geology*', *Isis*, **61** (1970): 5–33 and 'Uniformity and Progress: Reflections on the Structure of Geological Theory in the Age of Lyell', in Duane H. D. Roller (ed.), *Perspectives in the History of Science and Technology* (Norman: University of Oklahoma Press, 1971), pp. 209–27. Rudwick has also explored the context of Scrope's theory; see his 'Poulett Scrope on the Volcanoes of Auvergne: Lyellian Time and Political Economy', *British Journal for the History of Science*, **7** (1974): 202–42. Leonard G. Wilson's biography *Charles Lyell: The Years to 1841* (New Haven, Conn.: Yale University Press, 1972) covers the early part of Lyell's career; on the theory of climatic change, see Dov Ospovat, 'Lyell's Theory of Climate', *Journal of the History of Biology*, **10** (1977): 317–39. The challenge from the physicists is explored in Joe D. Burchfield, *Lord Kelvin and the Age of the Earth* (New York: Science History Publications, 1975).

The complex relationship between fieldwork and theoretical debate is explored by M. J. S. Rudwick, *The Great Devonian Controversy: The Shaping of Scientific Knowledge among Gentlemanly Specialists* (Chicago: University of Chicago Press, 1985); James Secord, *Controversy in Victorian Geology: The Cambrian–Silurian Debate* (Princeton, NJ: Princeton University Press, 1986); and D. R. Oldroyd, *The Highlands Controversy: Constructing Geological Knowledge through Fieldwork in Nineteenth-Century Britain* (Chicago: University of Chicago Press, 1990). Mott T. Greene, *Geology in the Nineteenth Century: Changing Views of a Changing World* (Ithaca, NY: Cornell University Press, 1982) is particularly valuable on continental and on American theories; on the latter, see also Ellen T. Drake and William M. Jordan (eds), *Geologists and Ideas: A History of North American Geology* (Boulder, Colo.: Geological Society of America, 1985). On the ice age theory, see the introduction to Carozzi's translation of Agassiz's *Studies on Glaciers* and Davies, *The Earth in Decay*. See also John and Katherine Imbrie, *Ice Ages: Solving the Mystery* (London: Macmillan, 1979), and on the impact of Croll's theory, Christopher Hamlin, 'James Geikie, James Croll and the Eventful Ice Age', *Annals of Science*, **39** (1982): 565–83.

On Darwin's alternative, see M. J. S. Rudwick, 'Darwin and Glen Roy: A "Great Failure" in Scientific Method?', *Studies in the History and Philosophy of Science*, **5** (1974): 97–185.

CHAPTER 7

The Darwin industry dominates the study of nineteenth-century natural history, with the result that it is far easier for the non-specialist to find suitable reading on Darwin than on any of the rival theorists. Reprinted editions of Darwin's *Journal of Researches into the Natural History and Geology of the Various Countries Visited by H.M.S. Beagle* and of *The Origin of Species* have always been widely available. There is a modern edition of *Charles Darwin's Beagle Diary*, ed. R. D. Keynes (Cambridge: Cambridge University Press, 1988); this is uniform with *The Correspondence of Charles Darwin*, ed. Frederick Burkhardt and Sydney Smith (Cambridge: Cambridge University Press, 1984–), which has reached volume 6 (1856–7) at the time of writing. On the discovery of natural selection, see *Charles Darwin's Theoretical Notebooks (1836–1844)*, ed. Paul H. Barrett *et al.* (Cambridge: Cambridge University Press, 1987). The 1842 sketch of the theory and the 1844 essay are reprinted along with the Darwin–Wallace papers of 1858 in Darwin and Wallace, *Evolution by Natural Selection* (Cambridge: Cambridge University Press, 1958). See also *Charles Darwin's Natural Selection; being the Second Part of his Big Species Book Written from 1856 to 1858*, ed. R. C. Stauffer (Cambridge: Cambridge University Press, 1975). Wallace's *The Malay Archipelago* has recently been reprinted (Singapore: Oxford University Press, 1989).

Outside the realm of Darwin studies, see the late edition of Cuvier's *The Animal Kingdom* (1863; reprinted New York: Kraus, 1969) and the *Recherches sur les ossemens fossiles de quadrupèdes* (four vols, 1812; reprinted Brussels: Culture et Civilization, 1969). There is a facsimile of Humboldt's *Essai sur la géographie des plantes* (London: Society for the Bibliography of Natural History, 1959). Owing to his position as the founding father of American zoology, Louis Agassiz's works have been widely reprinted and form a useful way of studying a totally non-evolutionary approach; see his *Essay on Classification,* ed. Edward Lurie (Cambridge, Mass.: Harvard University Press, 1962) and *Methods of Study in Natural History* (1863; reprinted New York: Arno, 1970). Lyell's puzzling over the species question is revealed in *Sir Charles Lyell's Journals on the Species Question,* ed. Leonard J. Wilson (New Haven, Conn.: Yale University Press, 1970). There is a reprint of Chambers' *Vestiges of the Natural History of Creation* (Leicester: Leicester University Press, 1969). Owen's *On the Archetypes and Homologies of the Vertebrate Skeleton* is available (1848; reprinted New York: AMS Press), as is Hugh Miller's *The Old Red Sandstone; or New Walks in an Old Field* (1857 edn; reprinted New York: Arno, 1977). See also Miller's *Footprints of the Creator; or the Asterolepis of Stromness* (1861 edn; reprinted Farnborough: Gregg, 1971). Baden Powell's views on 'creation by law' were expounded in his *Essays on the Spirit of the Inductive Philosophy* of 1855 (reprinted Farnborough: Gregg, 1969). Two valuable collections edited by Keir B. Sterling are *Selected Works in Nineteenth-Century North American Paleontology* (New York: Arno, 1975) and *Contributions to the History of American Natural History* (New York: Arno, 1975).

For a general survey of nineteenth-century biology stressing the importance of experimentalism, see William Coleman, *Biology in the Nineteenth Century: Problems of Form, Function and Transformation* (New York: Wiley, 1971). On the debates in French natural history in the early decades of the century, see Pietro Corsi, *The Age of Lamarck;* Toby Appel, *The Cuvier–Geoffroy Debate: French Biology in the Decades before Darwin* (Oxford: Oxford University Press, 1987); and Dorinda Outram, *Georges Cuvier: Vocation, Science, and Authority in Post-Revolutionary France* (Manchester: Manchester University Press, 1984). On German biology, see Timothy Lenoir, *The Strategy of Life: Teleology and Mechanics in Nineteenth-Century German Biology* (Dordrecht: Reidel, 1982). The social context of the pre-Darwinian debates in England is explored by Adrian Desmond, *The Politics of Evolution: Morphology, Medicine, and Reform in Radical London* (Chicago: University of Chicago Press, 1989). For more conventional accounts of the structure of British science, see Turrill, *The Royal Botanic Gardens, Kew* and Mea Allan, *The Hookers of Kew, 1785–1911* (London: Michael Joseph, 1967). On the role of museums, see Albert E. Gunther, *A Century of Zoology at the British Museum Through the Lives of Two Keepers, 1815–1914* (London: Dawson, 1975); and W. T. Stearn, *The Natural History Museum at South Kensington* (London: Heinemann, 1981). On zoos, see Wilfrid Blunt, *The Ark in the Park: The Zoo in the Nineteenth Century* (London: Hamish Hamilton, 1976). On American science, see Dupree, *Science in the Federal Government* and Jeffrey K. Barnes, *Asa Fitch and the Emergence of American Entomology*, Bulletin No. 461 (Albany: New York State Museum, 1988).

On changing attitudes towards Nature, see Allen, *The Naturalist in Britain*. On American naturalists, see J. Kastner, *A World of Naturalists* (London: John Murray, 1978). A great deal has been written recently on the impact of imperialism: see Harriet Ritvo, *The Animal Estate: The English and Other Creatures in the Victorian Age* (Cambridge, Mass.: Harvard University Press, 1987); John M. Mackenzie, *The Empire of Nature: Hunting, Conservation, and British Imperialism* (Manchester: Manchester University Press, 1988); and Lynn L. Merrill, *The Romance of Victorian Natural History* (Oxford: Oxford University Press, 1989).

On new developments in early-nineteenth-century natural history, see Philip F. Rehbock, *The Philosophical Naturalists: Themes in Early Nineteenth-Century British Biology* (Madison: University of Wisconsin Press, 1983). Cunningham and Jardine (eds), *Romanticism and the Sciences* contains articles on transcendental anatomy by Timothy Lenoir, by Evelleen Richards and by Rehbock. Another valuable collection is Alwyne Wheeler and James H. Price (eds), *From Linnaeus to Darwin: Commentaries on the History of Biology and Geology* (London: Society for the History of Natural History, British Museum (Natural History), 1985). On marine biology, see Rehbock's 'The Early Dredgers: "Naturalizing" in British Seas, 1830–1850', *Journal of the History of Biology*, **12** (1979): 293–368. For a more conventional account of British natural history stressing the role of natural theology, see Gillispie, *Genesis and Geology*. On classification, see Knight, *Ordering the World* and Mary P. Winsor, *Starfish, Jellyfish, and the Order of Life* (New Haven, Conn.: Yale University Press, 1976). On biogeography, see Browne, *The Secular Ark* and Nicholson, 'Alexander von Humboldt and the Geography of Vegetation'. See also Frank N. Egerton,

'Hewett C. Watson, Great Britain's First Phytogeographer', *Huntia*, **3** (1979): 87–102. Egerton also discusses changing views on the factors limiting population in a series of articles including 'Studies of Animal Populations from Lamarck to Darwin', *Journal of the History of Biology*, **1** (1968): 225–59 and 'Humboldt, Darwin, and Population', *ibid.*, **3** (1970): 325–60.

On paleontology, see Rudwick, *The Meaning of Fossils* and Rupke, *The Great Chain of History*; see also Peter J. Bowler, *Fossils and Progress: Paleontology and the Idea of Progressive Evolution in the Nineteenth Century* (New York: Science History Publications, 1976); Adrian Desmond, *Archetypes and Ancestors: Paleontology in Victorian London, 1850–1875* (London: Blond and Briggs / Chicago: University of Chicago Press, 1982); and Buffetaut, *A Short History of Vertebrate Paleontology*. On the debate between linear and cyclic models of progress, see Bowler, *The Invention of Progress*.

There is an immense literature on the rise of evolutionism; for a survey, see Bowler, *Evolution: The History of an Idea* and Mayr, *The Growth of Biological Thought*. For a survey that concentrates on the mid nineteenth century, see Michael Ruse, *The Darwinian Revolution: Science Red in Tooth and Claw* (Chicago: University of Chicago Press, 1979). Chambers' *Vestiges* is discussed by Milton Millhauser, *Just Before Darwin: Robert Chambers and Vestiges* (Middletown, Conn.: Wesleyan University Press, 1959) and more recently by James Secord, 'Behind the Veil: Robert Chambers and *Vestiges*', in J. R. Moore (ed.), *History, Humanity and Evolution*, pp. 165–94 (this collection includes a number of other valuable articles). On Owen's complex attitude to transmutation, see Evelleen Richards, 'A Question of Property Rights: Richard Owen's Evolutionism Reassessed', *British Journal for the History of Science*, **20** (1987): 129–72. On the development of a more liberal attitude towards transmutation, see Pietro Corsi, *Science and Religion: Baden Powell and the Anglican Debate, 1800–1860* (Cambridge: Cambridge University Press, 1988).

Biographies of Darwin include Gavin De Beer, *Charles Darwin* (London: Nelson, 1963) and more recently Peter J. Bowler, *Charles Darwin: The Man and His Influence* (Oxford: Blackwell, 1990). Adrian Desmond and James Moore's *Darwin*

(London: Michael Joseph, 1991) stresses the social background of Darwin's work. An extensive introduction to modern Darwin scholarship is David Kohn (ed.), *The Darwinian Heritage* (Princeton, NJ: Princeton University Press, 1985). A popular account of the *Beagle* voyage is Alan Moorehead, *Darwin and the Beagle* (London: Hamish Hamilton, 1969). There are many biographies of Wallace, some of which are spoiled by their author's determination to discredit Darwin; more balanced accounts include H. L. McKinney, *Wallace and Natural Selection* (New Haven, Conn.: Yale University Press, 1972) and Martin Fichman, *Alfred Russel Wallace* (Boston, Mass.: Twayne, 1981). See also Barbara G. Beddall, *Wallace and Bates in the Tropics* (London: Macmillan, 1969) and on the complex relationship between the two discoverers of the selection theory, Malcolm Kottler, 'Charles Darwin and Alfred Russel Wallace', in Kohn (ed.), *The Darwinian Heritage*, pp. 367–432. For an unconventional perspective on the role played by the selection theory in the emergence of nineteenth-century evolutionism, see P. J. Bowler, *The Non-Darwinian Revolution: Reinterpreting a Historical Myth* (Baltimore: Johns Hopkins University Press, 1988).

CHAPTER 8

Relatively few late-nineteenth-century scientific texts have been reprinted because the originals are available in many libraries. G. P. Marsh's *Man and Nature* has been reprinted ed. David Lowenthal (Cambridge, Mass.: Harvard University Press, 1965). The first edition of Darwin's *Origin of Species* is available in facsimile with an introduction by Ernst Mayr (Cambridge, Mass.: Harvard University Press, 1964); there are many reprints of the much-modified sixth edition of 1872. Most of Darwin's other books have been reprinted by the AMS Press of New York, including his studies of the fertilization of plants by insects, of climbing plants and of earthworms. David Hull has edited a valuable collection of reviews of the *Origin* under the title *Darwin and His Critics: the Reception of Darwin's Theory of Evolution by the Scientific Community* (Cambridge, Mass.: Harvard University Press, 1973). The nine volumes of T. H. Huxley's *Collected Essays* are widely available and were reprinted (Hildesheim: Georg Olms, 1970); see especially *Darwiniana* (vol. 2) and *Man's Place in Nature* (vol. 7). Asa Gray's *Darwiniana: Essays and Reviews Pertaining to Darwinism* has been reprinted ed. A. Hunter Dupree (Cambridge, Mass.: Harvard University Press, 1963). For anti-Darwinian ideas, see Owen's *Anatomy of the Vertebrates* of 1866–8 (reprinted New York: AMS Press, 1973, 3 vols). E. D. Cope's two books, *The Origin of the Fittest* (1887) and *The Primary Factors of Organic Evolution* (1896), have been reprinted together (New York: AMS Press, 1974). Sterling's *Selected Works in Nineteenth-Century North American Palaeontology* contains useful material from the later part of the century. William Coleman edited a useful selection of morphological works under the title *The Interpretation of Animal Form* (New York: Johnson, 1967), which includes E. Ray Lankester's discussion of evolutionary degeneration.

In the area of ecology, see Hermann Müller's *The Fertilization of Flowers* (1883; reprinted New York: Arno, 1977); the same press has reprinted Karl Semper's *Animal Life as Affected by the Natural Conditions of Existence* (1881), Alpheus Packard's *The Cave Fauna of North America* (1888), Oscar Drude's *Handbook of Phytogeography* (1890), Eugenius Warming's *Oecology of Plants* (1909), Roscoe Pound and Frederic Clements' *Phytogeography of Nebraska* (1900) and Clements' *Research Methods in Ecology* (1905). Useful collections of primary source readings include Frank N. Egerton (ed.), *American Plant Ecology, 1897–1917* and the same editor's *Early Marine Ecology* and *History of American Ecology* (all New York: Arno, 1977). See also Keir B. Sterling (ed.), *Selections from the Literature of American Biogeography* and the same editor's collection of C. Hart Merriam's *Selected Works* (both New York: Arno, 1974). See also Frank B. Golley (ed.), *Ecological Succession* (Stroudsberg, Pa.: Dowden, Hutchinson and Ross, 1977).

Several of the books cited above for chapter 7 are again useful on changing attitudes towards Nature in Britain, including David Allen's *The Naturalist in Britain*, Ritvo's *The Animal Estate* and Mackenzie's *The Empire of Nature*. Stearn's *The Natural History Museum at South Kensington* is strong on this period. Mea Allan's *The Hookers of Kew* and Brockway, *Science and Colonial Expansion* cover the imperialist role of botanical gardens; while on the impact of cheap quinine, see Henry Hobhouse, *Seeds of Change: Five Plants that Transformed Mankind* (London: Sidgwick and Jackson, 1985) and Daniel R. Headrick, *The Tools of Empire: Technology and European Imperialism in the Nineteenth Century* (New York: Oxford University Press, 1981), chap. 3. A useful survey of recent work on the British government's attitude to science is Peter Alter, *The Reluctant Patron: Science and the State in Britain, 1850–1920* (Oxford and Hamburg: Berg/New York: St Martin's Press, 1987). More generally see John M. Mackenzie (ed.), *Imperialism and the Natural World* (Manchester: Manchester University Press, 1990). On the exploitation of science by the state in America, see Dupree, *Science and the Federal Government* and Manning, *Government in Science*. Robert V. Bruce, *The Launching of Modern American Science, 1846–1876* (Ithaca, NY: Cornell University Press, 1988) is a more recent study.

On the origins of environmentalism, see the introductory chapters of John McCormick, *The Global Environment Movement: Reclaiming Paradise* (Bloomington: Indiana University Press/London: Belhaven, 1989). The conservation movement in Britain is described by John Sheal, *Nature in Trust: the History of Nature Conservation in Britain* (Glasgow: Blackie, 1976). There are several biographies of Patrick Geddes, the most recent of which is Helen Meller, *Patrick Geddes: Social Evolutionist and City Planner* (London: Routledge, 1989). More generally on Europe, see Anna Bramwell, *Ecology in the 20th Century: A History* (New Haven, Conn.: Yale University Press, 1989); note that 'ecology' here means environmentalism, not the science. See also Peter C. Gould, *Early Green Politics: Back to Nature, Back to the Land, and Socialism, 1880–1900* (London: Harvester Wheatsheaf, 1988). Juan Martinez-Alier's *Ecological Economics* (Oxford: Blackwell, 1987) charts the emergence of concern about non-renewable resources.

On American environmentalism, see Donald Worster's *Nature's Economy*, and the same author's *Rivers of Empire: Water, Aridity, and the Growth of the American West* (New York: Pantheon, 1985). On F. L. Olmsted, see Laura Wood Roper, *FLO: a Biography of Frederick Law Olmsted* (Baltimore: Johns Hopkins University Press, 1973) and more generally Michael L. Scott, *Pacific Visions:*

California Scientists and the Environment, 1850–1915 (New Haven, Conn.: Yale University Press, 1987). Classic discussions of changing attitudes towards Nature in America include Arthur A. Ekirch Jr, *Man and Nature in America* (New York: Columbia University Press, 1963); Hans Huth, *Nature and the American: Three Centuries of Changing Attitudes* (Berkeley: University of California Press, 1957); Roderick Nash, *Wilderness and the American Mind* (New Haven, Conn.: Yale University Press, 1957); and Peter Schmidt, *Back to Nature: Arcadian Myth in Urban America* (New York: Oxford University Press, 1969). On Canada, see W. A. Weiser, *The Field Naturalist: John Macoun, the Geological Survey and Natural Science* (Toronto: University of Toronto Press, 1989).

Coleman's *Biology in the Nineteenth Century* explores the later decades of the century, while Garland E. Allen's *Life Science in the Twentieth Century* (New York: Wiley, 1975) argues that the 'revolt against morphology' in the 1890s paved the way for the emergence of modern fields such as genetics. The attempt to create a unified science in the mid nineteenth century is explored by Joseph A. Caron, '"Biology" in the Life Sciences: a Historiographical Contribution', *History of Science*, **26** (1988): 223–68. On the critical developments in American biology at the turn of the century, see Ronald Rainger, Keith R. Benson and Jane Maienschein (eds), *The American Development of Biology* (Philadephia: University of Pennsylvania Press, 1988).

There are innumerable accounts of the 'Darwinian revolution' and its impact. Surveys already mentioned that deal with this issue are: Bowler, *Evolution: the History of an Idea*; Eiseley, *Darwin's Century*; Greene, *The Death of Adam*; Ruse, *The Darwinian Revolution*; and Kohn, *The Darwinian Heritage*. Bowler, *The Non-Darwinian Revolution* stresses the differences between nineteenth-century and modern Darwinism. An analysis of the religious debates that challenges the 'warfare' metaphor is James R. Moore, *The Post-Darwinian Controversies: a Study of the Protestant Struggle to Come to Terms with Darwin in Great Britain and America, 1870–1900* (New York: Cambridge University Press, 1979). For a modern evaluation of the famous confrontation at the 1860 meeting of the British Association, see J. Vernon Jensen, 'Return to the Wilberforce–Huxley Debate',

British Journal for the History of Science, **21** (1988): 161–80. On the ideological dimension of Darwinism, see John C. Greene, *Science, Ideology and World View: Essays in the History of Evolutionary Ideas* (Berkeley: University of California Press, 1981) and Robert M. Young, *Darwin's Metaphor: Nature's Place in Victorian Culture* (Cambridge: Cambridge University Press, 1985). For an account of the emergence of non-Darwinian evolutionism, see Peter J. Bowler, *The Eclipse of Darwinism: Anti-Darwinian Evolution Theories in the Decades around 1900* (Baltimore: Johns Hopkins University Press, 1983).

For a survey of the reaction to Darwinism in different countries, see Thomas F. Glick (ed.), *The Comparative Reception of Darwinism* (Austin: University of Texas Press, 1974). On France, see Y. Conry, *L'introduction de Darwinisme en France au XIXe siècle* (Paris: Presses Universitaires de France, 1972). On Germany, see Alfred Kelly, *The Descent of Darwinism: The Popularization of Darwinism in Germany, 1860–1914* (Chapel Hill, NC: University of North Carolina Press, 1981). On Russia, see Daniel P. Todes, *Darwin without Malthus: The Struggle for Existence in Russian Evolutionary Thought* (Oxford: Oxford University Press, 1989) and Alexander Vucinich, *Darwin in Russian Thought* (Berkeley: University of California Press, 1988).

On evolutionary morphology, see E. S. Russell, *Form and Function: a Contribution to the History of Animal Morphology* (London: John Murray, 1916). There are few modern accounts of this topic, but see P. J. Bowler, 'Development and Adaptation: Evolutionary Concepts in British Morphology, 1870–1914', *British Journal for the History of Science,* **22** (1989): 283–97. On zoological debates in this period, see Mary P. Winsor, *Reading the Shape of Nature: Comparative Zoology at the Agassiz Museum* (Chicago: University of Chicago Press, 1990). On the recapitulation theory, see Stephen Jay Gould, *Ontogeny and Phylogeny* (Cambridge, Mass.: Harvard University Press, 1978), part 1. Darwin's use of recapitulationism is stressed by Robert J. Richards, *The Meaning of Evolution: The Morphological Construction and Ideological Reconstruction of Darwin's Theory* (Chicago: University of Chicago Press, 1991), which also argues that historians have minimized the role of progressionism in Darwin's thinking.

Rudwick's *The Meaning of Fossils*, Desmond's *Archetypes and Ancestors* and Bowler's *Fossils and Progress* all deal with paleontology through into the 1870s but not beyond. Buffetaut's *A Short History of Vertebrate Palaeontology* is useful on the late nineteenth and twentieth centuries. There are several popular accounts of fossil hunting in the 'wild west', including Robert Plate, *The Dinosaur Hunters: Othniel C. Marsh and Edward D. Cope* (New York: D. McKay, 1964); Url Lanham, *The Bone Hunters* (New York: Columbia University Press, 1973); and Elizabeth Shor, *The Fossil Feud between E. D. Cope and O. C. Marsh* (Hicksville, NY: Exposition Press, 1974). Three valuable articles by Ronald Rainger are: 'The Continuation of the Morphological Tradition: American Paleontology, 1880–1910', *Journal of the History of Biology*, **14** (1981): 129–58; 'Just before Simpson: William Diller Matthew's Understanding of Evolution', *Proceedings of the American Philosophical Society*, **130** (1986): 453–74; and 'Vertebrate Paleontology as Biology: Henry Fairfield Osborn and the American Museum of Natural History', in Rainger, Benson and Maienschein (eds), *The American Development of Biology*, pp. 219–56. Rainger also has a book on Osborn, *An Agenda for Antiquity: Henry Fairfield Osborn and Vertebrate Paleontology at the American Museum of Natural History, 1890–1935* (Tuscaloosa: University of Alabama Press, 1991). On human origins, see Peter J. Bowler, *Theories of Human Evolution: a Century of Debate, 1844–1944* (Baltimore: Johns Hopkins University Press, 1986), and on the comparison with theories of history, Bowler, *The Invention of Progress*.

Janet Browne's *The Secular Ark* deals with biogeography through to the debates between Darwin and Hooker; on Wallace's contributions, see Martin Fichman, 'Wallace: Zoogeography and the Problem of Land Bridges', *Journal of the History of Biology*, **10** (1977): 45–63. Poulton's work is discussed in William C. Kimler, 'Advantage, Adaptiveness, and Evolutionary Ecology', *Journal of the History of Biology*, **19** (1986): 215–33. On biometry, particularly as it came into conflict with genetics, see William B. Provine, *The Origins of Theoretical Population Genetics* (Chicago: University of Chicago Press, 1971) and Bernard Norton, 'The Biometric Defence of Darwinism', *Journal of the History of Biology*, **6** (1973): 283–316.

Mayr, *The Growth of Biological Thought* is also strong on this period. Recent decades have seen substantial developments in historians' interpretation of the origins of genetics; for a survey of the new ideas, see Peter J. Bowler, *The Mendelian Revolution: the Emergence of Hereditarian Concepts in Modern Science and Society* (London: Athlone/Baltimore: Johns Hopkins University Press, 1989).

The origins of scientific ecology have been outlined in Donald Worster's *Nature's Economy* and in Robert P. McIntosh, *The Background of Ecology: Concept and Theory* (Cambridge: Cambridge University Press, 1985). Worster's book has been criticized for overstressing the role of environmentalism, while McIntosh concentrates on the science so zealously that his text becomes very heavy going for the non-specialist. Frank N. Egerton's 'Changing Concepts of the Balance of Nature' follows this theme through into the twentieth century. For surveys of the more technical historical literature, see the two parts of Egerton's 'The History of Ecology: Achievements and Opportunities', *Journal of the History of Biology*, **16** (1983): 259–310 and **18** (1985): 103–43 (part 1 deals with theoretical ecology, part 2 with practical applications). Edward J. Kormondy and J. Frank McCormick (eds), *Handbook of Contemporary Developments in World Ecology* (Westport, Conn.: Greenwood, 1981) contains a useful historical background.

The background to Warming's work has been described by William Coleman, 'Evolution into Ecology? The Strategy of Warming's Ecological Plant Geography', *Journal of the History of Biology*, **19** (1986): 181–96. On the rise of ecology in America, see Eugene Cittadino, 'Ecology and the Professionalization of Botany in America, 1890–1905', *Studies in the History of Biology*, **4** (1980): 171–98 and Ronald C. Tobey, *Saving the Prairies: the Life Cycle of the Founding School of American Plant Ecology, 1895–1955* (Berkeley: University of California Press, 1981). On the ideological dimensions of Clements' work, see Malcolm Nicolson, 'National Styles, Divergent Classifications: A Comparative Case Study from the History of French and American Plant Ecology', *Knowledge and Society*, **8** (1989): 139–86. On British ecology, see John Sheail, *Seventy-Five Years in Ecology: the British Ecological Society* (Oxford: Blackwell, 1987). On Germany, see Eugene Cittadino, *Nature as the Laboratory: Darwinian Plant Ecology in the German Empire, 1880–1900* (Cambridge: Cambridge University Press, 1991). For a detailed account of the origins of marine ecology, see Eric L. Mills, *Biological Oceanography: An Early History, 1870–1960* (Ithaca, NY: Cornell University Press, 1989). On limnology, see Stephen Bocking, 'Stephen Forbes, Jacob Reighard, and the Emergence of Aquatic Ecology in the Great Lakes Region', *Journal of the History of Biology*, **23** (1990): 461–98.

CHAPTER 9

A classic account of the expansion of resources invested in science is Derek De Solla Price's *Little Science, Big Science* (New York: Columbia University Press, 1965). On the structure and impact of modern science, see Steven Yearley's *Science, Technology and Social Change*. A. Hunter Dupree's *Science in the Federal Government* covers the period up to the Second World War and recounts the impact of war upon American science. On the British government and science up to 1920, see Peter Alter's *The Reluctant Patron*.

On the survival of geography as a discipline, see the chapters on modern geography in Preston E. James, *All Possible Worlds*; see also Robert E. Dickinson, *The Makers of Modern Geography* (London: Routledge and Kegan Paul, 1969). There are several books on the National Geographic Society, some supportive and others critical of its influence; see C. D. B. Bryan, *The National Geographic Society: One Hundred Years of Adventure and Discovery* (New York: Abrams, 1987) and Howard S. Abramson, *National Geographic: Behind America's Lens on the World* (New York: Crown, 1987). On geophysics, the American Geophysical Union is issuing a series of volumes collecting papers on historical matters; see C. Stewart Gillmor (ed.), *History of Geophysics* (vols 1 and 2, Washington, DC: American Geophysical Union, 1984, 1986). On modern developments in paleoclimatology, see John and Katherine Imbrie, *Ice Ages: Solving the Mystery*. Little has been written by historians on the emergence of meteorology, with the notable exception of Robert Marc Friedman, *Appropriating the Weather: Vilhelm*

Bjerknes and the Construction of Modern Meteorology (Ithaca, NY: Cornell University Press, 1989). On American meteorology, see James Rodger Fleming, *Meteorology in America, 1800–1870* (Baltimore: Johns Hopkins University Press, 1990) and Patrick Hughes, *A Century of Weather Service: A History of the Birth and Growth of the National Weather Service, 1870–1970* (New York: Gordon and Breach, 1970).

The range of topics addressed by early-twentieth-century geologists can be judged from Kirtley F. Mather (ed.), *Source Book in Geology, 1900–1950* (Cambridge, Mass.: Harvard University Press, 1967). The most accessible version of Wegener's *The Origin of Continents and Oceans* is the translation of the fourth German edition by John Biram (New York: Dover, 1966). There is a biography of Wegener by Martin Schwarzbach, *Alfred Wegener: the Father of Continental Drift* (Madison: Science Tech/Berlin: Springer-Verlag, 1986). Anthony Hallam's *A Revolution in the Earth Sciences: From Continental Drift to Plate Tectonics* (Oxford: Clarendon, 1973) offers a survey of these developments written by an earth scientist; there is also a chapter on this theme in Hallam's *Great Geological Controversies*. Joe D. Burchfield's *Lord Kelvin and the Age of the Earth* has a chapter on the impact of radioactivity on geology. Mott Greene's *Geology in the Nineteenth Century* notes the continuity between Wegener's theory and earlier ideas.

Robert Muir Wood's *The Dark Side of the Earth* (London: Allen and Unwin, 1985) stresses the impact of the new geophysical techniques of the 1960s and argues that the traditional framework of geology had to be overthrown for the revolution to become possible. On Holmes' theory, see Henry Frankel, 'Arthur Holmes and Continental Drift', *British Journal for the History of Science*, **11** (1978): 130–50. William Glen's *The Road to Jaramillo: Critical Years of the Revolution in Earth Science* (Stanford, Calif.: Stanford University Press, 1982) recounts the developments of the timescale of paleomagnetic reversals and its application to the theory of sea-floor spreading, with extensive quotes from interviews with the scientists involved.

Two accounts that use the revolution in the earth sciences to test various models of how science progresses are H. E. Le

Grand, *Drifting Continents and Shifting Theories: The Modern Revolution in Geology and Scientific Change* (Cambridge: Cambridge University Press, 1988) and John A. Stewart, *Drifting Continents and Colliding Paradigms: Perspectives on the Geoscience Revolution* (Bloomington: Indiana University Press, 1990).

Immanuel Velikovsky's *Worlds in Collision* (Garden City, NY: Doubleday, 1950) is still in print, along with a series of Velikovsky's later books; see A. De Grazia, *The Velikovsky Affair* (New Hyde Park, NY: University Books, 1966). For examples of modern scientific 'catastrophism', see Michael Allaby and James Lovelock, *The Great Extinction* (London: Secker and Warburg, 1983) and Victor Clube and Bill Napier, *The Cosmic Serpent* (London: Faber, 1982).

CHAPTER 10

The reader looking for original material on twentieth-century biology is better served than in the case of the earth sciences. Most of the classic contributions to the theory of plate tectonics took the form of papers in the technical literature, but many of the important ideas dealt with in this chapter were published in the form of books, which are still fairly easily available (some went through a number of later editions, and a few have been reprinted with historical commentaries). The anti-Darwinian literature is less well represented, although there are exceptions. William Bateson's *Problems of Genetics* (1913) has been reprinted with an introduction by G. Evelyn Hutchinson and Stan Rachootin (New Haven, Conn.: Yale University Press, 1979), while Stephen Jay Gould has performed the same function for Richard Goldschmidt's *The Material Basis of Evolution* (New Haven, Conn.: Yale University Press, 1982).

There is a later edition of R. A. Fisher's *Genetical Theory of Natural Selection* of 1930 (New York: Dover, 1958); classic works of the evolutionary synthesis should be readily available, including Theodosius Dobzhansky's *Genetics and the Origin of Species* (New York: Columbia University Press, 1937); Ernst Mayr's *Systematics and the Origin of Species* (New York: Columbia University Press, 1942); G. G. Simpson's *Tempo and Mode in Evolution* (New York: Columbia University Press, 1944); and

Julian Huxley's *Evolution: The Modern Synthesis* (London: Allen and Unwin, 1942). On the broader implications of the synthesis, see Simpson's *This View of Life* (New York: Harcourt, Brace and World, 1953) and Huxley's *Evolution in Action* (London: Chatto and Windus, 1953); see also the translation of Pierre Teilhard De Chardin's *The Phenomenon of Man* introduced by Huxley (London: Collins, 1959). For a rejection of Darwinian (and other forms of) materialism, see Arthur Koestler, *The Ghost in the Machine* (New York: Macmillan, 1967).

Several surveys were published in the early 1980s to gather together material on the post-synthesis debate on the evolutionary mechanism; see Jeremy Cherfas (ed.), *Darwin Up to Date* (London: IPC Magazines, 1982); J. Maynard Smith (ed.), *Evolution Now* (London: *Nature*/Macmillan, 1982); R. Milkman (ed.), *Perspectives on Evolution* (Sunderland, Mass.: Sinaur, 1982); Michael Ruse, *Darwinism Defended: A Guide to the Evolution Controversies* (Reading, Mass.: Addison-Wesley, 1982); and Mark Ridley, *The Problems of Evolution* (Oxford: Oxford University Press, 1985). On the rise of creationism, see for instance Dorothy Nelkin, *The Creation Controversy: Science or Scripture in the Public Schools* (New York: Norton, 1983).

In the area of ethology, Solly Zuckerman's *The Social Life of Monkeys and Apes* of 1932 has been reprinted with a postscript (London: Routledge and Kegan Paul, 1981). Konrad Lorenz's often amusing *King Solomon's Ring* is easily available (2nd edn, London: Methuen, 1960) and on a more serious note his *On Aggression* (London: Methuen, 1966). Nicolaas Tinbergen's *Curious Naturalists* (London: Penguin, 1972) is partly autobiographical. On primates, see Jane van Lawick-Goodall [her married name], *In the Shadow of Man* (London: Collins, 1970). On sociobiology, see E. O. Wilson's *Sociobiology: The New Synthesis* (Cambridge, Mass.: Harvard University Press, 1975) and Richard Dawkins, *The Selfish Gene* (Oxford: Oxford University Press, 1976).

Several of the more important biologists have written autobiographies, or have had selections of their private letters published. These include G. G. Simpson, *Concessions to the Improbable: An Unconventional Autobiography* (New Haven, Conn.: Yale University Press, 1978); L. F. Laporte (ed.),

Simple Curiosity: Letters from George Gaylord Simpson to his Family, 1921–1970 (Berkeley: University of California Press, 1987); Julian Huxley, *Memories* (London: Allen and Unwin, 1970); Solly (Lord) Zuckerman, *From Apes to Warlords: The Autobiography (1904–1946)* (London: Hamish Hamilton, 1978); and Theodosius Dobzhansky, *The Roving Naturalist: Travel Letters*, ed. Bentley Glass (Philadelphia: American Philosophical Society, 1980). There is also a collected work of specially written autobiographies edited by D. A. Dewsbury, *Leaders in the Study of Animal Behavior: Autobiographical Perspectives* (Lewisburg, Pa.: Bucknell University Press, 1985), reprinted as *Studying Animal Behavior: Autobiographies of the Founders* (Chicago: University of Chicago Press, 1989), including material by Konrad Lorenz, Niko Tinbergen, E. O. Wilson and V. C. Wynne-Edwards.

The alleged link between Haeckel's monist philosophy and Nazism is explored by Daniel Gasman, *The Scientific Origins of National Socialism: Social Darwinism in Ernst Haeckel and the Monist League* (New York: American Elsevier, 1971). Donna Haraway, *Primate Visions: Gender, Race and Nature in the World of Modern Science* (New York: Routledge, 1989) is invaluable on the ideological dimensions of primatology and related areas. There are many books on eugenics, of which the most accessible is Daniel Kevles, *In the Name of Eugenics: Genetics and the Uses of Human Heredity* (New York: Knopf, 1985; also a Pelican paperback).

General histories of biology valuable in this area include Mayr, *The Growth of Biological Thought* and Allen, *Life Science in the Twentieth Century*; Nordenskiöld's *History of Biology* should be avoided because it was written when Darwinism had still not begun the comeback that would lead to the modern synthesis. On evolution in the twentieth century, see Bowler, *Evolution: the History of an Idea*. The continuation of non-Darwinian evolutionism into the twentieth century is explored in Bowler, *The Eclipse of Darwinism* and *The Non-Darwinian Revolution*. On early-twentieth-century paleontology, see the articles and book by Ronald Rainger cited in chapter 8 above; also see Rainger's 'What's the Use: William King Gregory and the Functional Morphology of Fossil Vertebrates', *Journal of the History of Biology*, **22** (1989): 103–39. Stephen Jay Gould's *Wonderful Life: The Burgess Shale and the Meaning of History* (London:

Hutchinson Radius, 1989) combines a study of changing interpretations of the earliest fossils with a discussion of the implications of a Darwinian view of the history of life. The suspicion that Darwinism had contributed to a war mentality is explored by Gregg Mitman, 'Evolution as Gospel: William Patten, the Language of Democracy, and the Great War', *Isis*, **81** (1990): 446–63.

On human fossils, see J. Reader, *Missing Links: The Hunt for Earliest Man* (London: Collins, 1981), and on ideas about human origins, Bowler, *Theories of Human Evolution* and Misia Landau, *Narratives of Human Evolution: The Hero Story* (New Haven, Conn.: Yale University Press, 1991). For a popular account of the Scopes 'Monkey trial', see Ray Ginger, *Six Days or Forever* (Boston, Mass.: Beacon, 1958); L. Sprague De Camp, *The Great Monkey Trial* (Garden City, NY: Doubleday, 1968); and M. L. Settle, *The Scopes Trial* (New York: Frankin Watts, 1972). On later creationism, see Nelkin, *The Creation Controversy*.

For a survey of current thinking on the impact of genetics, see Bowler, *The Mendelian Revolution*, and on population genetics, see Provine, *The Origins of Theoretical Population Genetics*. On the evolutionary synthesis, see Ernst Mayr and W. Provine (eds), *The Evolutionary Synthesis: Perspectives on the Unification of Biology* (Cambridge, Mass.: Harvard University Press, 1980) and Marjorie Grene (ed.), *Dimensions of Darwinism: Themes and Counterthemes in Twentieth-Century Evolutionary Theory* (Cambridge: Cambridge University Press, 1983). For an insider's view of the controversies in post-synthesis evolution theory, see David L. Hull, *Science as a Process: An Evolutionary Account of the Social and Conceptual Development of Science* (Chicago: University of Chicago Press, 1988). The second edition of Hallam's *Great Geological Controversies* has an account of the debate over mass extinctions. Adrian Desmond's *The Hot-Blooded Dinosaurs: A Revolution in Palaeontology* (New York: Dial Press, 1976) popularized this particular controversy, and also provides useful historical background on how the dinosaurs have been interpreted. On the controversy over the 'molecular clock', see John Gribben and Jeremy Cherfas, *The Monkey Puzzle: A Family Tree* (London: Bodley Head, 1982).

Biographies of major figures in modern evolutionism include

Joan Fisher Box, *R. A. Fisher: The Life of a Scientist* (New York: Wiley, 1978); Ronald W. Clark, *JBS: The Life and Work of J. B. S. Haldane* (New York: Coward-McCann, 1969); and William Provine, *Sewall Wright and Evolutionary Biology* (Chicago: University of Chicago Press, 1986). See also John C. Greene, 'The Interaction of Science and World View in Sir Julian Huxley's Evolutionary Biology', and James F. Crow, 'Sewall Wright's Place in Twentieth-Century Biology', both in *Journal of the History of Biology*, **23** (1990): 39–55 and 57–89.

The surveys edited by Mayr and Provine, and by Greene, include discussions of how the synthesis was received in different countries; Jan Sapp's *Beyond the Gene: Cytoplasmic Inheritance and the Struggle for Authority in Genetics* (New York: Oxford University Press, 1987) deals explicitly with the European opposition to genetics and Darwinism. On a less academic note, see also Arthur Koestler's account of the Kammerer affair, *The Case of the Midwife Toad* (London: Hutchinson, 1971). The link between Kammerer's defender, E. W. MacBride, and race theory is discussed in P. J. Bowler, 'E. W. MacBride's Lamarckian Eugenics', *Annals of Science*, **41** (1984): 245–60. There are several studies of the Lysenko affair, including David Joravksy, *The Lysenko Affair* (Cambridge, Mass.: Harvard University Press, 1970) and Z. Medvedev, *The Rise and Fall of T. D. Lysenko* (New York: Columbia University Press, 1969); for an alternative that explores Lysenko's links with other non-Darwinian theories, see Nils Roll-Hansen, 'A New Perspective on Lysenko?', *Annals of Science*, **42** (1985): 261–76.

For a detailed study of the role played by materialism in early-twentieth-century American biology, see Philip J. Pauly, *Controlling Life: Jacques Loeb and the Engineering Ideal in Biology* (Berkeley: University of California Press, 1987). On the relationship between biology, psychology and the social sciences, see Hamilton Cravens, *The Triumph of Evolution: American Scientists and the Heredity–Environment Controversy, 1900–1941* (Philadelphia: University of Pennsylvania Press, 1978). The best account of theories of the origin of mind and instinct is Robert J. Richards, *Darwin and the Emergence of Evolutionary Theories of Mind and Behavior* (Chicago: University of Chicago Press, 1987).

Popular accounts of the development of ethology include W. H. Thorpe, *The Origins and Rise of Ethology* (New York: Praeger, 1979) and John Sparks, *The Discovery of Animal Behaviour* (London: Collins/BBC, 1982). Two shorter but more serious accounts are John Durrant, 'Innate Character in Animals and Man: A Perspective on the Origins of Ethology', in Charles Webster (ed.), *Medicine and Society, 1840–1940* (Cambridge: Cambridge University Press, 1981), pp. 157–92; and Richard W. Burkhardt, 'The Development of an Evolutionary Ethology', in D. W. Bendall (ed.), *Evolution from Molecules to Men* (Cambridge: Cambridge University Press, 1983), pp. 429–44. More specialized studies include M. A. and H. E. Evans, *William Morton Wheeler, Biologist* (Cambridge, Mass.: Harvard University Press, 1970) and Alec Nisbet, *Konrad Lorenz* (London: Dent, 1972). On Lorenz and Nazism, see Theodora J. Kalikow, 'Konrad Lorenz's Ethological Theory: Explanation and Ideology', *Journal of the History of Biology*, **16** (1983): 39–73.

Haraway's *Primate Visions* is strong on the studies of ape behaviour. There are innumerable books and articles on the sociobiology controversy; a useful collection of early literature is Arthur L. Caplan (ed.), *The Sociobiology Debate* (New York: Harper and Row, 1978). A sympathetic study is Michael Ruse, *Sociobiology; Sense or Nonsense?* (Dordrecht: Reidel, 1979), while wholesale opposition to hereditarian ideas is expressed in S. Rose, L. J. Kamin and R. C. Lewontin, *Not in Our Genes: Biology, Ideology and Human Nature* (Harmondsworth: Penguin, 1984).

CHAPTER 11

Easily available works on ecology include F. E. Clements, *Plant Succession* (Washington: Carnegie Institution, 1916); Charles Elton, *Animal Ecology* (London: Sidgwick and Jackson, 1927); W. C. Allee, *Animal Aggregations* (Chicago: University of Chicago Press, 1931); W. C. Allee *et al.*, *Principles of Animal Ecology* (Philadelphia: W. B. Saunders, 1949); David Lack, *Darwin's Finches* (Cambridge: Cambridge University Press, 1947); and G. Evelyn Hutchinson, *The Ecological Theatre and the Evolutionary Play* (New Haven, Conn.: Yale University Press, 1965) and *An Introduction to Population Ecology* (New Haven, Conn.:

Yale University Press, 1978). Hutchinson has also written an autobiography, *The Kindly Fruits of the Earth: Recollections of an Embryo Ecologist* (New Haven, Conn.: Yale University Press, 1979). A useful collection of original papers is Leslie A. Real and James H. Brown (eds), *Foundations of Ecology: Classic Papers with Commentaries* (Chicago: University of Chicago Press, 1991). See also James Lovelock, *Gaia: A New Look at Life on Earth* (new edn, Oxford: Oxford University Press, 1987); see also Lawrence E. Joseph, *Gaia: The Growth of an Idea* (London: Penguin, 1991).

Most of the sources cited for the history of environmentalism in chapter 8 above are still useful for the mid twentieth century; see especially Bramwell, *Ecology in the Twentieth Century* on Europe; Sheail, *Nature in Trust* and *Seventy-Five Years in Ecology* on Britain; and Worster, *Nature's Economy* on America. For Russia, see Douglas R. Weiner, *Models of Nature: Ecology, Conservation, and Cultural Revolution in Soviet Russia* (Bloomington: Indiana University Press, 1988) and Kendall Bailes, *Science and Russian Culture in an Age of Revolution: V. I. Vernadsky and his Scientific School (1863–1945)* (Bloomington: Indiana University Press, 1990). McCormick's *The Global Environment Movement* comes into its own for the post-war period; see also Andrew Dobson, *Green Political Thought: An Introduction* (London: Unwin Hyman, 1990).

Part 2 of Egerton's 'The History of Ecology' deals with practical applications; see also Paolo Palladino, 'Ecological Theory and Pest Control Practice: A Study of the Institutional and Conceptual Dimensions of a Scientific Debate', *Social Studies of Science*, **20** (1990): 255–81. On fisheries policy, see Arthur F. McEvoy, *The Fisherman's Problem: Ecology and Law in the California Fisheries, 1850–1980* (Cambridge: Cambridge University Press, 1986). There are several useful papers in Donald Worster (ed.), *The Ends of the Earth: Perspectives on Modern Environmental History* (Cambridge: Cambridge University Press, 1988). On the International Biological Programme, see E. B. Worthington (ed.), *The Evolution of the International Biological Programme* (Cambridge: Cambridge University Press, 1975).

General works on the history of ecological theory already cited in chapter 8 include those by Worster, *Nature's Economy*;

McIntosh, *The Background of Ecology*; Egerton, 'The History of Ecology'; and Tobey, *Saving the Prairies*. On the contrast between American ecology and the 'Sigmatist' group, see Malcolm Nicolson, 'National Styles, Divergent Classifications: A Comparative Case Study from the History of French and American Plant Ecology', *Knowledge and Society*, **8** (1989): 139–86. On individualist ecology, see Nicolson's 'Henry Allan Gleason and the Individualist Hypothesis: The Structure of a Botanist's Career', *Botanical Review*, **56** (1990): 91–160. See also Joel B. Hagen, 'Ecologists and Taxonomists: Divergent Traditions in Twentieth-Century Plant Geography', *Journal of the History of Biology*, **19** (1986): 197–214. There are two useful accounts of experimental taxonomy: John Dean, 'Controversies over Classification: A Case Study from the History of Botany', in B. Barnes and S. Shapin (eds), *Natural Order: Historical Studies of Scientific Culture* (Beverly Hills: Sage, 1979), pp. 211–30; and Joel B. Hagen, 'Experimentalists and Naturalists in Twentieth-Century Botany: Experimental Taxonomy, 1920–1950', *Journal of the History of Biology*, **17** (1984): 249–70.

Turning to animal ecology, there is a study of Charles Elton's influence by Peter Crowcroft, *Elton's Ecologists: A History of the Bureau of Animal Population* (Chicago: University of Chicago Press, 1991). There is a detailed study of population ecology by Sharon E. Kingsland, *Modelling Nature: Episodes in the History of Population Ecology* (Chicago: University of Chicago Press, 1985). See also William C. Kimler, 'Advantage, Adaptiveness, and Evolutionary Ecology', *Journal of the History of Biology*, **19** (1986): 215–33. On the Chicago school, see Edwin M. Banks, 'Warder Clyde Allee and the Chicago School of Animal Behavior', *Journal of the History of the Behavioral Sciences*, **22** (1985); 345–53; and Gregg Mitman, 'From the Population to Society: The Cooperative Metaphors of W. C. Allee and A. E. Emerson', *Journal of the History of Biology*, **21** (1988): 173–94. On H. T. Odum, see Peter J. Taylor, 'Technocratic Optimism, H. T. Odum, and the Partial Transformation of Ecological Metaphor after World War II', also in the *Journal of the History of Biology*, **21** (1988): 213–44. On evolutionary ecology, see James P. Collins, 'Evolutionary Ecology and the Use of Natural Selection in Ecological Theory', *Journal of the History of Biology*, **19** (1986):

257–88. See also Paolo Palladino, 'Defining Ecology: Ecological Theories, Mathematical Models, and Applied Biology in the 1960s and 1970s', *ibid.*, **24** (1991): 223–43. The ecology of the oceans is covered by Mills, *Biological Oceanography*, which deals with the British school in the inter-war years and the later developments in America.

INDEX

Fontana Modern Masters
Editor: Frank Kermode

Einstein
Second Edition

Jeremy Bernstein

Future generations may well refer to the first half of the twentieth century as 'The Age of Einstein' – so outstanding was his contribution to modern thought. In this book Jeremy Bernstein approaches his career through the three basic themes in his work: the special theory of relativity, the general theory of relativity and gravitation, and the quantum theory. Studying these, the author traces the stages of both Einstein's own life in science and the growth of modern scientific awareness. His account is both lucid and evocative, and goes a long way towards helping the layman to a fuller understanding of Albert Einstein and the central issues of modern physics.

'Jeremy Bernstein's book is good value for money; it devotes itself to an attempt to translate the scientific theories into a more mundane and ambiguous language. Biographical details occupy only a small fraction of the contents, but present the essential elements of Einstein's professional career.'　　　*Economist*

'Bernstein concentrates on Einstein's physics, ably relating it to the immediate pre-Einsteinian situation, and to its spectacular confirmation.'　　　*Times Literary Supplement*

Fontana Press

Fontana Press

Fontana Press is a leading paperback publisher of non-fiction. Below are some recent titles.

You can buy Fontana Paperbacks at your local bookshops or newsagents. Or you can order them from Fontana, Cash Sales Department, Box 29, Douglas, Isle of Man. Please send a cheque, postal or money order (not currency) worth the price plus 24p per book for postage (maximum postage required is £3.00 for orders within the UK).

NAME (Block letters)_____

ADDRESS_____

_____